Lagrangian mechanics of nonconservative nonholonomic systems

Monographs and textbooks on mechanics of solids and fluids

editor-in-chief: G. Æ Oravas

Mechanics: Dynamical systems

editor: L. Meirovitch

1. E. H. Dowell
   Aeroelasticity of plates and shells

2. D. G. B. Edelen
   Lagrangian mechanics of nonconservative nonholonomic systems

Lagrangian mechanics of nonconservative nonholonomic systems

Dominic G. B. Edelen

Center for the Application of Mathematics
and Department of Mathematics
Lehigh University
Bethlehem, Pennsylvania

NOORDHOFF INTERNATIONAL PUBLISHING LEYDEN

Softcover reprint of the hardcover 1st edition 1977

A division of A. W. Sijthoff International Publishing Company B.V., Leyden, The Netherlands.

ISBN 978-90-481-8464-4

To

JAMES F. BELL

who first opened for me
the pages of Whittaker

PREFACE

Real problems in the real world rarely exhibit themselves in those pleasant and pristine forms wherein one can model them in terms of systems with conservative forces and holonomic constraints. Indeed, in a deeper sense, that ultimate democratic principle of natural science, referred to in its phenomenological aspect as the second law of thermodynamics, tells us that such conservative, holonomic representations must ultimately degenerate or excise themselves from the domain of the real into ethereal flights of fancy. It is thus altogether surprising to discover that many, if not almost all recent applications of E. Cartan's exterior calculus to mechanics confine the expositions to problems of the pleasant and pristine variety. As attested by the title, this tract addresses itself to the daily problems of real mechanics. The mode of exposition is through the exterior calculus, for this calculus knows nothing about, nor takes cognizance of the intrinsic limitations imposed by the monads of the conservative or the holonomic. It also adheres to the Lagrangian formulation throughout, for this is the natural formulation, the Hamiltonian formulation being dependent on Legendre transformations whose very existence can fail to obtain in real problems. My one wish is that the reader derive as much pleasure from the reading as the writer derived from the writing.

I wish to acknowledge a debt of gratitude to Mrs. Fern Sotzing and Mrs. Lois Nakata for typing the several drafts of this tract, and to Mrs. Fern Sotzing for her patience and endurance in the preparation of the camera-ready copy.

Bethlehem, PA
November, 1976

Dominic G. B. Edelen

CONTENTS

CHAPTER I

KINEMATIC SPACE - THE ARENA OF NEWTONIAN MECHANICS

1. PRELIMINARY REMARKS AND NOTATION

As is well known, the intrinsic utility of a mathematical development of a physical discipline is determined almost solely by the care with which one constructs the underlying space or arena. Any old space or arena will do. The basic structural assumptions of the physical discipline must appear as structural assumptions for the underlying space if the resulting mathematical structure is to yield direct and simple expressions for the quantification of the significant physical variables. The purpose of this chapter is to construct an intrinsic underlying space for Newtonian Mechanics. We do this by making a serious attempt to extract all of the meaning and structural significance that is contained in the assumption of the existence of an inertial frame of reference.

Our method of approach is primarily a geometric one. Full use is therefore made of the Cartan calculus of exterior forms. We summarize here the standard notation and refer the reader not overly familiar with the exterior calculus to the Appendix. The collection of all tangent vector fields on a space K is denoted by $T(K)$. The collection of all exterior forms on K is denoted by $\Lambda(K)$, while $\Lambda^k(K)$ denotes the collection of all exterior forms of degree k on K . Thus $\Lambda^0(K)$ consists of scalar valued functions on K and $\Lambda^1(K)$ consists of differential forms on K . In order to eliminate the need for specifying the continuity class of each and every function that arises, we simply assume throughout this tract that all functions are C^∞-functions. In most cases, we will actually only need C^1 or C^2 functions, but this should be apparent to the reader from the context. The following notation is used throughout:

$\wedge$ for exterior product,

d for exterior differentiation,

$\lrcorner$ for inner multiplication,

$\pounds$ for Lie differentiation.

These symbols are related by the basic formula

$$\pounds_v \omega = v \lrcorner d\omega + d(v \lrcorner \omega) ,$$

where ω is any exterior form and v is any tangent vector field.

2. INERTIAL FRAME OF REFERENCE AND NEWTONIAN MOTIONS

We start with the basic tenet of Newtonian mechanics: *There exists an inertial frame of reference.* There is almost universal agreement as to the meaning of this fundamental assumption even though a careful mathematical statement involves a number of additional primitives in the logical sense. For the purposes of this tract, we take the existence of an inertial frame to mean the existence of a three-dimensional Euclidean space E_3 whose points are quantified by a fixed global Cartesian coordinate cover (x, y, z), and the existence of an invariable time measure, t.

If E_3 is populated by N particles of masses $m_1, m_2, \ldots, m_N$ (the concept and quantification of "mass" are taken to be primitive) there are $3N$ functions $\{\phi^\alpha(t)\}$ that quantify the positions of the particles at any given time t in E_3 by the relations

$$\phi^{3i-2}(t)=x_i(t), \quad \phi^{3i-1}(t)=y_i(t), \quad \phi^{3i}(t)=z_i(t), \quad i=1,\ldots,N . \tag{2.1}$$

As the time t varies, the functions $\{\phi^\alpha(t)\}$ define a motion of the N particles in E_3 . Such motions are *Newtonian Motions under the given forces* $\{F_{x_i}, F_{y_i}, F_{z_i}\}$ (the concept and quantification of "force" are taken to be primitive) if and only if the functions $\{\phi^\alpha(t)\}$ satisfy the system of differential equations

$$(2.2) \qquad m_i \frac{d^2x_i(t)}{dt^2} = F_{x_i}, \; m_i \frac{d^2y_i(t)}{dt^2} = F_{y_i}, \; m_i \frac{d^2z_i(t)}{dt^2} = F_{z_i}$$

under the identifications given by the relations (2.1).

The following sections of this chapter develop the "spatial structures" whereby the above skeletal concepts can be clothed with greater intrinsic meaning. For the time being, however, it is sufficient for our present purposes to deal with the fundamental notion of the existence of an inertial frame, and, in particular, the existence of an invariable time measure.

3. CONFIGURATION SPACE, EVENT SPACE

We assume from now on that the number N of particles that populate E_3 is fixed. Since the quantification of the positions of these N particles at any time requires 3N numbers, an obvious simplification can be achieved by the introduction of *Configuration Space*,

$$C = E_{3N} = E_3 \times E_3 \times \ldots \times E_3 \quad \text{(N-fold Cartesian product)},$$

with the global coordinate cover (q^α) $(\alpha=1,\ldots,3N)$ that is induced by the given coordinate cover (x, y, z) of E_3 through

$$(3.1) \qquad q^{3i-2} = x_i, \quad q^{3i-1} = y_i, \quad q^{3i} = z_i, \quad i=1,\ldots,N,$$

and (x_i, y_i, z_i) is the given coordinate cover to the ith replica of E_3 .

As time varies, the N particles move about in E_3 in accordance with the relations (2.2). This induces a motion of the *configuration point* in C that is quantified by the relations

$q^\alpha = \phi^\alpha(t)$. A motion of the N particles in E_3 is thus equivalent to a map from $\mathbb{R}$ into C . If we were to restrict our attention just to C , an essential aspect of Newtonian mechanics would then be lost, for we would not be able to include the requirement of the existence of an invariable time. It is therefore logical to introduce *Event Space*,

$$E = \mathbb{R} \times C ,$$

in which the first factor allows us to account for the invariable time measure. A motion of the N particles in E_3 with respect to the invariable time measure t can now be described by a map

$$(3.2) \qquad \phi\colon J \subset \mathbb{R} \to E \,|\, \{id;\ \phi^\alpha(\cdot)\}$$

with respect to the given coordinate cover $(t;\ q^\alpha)$ of E . It is customary to refer to $\phi(t) = \{t,\ \phi^\alpha(t)\}$ as the *event at time* t .

Let π denote the projection of E onto C that is defined by

$$(3.3) \qquad \pi\colon E \to C \,|\, \{t,\ q^\alpha\} \rightarrowtail \{q^\alpha\} ,$$

then

$$\pi\circ\phi(t) \ |\ \{t,\ \phi^\alpha(t)\} \rightarrowtail \{\phi^\alpha(t)\}$$

gives the *configuration* of the system at time t under the motion represented by ϕ . On the other hand, if we define $\pi\circ\phi$ by

$$(3.4) \qquad \pi\circ\phi\colon J \to C \,|\, q^\alpha = \bigcup_{t\in J} \phi^\alpha(t) ,$$

i.e. $\pi\circ\phi = \bigcup_{t\in J} \pi\circ\phi(t)$, then $\pi\circ\phi$ gives the *orbit* of the configuration point in configuration space that is determined by the map ϕ .

The tangent vector field to the orbit $\pi\circ\phi$ of ϕ in C is given by the mapping†

(3.5) $\pi \circ D\phi : J \to T_{\pi\circ\phi}(C) \,|\, \{D\phi^{\alpha}(\cdot)\}$

where the entries of $\{D\phi^{\alpha}(\cdot)\}$ are given with respect to the natural basis $\{\partial/\partial q^{\alpha}\}$ of $T(C)$, and the operator D is defined by

(3.6) $D\phi^{\alpha}(t) \;=\; d\phi^{\alpha}(t)/dt$.

This tangent field to $\pi\circ\phi$ in C lifts in a trivial fashion to a tangent vector field to ϕ in E that is given by the mapping

(3.7) $D\phi : J \to T_{\phi}(E) \,|\, \{1, D\phi^{\alpha}(\cdot)\}$,

where the entries of $\{1, D\phi^{\alpha}(\cdot)\}$ are given with respect to the natural basis $\{\partial/\partial t, \partial/\partial q^{\alpha}\}$ of $T(E)$. We take particular note of the fact that the first entry in the representation of $D\phi$ is exactly unity. This has not come about by happenstance; rather, it can be interpreted as a restatement of the invariable nature of the time measure that is subsumed in Newtonian mechanics. A simple comparison of (3.5) and (3.7) immediately shows that an analysis based only on configuration space, C , would miss this essential ingredient of Newtonian mechanics, while an analysis based on event space, E , contains the invariability of the Newtonian time as a basic ingredient.

4. KINEMATIC SPACE

Now, as is well known to any student of mechanics, we have to replace event space by a larger space. Forces can depend both on position and velocity, and, more importantly, acceleration, the essential kinematic ingredient of the equations of motion (2.2), is the time derivative of velocity. Thus, in order to pursue

---

†We use $T_P(K)$ to denote the vector space of tangents that is attached to the space K at the point P .

Newtonian mechanics in an orderly manner, the velocity variables should be available to us as part of the underlying domain space. It turns out that it is sufficient to consider *Kinematic space*, K . This is a space of $(6N+1)$-dimensions whose points P are labeled with respect to a global coordinate cover by

$$(4.1) \qquad P(K) = (t;\, q^{\alpha};\, y^{\alpha}) \;, \quad \alpha=1,\dots,3N$$

when E is referred to the coordinate cover $(t;\, q^{\alpha})$. Thus K has the generic structure

$$(4.2) \qquad K = \mathbb{R}\times C\times S = E\times S \;,$$

where the factor S is 3N-dimensional.

We note that the above statement does not define K ; rather, it assigns a dimension to K and a product structure, but leaves the properties of the factor space S unspecified. We thus have the problem of assigning an appropriate structure to K (to S), where the criteria for appropriateness should rest on the intrinsic properties associated with the underlying notion of an inertial frame.

If we have a motion

$$(4.3) \qquad \phi\colon J \to E|\{\mathrm{id};\, \phi^{\alpha}(\cdot)\} \;,$$

an abvious first requirement is that the map ϕ should extend to a map Φ of J into K such that

$$(4.4) \qquad \Phi\colon J \to K|\{\mathrm{id};\, \phi^{\alpha}(\cdot);\, D\phi^{\alpha}(\cdot)\} \;;$$

that is, the action of Φ gives the assignment

$$(4.5) \qquad \Phi(t) = t, \quad \Phi(q^{\alpha}) = \phi^{\alpha}(t), \quad \Phi(y^{\alpha}) = D\phi^{\alpha}(t) \;.$$

This certainly places a structure on the factor space S so that its elements give the velocity associated with the given motion ϕ in E . In order to place this requirement in a setting that does not depend on the given motion ϕ in E, we proceed as follows. We first note that $\Psi^*(dq^\alpha) = D\psi^\alpha\, dt$ for any map

$$\Psi: J \to K | \Psi(t) = t, \quad \Psi(q^\alpha) = \psi^\alpha(t).$$

Since $y^\alpha dt$ is a 1-form on K for each value of the index α, we can accordingly consider $\Psi^*(y^\alpha dt) = \Psi^*(y^\alpha)\Psi^*(dt) = (y^\alpha \circ \Psi)dt$, which is a 1-form on the 1-dimensional space $\mathbb{R}$. The condition (4.4) can then be made universal by the requirement that

$$\Psi^*(dq^\alpha - y^\alpha dt) = 0, \quad \alpha = 1, \ldots, 3N,$$

for all maps $\Psi: J \to K$ such that $\Psi(t) = t$; that is $D\psi^\alpha(t)dt = (y^\alpha \circ \Psi)dt \ \forall\ \Psi | \Psi(t) = t$. The sole restriction on the maps Ψ, namely $\Psi(t) = t$, is a direct reflection of the previous requirement of the existence of an absolute time scale. The condition (4.6) states that the structure of S is such that $\Psi^*(y^\alpha)$ assigns the αth component of the velocity with respect to the fixed time measure t for each curve $\psi: J \to E | \{id; \psi^\alpha(\cdot)\}$ in E. Thus, all maps $\psi: J \to E | \psi(t) = t$, $\psi(q^\alpha) = \psi^\alpha(t)$ extend to maps $\Psi: J \to K | \Psi(t) = t$, $\Psi(q^\alpha) = \psi^\alpha(t)$, $\Psi(y^\alpha) = D\psi^\alpha(t)$, with $D = d/dt$. This is not necessarily true for maps with a domain space of dimension greater than one.

We can actually go one step further and relax the condition that $\Psi(t) = t$. Consider an arbitrary map

$$\Psi: \mathbb{R} \to K | \Psi(t) = \psi_0(\sigma), \quad \Psi(q^\alpha) = \psi_1^\alpha(\sigma), \quad d\psi_0(\sigma)/d\sigma \neq 0.$$

We then have

$$\Psi^*(dt) = \frac{d\psi_0(\sigma)}{d\sigma} d\sigma, \quad \Psi^*(dq^\alpha) = \frac{d\psi_1^\alpha(\sigma)}{d\sigma} d\sigma,$$

and hence

$$\begin{aligned}\Psi^*(dq^\alpha - y^\alpha dt) &= \Psi^*(dq^\alpha) - \Psi^*(y^\alpha)\Psi^*(dt)\\ &= \left(\frac{d\psi_1^\alpha(\sigma)}{d\sigma} - \Psi^*(y^\alpha)\frac{d\psi_0(\sigma)}{d\sigma}\right) d\sigma.\end{aligned}$$

If we require

$$\Psi^*(dq^\alpha - y^\alpha dt) = 0 \ , \ \alpha=1,\ldots,3N$$

for all maps $\Psi: \mathbb{R} \to K | \Psi^*(dt) \neq 0$, we obtain

$$\Psi^*(y^\alpha) = (y^\alpha \circ \Psi)(\sigma) = \frac{d\psi_1^\alpha(\sigma)}{d\sigma} \Big/ \frac{d\psi_0(\sigma)}{d\sigma} = \frac{dq^\alpha(\sigma)}{dt(\sigma)}$$

because $\Psi^*(dt) = d\psi_0(\sigma) \neq 0$ guarantees that we can solve $\Psi(t) = \psi_0(\sigma)$ for σ as a function of t . This accomplishes the desired purpose of assigning to (y^α) the derivative of $q^\alpha(\sigma)$ with respect to $t(\sigma)$ along any curve in K for which $t(\sigma)$ is strictly monotone. The condition (4.4) can thus be replaced by the *universal requirement*†

$$(4.6) \qquad \Psi^*(dq^\alpha - y^\alpha dt) = 0 \qquad \forall\ \Psi: \mathbb{R} \to K | \Psi^*(dt) \neq 0 \ .$$

Since $\Phi: J \to K | \Phi(t) = t$ defines a curve in K , it has a tangent vector in $T_\Phi(K)$. A second natural requirement is that $D\phi: J \to T_\phi(E)$ should extend to a map $D\Phi$ of J into $T_\Phi(K)$ such that

$$(4.7) \qquad D\Phi: J \to T_\Phi(K) | \{1;\ D\phi^\alpha(\cdot);\ D^2\phi^\alpha(\cdot)\} \ .$$

This requirement can be stated in a universal form by

$$(4.8) \qquad D\Psi: J \to T_\psi(K) | \{1;\ D\psi^\alpha(\cdot);\ D^2\psi^\alpha(\cdot)\} \qquad \forall\ \Psi: J \to K |$$

$$\Psi(t) = t \ , \quad \Psi(q^\alpha) = \psi^\alpha(t)$$

and gives rise to the acceleration, $D^2\phi^\alpha(t)$, with respect to the absolute time measure t in a natural manner.

5. THE STRUCTURE OF KINEMATIC SPACE

We now turn to the basic problem of characterizing the

---

†The author is indebted to Mr. E. Fahy for pointing out an error in the first draft of these considerations.

structure of Kinematic Space, K . Although, at first glance, the second and third factor spaces, $C \times S$, of K may appear to be the tangent space of C , this can not be the case. The tangent space of a space can use any parameterization of the families of curves that give rise to the tangent vector fields, while we are only permitted to use the absolute time, t , as a parameter in view of the conditions (4.6) and (4.8). This apparent, but deceptive similarity suggests, however, that we can determine the structure of K by first determining the structure of the tangent vector fields on K . We denote the collection of all (C^{∞}) tangent vector fields on K by $T(K)$.

Let f be a mapping with domain K and with range contained in $\mathbb{R}$, then, for the given coordinate cover $(t; q^{\alpha}; y^{\alpha})$ of K ,

$$df(t; q^{\alpha}; y^{\alpha}) = \frac{\partial f}{\partial t} dt + \frac{\partial f}{\partial q^{\alpha}} dq^{\alpha} + \frac{\partial f}{\partial y^{\alpha}} dy^{\alpha}$$

is scalar valued. Thus, $\{dt; dq^{\alpha}; dy^{\alpha}\}$ is a basis for $\Lambda^1(K)$ and $\{\partial/\partial t; \partial/\partial q^{\alpha}; \partial/\partial y^{\alpha}\}$ is a basis for $T(K)$. We can generate a tangent vector field on E by means of the differential equations

$$(5.1) \qquad \frac{dt}{d\sigma} = v_0(t; q^{\beta}) , \quad \frac{dq^{\alpha}}{d\sigma} = v_1^{\alpha}(t; q^{\beta}) ,$$

with the solutions $t = T(\sigma; t_0; q_0^{\beta})$, $q^{\alpha} = Q^{\alpha}(\sigma; t_0; q_0^{\beta})$ for the initial data $t(0) = t_0$, $q^{\alpha}(0) = q_0^{\alpha}$; we generate vector fields as fields of tangents to the orbits of 1-parameter groups of point transformations in E . Since the condition (4.6) must hold for all maps $\Psi: \mathbb{R} \to K | \Psi^*(dt) \neq 0$, it must hold for those maps Ψ that are generated by the orbits of the point transformations (5.1) for $v_0 \neq 0$. We must thus require that

$$(5.2) \qquad \Psi^*(dq^{\alpha} - y^{\alpha} dt) = 0 , \quad \alpha=1,\dots,3N$$

for all maps $\Psi: \mathbb{R} \to K | dt/d\sigma = v_0 \neq 0$, $dq^{\alpha}/d\sigma = v_1^{\alpha}$. These conditions are satisfied if and only if $(dq^{\alpha}/d\sigma - y^{\alpha}(\sigma) dt/d\sigma) d\sigma = 0$ holds on each orbit of the group of point transformations generated by (5.1) with $v_o \neq 0$; that is

$$(5.3) \qquad y^\alpha(\sigma) = v_1^\alpha(t(\sigma);\ q^\beta(\sigma))/v_o(t(\sigma);\ q^\beta(\sigma)) \ , \quad v_o \neq 0 \ .$$

A straightforward differentiation with respect to σ and use of $y^\alpha(\sigma) = v_1^\alpha/v_o$ gives us

$$(5.4) \qquad \frac{dy^\alpha(\sigma)}{d\sigma} = Z\left(v_1^\alpha(t;\ q^\beta) - y^\alpha\, v_o(t;\ q^\beta)\right) \ ,$$

where Z is a linear operator that is defined by

$$(5.5) \qquad Zf = \left(\frac{\partial}{\partial t} + y^\beta \frac{\partial}{\partial q^\beta}\right) f \ .$$

We have therefore established the following result.

Every element of $T(K)$ *has the form*

$$(5.6) \qquad V(t;\ q^\beta;\ y^\beta) = v_o \frac{\partial}{\partial t} + v_1^\alpha \frac{\partial}{\partial q^\alpha} + v_2^\alpha \frac{\partial}{\partial y^\alpha} \ ,$$

where $v_o = v_o(t;\ q^\beta)$, $v_1^\alpha = v_1^\alpha(t;\ q^\beta)$ *are* C^∞ *functions on* E ,

$$(5.7) \qquad v_2^\alpha = v_2^\alpha(t;\ q^\beta;\ y^\beta) = Z\left(v_1^\alpha(t;\ q^\beta) - y^\alpha\, v_o(t;\ q^\beta)\right) \ ,$$

Z *is the linear operator*

$$(5.8) \qquad Z = \frac{\partial}{\partial t} + y^\beta \frac{\partial}{\partial q^\beta} \ ,$$

and the orbits of $V(t;\ q^\beta;\ y^\beta)$ *consist of all solutions of*

$$(5.9) \qquad \frac{dt}{d\sigma} = v_o(t;\ q^\beta) \ ,$$

$$(5.10) \qquad \frac{dq^\alpha}{d\sigma} = v_1(t;\ q^\beta)$$

$$(5.11) \qquad \frac{dy^\alpha}{d\sigma} = v_2^\alpha(t;\ q^\beta;\ y^\beta) = Z(v_1^\alpha - y^\alpha v_o) \ .$$

Since this result allows us to obtain an explicit characterization of K , we pause at this point to do so and take up the further properties of $T(K)$ in succeeding paragraphs. We first note that (5.9) and (5.10) can be considered as the equations for a 1-parameter group G of (point) transformations acting on E with the associated operator

$$(5.12) \qquad X = \upsilon_o(t;\, q^\beta)\frac{\partial}{\partial t} + \upsilon_1^\alpha(t;\, q^\beta)\frac{\partial}{\partial q^\alpha}\ .$$

We know that for any $\Psi\colon J \to K|\Psi(t) = t$, $\Psi(q^\alpha) = \psi^\alpha(t)$, the corresponding y^α in K is $\Psi(y^\alpha) = D\psi^\alpha(t)$. This is, however, exactly the situation that occurs in extending the group G with respect to all relations $q^\alpha = \psi^\alpha(t)$†; in fact,

$$\upsilon_2^\alpha(t;\, q^\beta;\, y^\beta) = Z(\upsilon_1^\alpha - y^\alpha\upsilon_o)$$

is exactly the coefficient that arises in the operator

$$V = X + \upsilon_2^\alpha \frac{\partial}{\partial y^\alpha}$$

of the first extension, $\bar{G}$, of the group G with respect to all relations $q^\alpha = \psi^\alpha(t)$. These observations establish the following result.

Kinematic Space is the domain space of the first extension of all groups of (point) transformations on Event Space with respect to all maps $\psi\colon J \to E|\psi(t) = t$.

Thus, $C\times S$ is indeed distinct from the tangent bundle of C, and the careful reader will perceive that this distinction arises because of the demand that t be an absolute invariable time measure.

We now return to the study of $T(K)$. The fact that K is the domain space of first extensions of groups on E allows us to conclude that $V_1 \varepsilon T(K)$, $V_2 \varepsilon T(K)$ implies

$$(5.13) \qquad [V_1, V_2] = \pounds_{V_1} V_2 = -\,\pounds_{V_2} V_1$$

belongs to $T(K)$, where $\pounds$ denotes Lie differentiation. This follows from noting that $X_1 \varepsilon T(E)$, $X_2 \varepsilon T(E)$ implies $[X_1, X_2] \varepsilon T(E)$

† Eisenhart (Ref. 10), Section 26; Ovsjannikov (Ref. 11), Section 9.

and the fact that $V = X+v_2^\alpha \frac{\partial}{\partial y^\alpha}$, $v_2^\alpha = Z(v_1^\alpha - y^\alpha v_o)$. In fact, standard results from group extension theory[†] give the following results.

If G_r *is an* r*-parameter Lie group on* E *with structure constants* c_{ij}^k *and operators* X_i , $i=1,\ldots,r$,

$$[X_i, X_j] = c_{ij}^k X_k , \tag{5.14}$$

then the first extension, $\bar{G}_r$, *of* G *on* K *has the same structure constants* c_{ij}^k *with the operators* $V_i = X_i + v_{2i}^\alpha \frac{\partial}{\partial y^\alpha}$ *and*

$$[V_i, V_j] = c_{ij}^k V_k . \tag{5.15}$$

Thus, in particular, we have

$$[V_i, V_j] = [V_i, V_j]_o \frac{\partial}{\partial t} + [V_i, V_j]_1^\alpha \frac{\partial}{\partial q^\alpha} + [V_i, V_j]_2^\alpha \frac{\partial}{\partial y^\alpha} \tag{5.16}$$

and

$$[V_i, V_j]_2^\alpha = Z\Big([V_i, V_j]_1^\alpha - y^\alpha [V_i, V_j]_o\Big) . \tag{5.17}$$

If we define addition in $T(K)$ by (pointwise addition)

$$(V_1+V_2)(t;\, q^\beta;\, y^\beta) = V_1(t;\, q^\beta;\, y^\beta) + V_2(t;\, q^\beta;\, y^\beta) ,$$

then $T(K)$ is closed under the operation $+$. The standard pointwise definition of multiplication by the reals shows that $T(K)$ is closed under multiplication by elements of $\mathbb{R}$. It is then a trivial matter to conclude that $T(K)$ *forms a vector space over* $\mathbb{R}$, since multiplication by real numbers commutes with the operator Z (see (5.7)). If we simply replace the operation of multiplication by numbers by the operation of multiplication by C^∞-functions on E , then $T(K)$ does not form a module over the associative algebra of C^∞-functions on E because Z does not

---

[†] Ovsjannikov (Ref. 11), Chapter 2.

commute with multiplication by such functions. We can, however, modify the definition of "scalar multiplication" by C^∞-functions on E so as to make $T(K)$ into a module over the associative algebra of C^∞-functions on E.

Let $C^\infty(E)$ denote the associative algebra of C^∞-functions on E. Define the operation $\uparrow$ for any $\gamma(t;\, q^\beta)\varepsilon C^\infty(E)$ and any $V(t;\, q^\beta;\, y^\beta)\varepsilon T(K)$ by

$$\gamma(t;\, q^\beta)\uparrow V(t;\, q^\beta;\, y^\beta) = \gamma\upsilon_o \frac{\partial}{\partial t} + \gamma\upsilon_1^\alpha \frac{\partial}{\partial q^\alpha} + \left(\gamma\upsilon_2^\alpha + (\upsilon_1^\alpha - y^\alpha\upsilon_o)Z(\gamma)\right)\frac{\partial}{\partial y^\alpha} . \tag{5.18}$$

It follows immediately from the following calculation and (5.7) that $\uparrow$ is on $C^\infty(E)\times T(K)$ to $T(K)$:

$$Z\left((\gamma\upsilon_1^\alpha) - y^\alpha(\gamma\upsilon_o)\right) = \gamma Z(\upsilon_1^\alpha - y^\alpha\upsilon_o) + (\upsilon_1^\alpha - y^\alpha\upsilon_o)Z(\gamma) .$$

Since (5.18) is bilinear in γ and V, we see that, for any $a\varepsilon\mathbb{R}$, $\gamma\uparrow(aV) = (a\gamma)\uparrow V = a(\gamma\uparrow V)$, $\gamma\uparrow(V_1+V_2) = \gamma\uparrow V_1 + \gamma\uparrow V_2$, $(\gamma_1+\gamma_2)\uparrow V = \gamma_1\uparrow V + \gamma_2\uparrow V$. Thus, the only condition that remains to be verified in order to establish the modular structure[†] of $T(K)$ is $\gamma_1\uparrow(\gamma_2\uparrow V) = (\gamma_1\gamma_2)\uparrow V$. Now, (5.18) gives

$$\begin{aligned}
\gamma_1\uparrow(\gamma_2\uparrow V) &= \gamma_1\uparrow\left\{\gamma_2\upsilon_o \frac{\partial}{\partial t} + \gamma_2\upsilon_1^\alpha \frac{\partial}{\partial q^\alpha} + \left(\gamma_2\upsilon_2^\alpha + (\upsilon_1^\alpha - y^\alpha\upsilon_o)Z(\gamma_2)\right)\frac{\partial}{\partial y^\alpha}\right\} \\
&= \gamma_1\gamma_2\upsilon_o \frac{\partial}{\partial t} + \gamma_1\gamma_2\upsilon_1^\alpha \frac{\partial}{\partial q^\alpha} + \gamma_1\left(\gamma_2\upsilon_2^\alpha + (\upsilon_1^\alpha - y^\alpha\upsilon_o)Z(\gamma_2)\right)\frac{\partial}{\partial y^\alpha} \\
&\qquad + (\gamma_2\upsilon_1^\alpha - y^\alpha\gamma_2\upsilon_o)Z(\gamma_1)\frac{\partial}{\partial y^\alpha} \\
&= \gamma_1\gamma_2\upsilon_o \frac{\partial}{\partial t} + \gamma_1\gamma_2\upsilon_1^\alpha \frac{\partial}{\partial q^\alpha} + \left(\gamma_1\gamma_2\upsilon_2^\alpha + (\upsilon_1^\alpha - y^\alpha\upsilon_o)Z(\gamma_1\gamma_2)\right)\frac{\partial}{\partial y^\alpha} \\
&= (\gamma_1\gamma_2)\uparrow V_1 ,
\end{aligned}$$

and hence we have the following result.

[†]Matsushima (Ref. 8), p. 72.

The space T(K) *is a module over the associative algebra* $C^\infty(E)$ *of* C^∞-*functions on* E *with respect to the operations of pointside addition and scalar multiplication defined by*

$$(5.19)\qquad \uparrow: C^\infty(E)\times T(K) \to T(K) \mid \gamma\uparrow V = \gamma v_o \frac{\partial}{\partial t} + \gamma v_1^\alpha \frac{\partial}{\partial q^\alpha} + \left(\gamma v_2^\alpha + (v_1^\alpha - y^\alpha v_o) Z(\gamma)\right)\frac{\partial}{\partial y^\alpha} .$$

Notice that we can not replace the algebra $C^\infty(E)$ by the associative algebra of C^∞-functions on K because (5.19) would give $\gamma\uparrow V = \gamma(t;\ q^\beta;\ y^\beta) v_o(t;\ q^\beta)\frac{\partial}{\partial t} + \gamma(t;\ q^\beta;\ y^\beta) v_1^\alpha(t;\ q^\beta)\frac{\partial}{\partial q^\alpha} + (\ )\frac{\partial}{\partial y^\alpha}$ and this does not belong to T(K) because the coefficients of $\partial/\partial t$ and $\partial/\partial q^\alpha$ are no longer independent of the arguments $\{y^\beta\}$.

It is instructive to look more closely at the case $\Phi: J \to K \mid \Phi(t) = t,\ \Phi(q^\alpha) = \phi^\alpha(t),\ \Phi(y^\alpha) = D\phi^\alpha(t)$. Since Φ defines a curve in K , it has a tangent vector "field" that belongs to $T_\Phi(K)$. The first and second set of entries in this "tangent field" give

$$v_o(t;\ q^\beta) = 1 ,\quad v_1^\alpha(t;\ q^\beta) = D\phi^\alpha(t) ,$$

and accordingly, (5.7) and (5.8) yield

$$v_2^\alpha = Z(v_1^\alpha - y^\alpha v_o) = \left(\frac{\partial}{\partial t} + y^\beta \frac{\partial}{\partial q^\beta}\right)(D\phi^\alpha(t) - y^\alpha) = D^2\phi^\alpha(t) .$$

Thus, the structure already imposed on K and T(K) shows that

$$D\Phi: J \to T_\Phi(K) \mid \{1;\ D\phi^\alpha(t);\ D^2\phi^\alpha(t)\} .$$

Since this clearly holds for any map $\Psi: J \to K \mid \Phi(t) = t$, we see that the structure already imposed on K and T(K) is such that the second condition (4.8) is met. Thus, in effect, we have shown that satisfaction of the condition (4.6), namely $\Psi^*(dq^\alpha - y^\alpha dt) = 0$ for all $\Psi: J \to K \mid \Psi(t) = t$ implies satisfaction of the condition (4.8). Accordingly, we may conclude that *the structure of*

Kinematic Space is uniquely determined by the condition $\Psi^*(dq^\alpha - y^\alpha dt) = 0$ *for all* $\Psi: J \to K|\Psi(t) = t$.

A vector field of the form $\{1;\ D\psi^\alpha(\cdot);\ D^2\psi^\alpha(\cdot)\}$ is called a *vertical vector field*. Thus, *every map* $\Psi: J \to K|\Psi(t) = t$, $\Psi(q^\alpha) = \psi^\alpha(t)$ *generates a vertical vector field* $\{1;\ D\psi^\alpha(\cdot);\ D^2\psi^\alpha(\cdot)\}$ *on* Ψ *in* K . Careful note should be taken of the fact that vertical vector fields are not defined on all of $T(K)$; rather, they are defined only on the 1-dimensional set $T_\Psi(K)$.

There is a vector field on all of E , called a *rectilinear field* that is closely related to the class of vertical vector fields and is of particular importance in mechanics. This vector field is generated by the group of rectilinear motions on E with equations

$$\frac{dt}{d\sigma} = 1\ ,\quad \frac{dq^\alpha}{d\sigma} = y^\alpha \tag{5.20}$$

and orbits

$$t(\sigma) = t_o + \sigma\ ,\qquad q^\alpha(\sigma) = q^\alpha_o + \sigma y^\alpha\ .$$

Extension to K is accomplished by identifying the "y's" in (5.20) with the y-coordinates in K and using (5.7):

$$v_o = 1\ ,\quad v^\alpha_1 = y^\alpha\ ,\quad v^\alpha_2 = Z(v^\alpha_1 - y^\alpha v_o) = 0\ . \tag{5.21}$$

The *rectilinear field* on K is the vector field

$$t(t;\ q^\beta;\ y^\beta) = \frac{\partial}{\partial t} + y^\alpha \frac{\partial}{\partial q^\alpha}\ . \tag{5.22}$$

The orbits of this vector field on K are

$$t(\sigma) = t_o + \sigma\ ,\quad q^\alpha(\sigma) = q^\alpha_o + y^\alpha_o\sigma\ ,\quad y^\alpha(\sigma) = y^\alpha_o \tag{5.23}$$

and hence $t(t;\ q^\alpha;\ y^\alpha)$ is also a tangent to a rectilinear motion in K with constant "velocity coordinates" $\{y^\alpha_o\}$. We also note that this field is naturally associated with the condition $\Psi^*(dq^\alpha - y^\alpha dt) = 0$, for (5.20) yield

$$dq^\alpha - y^\alpha dt = y^\alpha d\sigma - y^\alpha d\sigma = 0$$

on any orbit of $\mathcal{t}(t;\ q^\beta;\ y^\beta)$. We also note that the operator $Z = \frac{\partial}{\partial t} + y^\beta \frac{\partial}{\partial q^\beta}$ is exactly the rectilinear field $\mathcal{t}(t;\ q^\beta;\ y^\beta)$ acting as a linear operator. Thus, if we denote this action by $\mathcal{t}[\ ]$, we have

$$(5.24) \qquad \mathcal{t}[\mathfrak{f}] \equiv Z(\mathfrak{f}) ,$$

which ties everything together very nicely. As a final remark, we note that $\mathcal{t}$ is not an element of $T(K)$ since the coefficient of $\frac{\partial}{\partial q^\alpha}$ is a function of the y's, a situation not allowed for elements of $T(K)$. The space K is thus strange since one can define vector fields on K that are not tangent vector fields.

6. BEHAVIOR UNDER MAPPINGS

Now that we know that K is the domain space of the first extension of groups acting on E with respect to all maps $q^\alpha = \psi^\alpha(t)$, we can examine what happens when K is subjected to mappings. This is clearly necessary since the analysis up to this point has been confined to considerations in which K is referred to the single global inertial coordinate cover that is the extension of the single global inertial coordinate cover of E we postulated at the beginning.

We start with the global inertial coordinate cover $(t;\ q^\beta;\ y^\beta)$ of K that is the extension of the global inertial coordinate cover $(t;\ q^\alpha)$ of E . Since K is the domain space of the first extension of groups acting on E with respect to all maps $\Psi(q^\alpha) = \psi^\alpha(t)$, the transformation properties of K may be expected to be determined by those of E . We therefore begin with the invertible map

$$(6.1) \qquad \gamma\colon E \to {}^{`}E \mid {}^{`}t = \gamma(t;\ q^\beta) , \quad {}^{`}q^\alpha = \gamma^\alpha(t;\ q^\beta) ,$$

and the equations

$$\text{(6.2)} \qquad \frac{dt}{d\sigma} = \upsilon_o(t;\ q^\beta)\ ,\quad \frac{dq^\alpha}{d\sigma} = \upsilon_1^\alpha(t;\ q^\beta)$$

of a 1-parameter group of point transformations on E. It is then immediate that

$$\text{(6.3)} \qquad `\upsilon_o(`t;`q^\beta) = \frac{d`t}{d\sigma} = \left(\frac{\partial `t}{\partial t}\upsilon_o + \frac{\partial `t}{\partial q^\beta}\upsilon_1^\beta\right)\circ \overset{-1}{\gamma}\ ,$$

$$\text{(6.4)} \qquad `\upsilon_1^\alpha(`t;`q^\beta) = \frac{d`q^\alpha}{d\sigma} = \left(\frac{\partial `q^\alpha}{\partial t}\upsilon_o + \frac{\partial `q^\alpha}{\partial q^\beta}\upsilon_1^\beta\right)\circ \overset{-1}{\gamma}\ ,$$

where $(\quad)\circ\overset{-1}{\gamma}$ denotes the fact that we must express the arguments $(t,\ q^\beta)$ in terms of $(`t,\ `q^\beta)$ by $\overset{-1}{\gamma}\colon `E \to E|t = \overset{-1}{\gamma}(`t,\ `q^\beta)\ ,\ q^\alpha = \overset{-1}{\gamma}{}^\alpha(`t,\ `q^\beta)$. Therefore, we have

(6.5)

$$\gamma_*\colon T(E) \to T(`E)|\gamma_*\left(\upsilon_o\frac{\partial}{\partial t} + \upsilon_1^\alpha\frac{\partial}{\partial q^\alpha}\right) = `\upsilon_o\frac{\partial}{\partial `t} + `\upsilon_1^\alpha\frac{\partial}{\partial `q^\alpha}\ .$$

Since $y^\alpha(\sigma) = (dq^\alpha(\sigma)/d\sigma)/(dt(\sigma)/d\sigma) = \upsilon_1^\alpha/\upsilon_o$, we see that

$$`y^\alpha(\sigma) = \frac{d`q^\alpha(\sigma)/d\sigma}{d`t(\sigma)/d\sigma} = \frac{\left(\frac{\partial `q^\alpha}{\partial t}\upsilon_o + \frac{\partial `q^\alpha}{\partial q^\beta}\upsilon_1^\beta\right)}{\left(\frac{\partial `t}{\partial t}\upsilon_o + \frac{\partial `t}{\partial q^\beta}\upsilon_1^\beta\right)}$$

$$= \frac{\frac{\partial `q^\alpha}{\partial t} + \frac{\partial `q^\alpha}{\partial q^\beta}y^\beta}{\frac{\partial `t}{\partial t} + \frac{\partial `t}{\partial q^\beta}y^\beta} = \frac{Z(`q^\alpha)}{Z(`t)}\ .$$

We have thus established the following result.

A map $\gamma\colon E \to `E|`t = \gamma(t;\ q^\beta)\ ,\ `q^\alpha = \gamma_1^\alpha(t;\ q^\beta)$ *induces the map*

(6.6)

$$\Gamma\colon K \to `K|`t{=}\gamma(t;\ q^\beta),\quad `q^\alpha{=}\gamma_1^\alpha(t;\ q^\beta),\quad `y^\alpha{=}\gamma_2^\alpha(t;\ q^\beta;\ y^\beta) = \frac{Z(\gamma_1^\alpha)}{Z(\gamma)}\ .$$

It is clear from (6.6) that Kinematic Space possesses some fearsome transformation properties whenever the transformation function $\gamma(t;\ q^\beta)$ of time is such that $Z(\gamma(t;\ q^\beta))$ can vanish at one or more points of K. In fact, it is clear that a map $\Gamma: K \to {}^\backprime K$ is *regular* (send points with finite coordinates into points with finite coordinates) only on its *domain of regularity*

$$R(\Gamma) = \{(x^i;\ q^\alpha;\ y^\alpha) \mid y^\beta \partial\gamma(t;\ q^\alpha)/\partial q^\beta \neq -\partial\gamma(t;\ q^\alpha)/\partial t\} \subset K .$$

Further, we do not know at this point that the image of $R(\Gamma)$ in ${}^\backprime K$ has the properties of a Kinematic Space, for it has not been shown that a vector field on K will transform into a vector field on ${}^\backprime K$ (i.e., that ${}^\backprime v_2^\alpha = {}^\backprime Z({}^\backprime v_1^\alpha - {}^\backprime y^\alpha\, {}^\backprime v_o)$ for ${}^\backprime Z = \partial/\partial{}^\backprime t + {}^\backprime y^\alpha \partial/\partial{}^\backprime q^\alpha$).

The above result is not unexpected, for we have determined the properties of K from the underlying requirement of the existence of an immutable time measure t. Accordingly, deformations of the time variable should lead to quite complex situations, as they indeed have. It is therefore natural to restrict attention to maps Γ that preserve the time measure t.

A map $\gamma: E \to {}^\backprime E$ is called *admissible* iff $\gamma(t;\ q^\beta) = t$. There is a subclass of admissible maps that will prove to be important, so we may as well define it now.

A map $\gamma: E \to {}^\backprime E$ is called a *factor* map iff γ is admissible and $\gamma(t;\ q^\beta) = \gamma_1^\alpha(q^\beta)$.

Thus, a factor map is the lift to E of a map of C and we have $\pi\circ\gamma = \gamma\circ\pi$. The following results are now immediate consequences of (6.6).

If $\gamma: E \to {}^\backprime E$ *is admissible, then* $\Gamma: K \to {}^\backprime K$ *is* t-*preserving,* $R(\Gamma) = K$, *and* $\{y^\alpha\}$ *maps linearly:*

$$(6.7) \qquad {}^\backprime y^\alpha = Z(\gamma_1^\alpha) = \frac{\partial{}^\backprime q^\alpha}{\partial t} + \frac{\partial{}^\backprime q^\alpha}{\partial q^\beta} y^\beta .$$

If γ *is a factor map, then* Γ *is a factor map,* $R(\Gamma) = K$, *and* $\{y^\alpha\}$ *maps linearly and homogeneously:*

$$(6.8)\qquad {}^{\backprime}y^{\alpha} = Z(\gamma_1^{\alpha}) = \frac{\partial {}^{\backprime}q^{\alpha}}{\partial q^{\beta}}\, y^{\beta} .$$

In this latter case, $\{y^{\alpha}\}$ *maps as the components of a tangent vector field on* C *:* $\Gamma(y^{\alpha}) \varepsilon \gamma_{*} T(C)$.

It is obvious from the above that admissible maps preserve the Newtonian absolute time, but lead, in general, to noninertial frames of reference on ${}^{\backprime}C$ $({}^{\backprime}q^{\alpha} = \gamma_1^{\alpha}(t;\, q^{\beta}))$. On the other hand, a factor map preserves the absolute nature of time and maps the inertial frame on C into an inertial frame on ${}^{\backprime}C$, where the inertial frame on ${}^{\backprime}C$ may be referred to general curvilinear coordinates. We should thus expect that factor maps can be utilized to recast Newtonian mechanics in terms of general curvilinear coordinate covers, while admissible, but not factor maps will give the results that obtain from the use of noninertial frames of reference with a fixed absolute time measure. The case of a general map Γ on its domain of regularity $R(\Gamma) \subset K$ represents a very complex situation. Here, we encounter transformations to curvilinear, noninertial frames of reference where there are deformations of the time measure that depend upon the instantaneous event in question (i.e., depend on $(t;\, q^{\alpha})$).

We now turn to the transformation properties of $T(K)$ that are induced by a general map $\Gamma\colon K \to {}^{\backprime}K$ that obtains from the lifting of a general, invertible map $\gamma\colon E \to {}^{\backprime}E$. It is clear at the outset that we have to restrict attention to the domain of regularity of Γ , and hence we shall study $T(R(\Gamma))$. We start by noting that (6.3) and (6.4) yield

$$(6.9)\qquad v_1^{\beta} \frac{\partial {}^{\backprime}q^{\alpha}}{\partial q^{\beta}} = {}^{\backprime}v_1^{\alpha} - \frac{\partial {}^{\backprime}q^{\alpha}}{\partial t} v_o ,\qquad v_1^{\beta} \frac{\partial {}^{\backprime}t}{\partial q^{\beta}} = {}^{\backprime}v_o - \frac{\partial {}^{\backprime}t}{\partial t} v_o .$$

Now, by definition,

$${}^{\backprime}v_2^{\alpha} = \frac{d\,{}^{\backprime}y^{\alpha}(\sigma)}{d\sigma} = \frac{\partial {}^{\backprime}y^{\alpha}}{\partial t} v_o + \frac{\partial {}^{\backprime}y^{\alpha}}{\partial q^{\beta}} v_1^{\beta} + \frac{\partial {}^{\backprime}y^{\alpha}}{\partial y^{\mu}} v_2^{\mu} .$$

Thus, since $v_2^{\alpha} = Z(v_1^{\alpha} - y^{\alpha} v_o)$ and ${}^{\backprime}y^{\alpha} = Z({}^{\backprime}q^{\alpha})/Z({}^{\backprime}t)$, we obtain

$$`v_2^\alpha = v_o \frac{\partial}{\partial t}\left[\frac{Z(`q^\alpha)}{Z(`t)}\right] + v_1^\beta \frac{\partial}{\partial q^\beta}\left[\frac{Z(`q^\alpha)}{Z(`t)}\right]$$

$$+ Z(v_1^\beta - y^\beta v_o)\frac{\partial}{\partial y^\beta}\left[\frac{Z(`q^\alpha)}{Z(`t)}\right]$$

on the image of $R(\Gamma)$ in $`K$. However, it is easily seen that

$$\frac{\partial}{\partial t} Z = Z\frac{\partial}{\partial t}, \quad \frac{\partial}{\partial q^\alpha} Z = \frac{\partial}{\partial q^\alpha}, \quad \frac{\partial}{\partial y^\beta} Z(`t) = \frac{\partial `t}{\partial q^\beta}, \tag{6.10}$$

$$\frac{\partial}{\partial y^\beta} Z(`q^\alpha) = \frac{\partial `q^\alpha}{\partial q^\beta},$$

and hence

$$`v_2^\alpha = \frac{1}{Z(`t)}\left\{Z\left(v_1^\beta \frac{\partial `q^\alpha}{\partial q^\beta}\right) - `y^\alpha Z\left(v_1^\beta \frac{\partial `t}{\partial q^\beta}\right)\right.$$

$$+ v_o\left[Z\left(\frac{\partial `q^\alpha}{\partial t}\right) - `y^\alpha Z\left(\frac{\partial `t}{\partial t}\right)\right]$$

$$\left. - Z(v_o)y^\beta\left[\frac{\partial `q^\alpha}{\partial q^\beta} - `y^\alpha \frac{\partial `t}{\partial q^\beta}\right]\right\}$$

on using $`y^\alpha = Z(`q^\alpha)/Z(`t)$. This establishes the following result.

If $\Gamma: K \to `K$ *is the lift to* $K$ *of an invertible map* $\gamma: E \to `E$, *then*

$$\Gamma_*\left(v_o \frac{\partial}{\partial t} + v_1^\alpha \frac{\partial}{\partial q^\alpha} + v_2^\alpha \frac{\partial}{\partial y^\alpha}\right) \tag{6.11}$$

$$= `v_o \frac{\partial}{\partial `t} + `v_1^\alpha \frac{\partial}{\partial `q^\alpha} + `v_2^\alpha \frac{\partial}{\partial `y^\alpha}$$

on $\Gamma(R(\Gamma))$ *with*

$$\begin{aligned} `v_o &= \left(\frac{\partial `t}{\partial t} v_o + \frac{\partial `t}{\partial q^\beta} v_1^\beta\right)\circ \overset{-1}{\gamma} \\ `v_1^\alpha &= \frac{\partial `q^\alpha}{\partial t} v_o + \frac{\partial `t}{\partial q^\beta} v_1 \circ \overset{-1}{\gamma} \end{aligned} \tag{6.12}$$

and

$$(6.13)\qquad {}^{\backprime}\upsilon_2^\alpha = \left(\frac{1}{Z({}^{\backprime}t)}\right)\circ\overset{-1}{\Gamma}\left\{Z\left(\upsilon_1^\beta\frac{\partial{}^{\backprime}q^\alpha}{\partial q^\beta}\right) - {}^{\backprime}y^\alpha Z\left(\upsilon_1\frac{\partial{}^{\backprime}t}{\partial q^\beta}\right)\right.$$

$$+ \upsilon_o\left[Z\left(\frac{\partial{}^{\backprime}q^\alpha}{\partial t}\right) - {}^{\backprime}y^\alpha Z\left(\frac{\partial{}^{\backprime}t}{\partial t}\right)\right]$$

$$\left. - Z(\upsilon_o)y^\beta\left[\frac{\partial{}^{\backprime}q^\alpha}{\partial q^\beta} - \partial y^\alpha\frac{\partial{}^{\backprime}t}{\partial q^\beta}\right]\right\}\circ\overset{-1}{\Gamma}\ .$$

It now remains to show that $\Gamma_*: T(R(\Gamma)) \to T(\Gamma(R(\Gamma)))$; that is, that ${}^{\backprime}\upsilon_2^\alpha = {}^{\backprime}Z({}^{\backprime}\upsilon_1^\alpha - {}^{\backprime}y^\alpha{}^{\backprime}\upsilon_o)$ on the image of $R(\Gamma)$. If we substitute from (6.9) into (6.13), we obtain

$${}^{\backprime}\upsilon_2^\alpha = \frac{1}{Z({}^{\backprime}t)}\left\{Z\left({}^{\backprime}\upsilon_1^\alpha - \frac{\partial{}^{\backprime}q^\alpha}{\partial t}\upsilon_o\right) - {}^{\backprime}y^\alpha Z\left({}^{\backprime}\upsilon_o - \frac{\partial{}^{\backprime}t}{\partial t}\upsilon_o\right)\right.$$

$$+ \upsilon_o\left[Z\left(\frac{\partial{}^{\backprime}q^\alpha}{\partial t}\right) - {}^{\backprime}y^\alpha Z\left(\frac{\partial{}^{\backprime}t}{\partial t}\right)\right]$$

$$\left. - Z(\upsilon_o)y^\beta\left[\frac{\partial{}^{\backprime}q^\alpha}{\partial q^\beta} - {}^{\backprime}y^\alpha\frac{\partial{}^{\backprime}t}{\partial q^\beta}\right]\right\}$$

$$= \frac{1}{Z({}^{\backprime}t)}\left\{Z({}^{\backprime}\upsilon_1^\alpha) - {}^{\backprime}y^\alpha Z({}^{\backprime}\upsilon_o)\right.$$

$$+ Z(\upsilon_o)\left[-\frac{\partial{}^{\backprime}q^\alpha}{\partial t} + {}^{\backprime}y^\alpha\frac{\partial{}^{\backprime}t}{\partial t} - y^\beta\frac{\partial{}^{\backprime}q^\alpha}{\partial q^\beta}\right.$$

$$\left.\left. + {}^{\backprime}y^\alpha y^\beta\frac{\partial{}^{\backprime}t}{\partial q^\beta}\right]\right\}$$

$$= \frac{1}{Z({}^{\backprime}t)}\{Z({}^{\backprime}\upsilon_1^\alpha) - {}^{\backprime}y^\alpha Z({}^{\backprime}\upsilon_o)$$

$$+ Z(\upsilon_o)[{}^{\backprime}y^\alpha Z({}^{\backprime}t) - Z({}^{\backprime}q^\alpha)]\}$$

$$= \frac{1}{Z({}^{\backprime}t)}\left(Z({}^{\backprime}\upsilon_1^\alpha) - {}^{\backprime}y^\alpha Z({}^{\backprime}\upsilon_o)\right)$$

$$+ Z(\upsilon_o)\left[{}^{\backprime}y^\alpha - \frac{Z({}^{\backprime}q^\alpha)}{Z({}^{\backprime}t)}\right]$$

$$= \frac{1}{Z({}^{\backprime}t)}\left(Z({}^{\backprime}\upsilon_1^\alpha) - {}^{\backprime}y^\alpha Z({}^{\backprime}\upsilon_o)\right) ,$$

since ${}^{\backprime}y^\alpha = Z({}^{\backprime}q^\alpha)/Z({}^{\backprime}t)$. However, $Z = \frac{\partial}{\partial t} + y^\alpha \frac{\partial}{\partial q^\beta}$ yields

$$Z = \left(\frac{\partial{}^{\backprime}t}{\partial t} + y^\beta \frac{\partial{}^{\backprime}t}{\partial q^\beta}\right)\frac{\partial}{\partial{}^{\backprime}t} + \left(\frac{\partial{}^{\backprime}q^\rho}{\partial t} + y^\beta \frac{\partial{}^{\backprime}q^\rho}{\partial q^\beta}\right)\frac{\partial}{\partial{}^{\backprime}q^\rho}$$

$$= Z({}^{\backprime}t)\frac{\partial}{\partial{}^{\backprime}t} + Z({}^{\backprime}q^\rho)\frac{\partial}{\partial{}^{\backprime}q^\rho} = Z({}^{\backprime}t)\left[\frac{\partial}{\partial t} + \frac{Z({}^{\backprime}q^\rho)}{Z({}^{\backprime}t)}\frac{\partial}{\partial{}^{\backprime}q^\rho}\right]$$

$$= Z({}^{\backprime}t)\left[\frac{\partial}{\partial t} + {}^{\backprime}y^\rho \frac{\partial}{\partial{}^{\backprime}q^\rho}\right] = Z({}^{\backprime}t){}^{\backprime}Z ,$$

on $R(\Gamma)$, so that

$$Z(R(\Gamma)) = Z({}^{\backprime}t)Z(\Gamma(R(\Gamma))) . \tag{6.14}$$

Thus, we see that

$${}^{\backprime}v_2^\alpha = {}^{\backprime}Z({}^{\backprime}v_1^\alpha) - {}^{\backprime}y^{\alpha}{}^{\backprime}Z({}^{\backprime}v_o) = {}^{\backprime}Z({}^{\backprime}v_1^\alpha - {}^{\backprime}y^{\alpha}{}^{\backprime}v_o) .$$

This establishes the following result.

The map Γ_* *, that is induced by a map* $\Gamma\colon K \to {}^{\backprime}K$ *that obtains from the lift to* K *of an invertible map* $\gamma\colon E \to {}^{\backprime}E$ *, is a map from* $T(R(\Gamma))$ *to* $T(\Gamma(R(\Gamma)))$.

The importance of this theorem is that it establishes the following useful fact.

The image of $R(\Gamma)$ *under a map* Γ *, that obtains from the lift to* K *of an invertible map* $\gamma\colon E \to {}^{\backprime}E$ *, is a Kinematic Space.*

Upon restriction to admissible maps, we thus see that *the map* Γ_* *, that is induced by an admissible map* $\Gamma\colon K \to {}^{\backprime}K$ *, is a map from* $T(K)$ *to* $T({}^{\backprime}K)$ *and* ${}^{\backprime}K$ *is a Kinematic Space.*

We also note the following:

If $\Gamma\colon K \to {}^{\backprime}K$ *is a factor map, then* Γ_* *yields*

$${}^{\backprime}v_o = v_o \circ \overset{-1}{\Gamma} , \quad {}^{\backprime}v_1^\alpha = \left(\frac{\partial{}^{\backprime}q^\alpha}{\partial q^\beta} v_1^\beta\right)\circ \overset{-1}{\Gamma} \tag{6.15}$$

$${}^{\backprime}v_2^\alpha = {}^{\backprime}Z\left[\left(\frac{\partial{}^{\backprime}q^\alpha}{\partial q^\beta}(v_1^\beta - y^\beta v_o)\right)\circ \overset{-1}{\Gamma}\right]. \tag{6.16}$$

The mapping properties of $\Lambda^1(K)$ follow almost directly from the definition of a map Γ and (6.7). We therefore simply record the results.

If $\Gamma: K \to {}^{\backprime}K$ *is admissible then* $\Gamma^*: \Lambda^1({}^{\backprime}K) \to \Lambda^1(K)$ *and*

$$(6.17)\qquad \Gamma^*(d^{\backprime}t) = dt\ ,\quad \Gamma^*(d^{\backprime}q^\alpha) = \frac{\partial^{\backprime}q^\alpha}{\partial t}dt + \frac{\partial^{\backprime}q^\alpha}{\partial q^\beta}dq^\beta$$

$$(6.18)\qquad \Gamma^*(d^{\backprime}y^\alpha) = dZ({}^{\backprime}q^\alpha) = d\left(\frac{\partial^{\backprime}q^\alpha}{\partial t} + \frac{\partial^{\backprime}q^\alpha}{\partial q^\beta}y^\beta\right)$$

$$= \left(\frac{\partial^{2\backprime}q^\alpha}{\partial t^2} + \frac{\partial^{2\backprime}q^\alpha}{\partial q^\beta\partial t}y^\beta\right)dt + \left(\frac{\partial^{2\backprime}q^\alpha}{\partial t\partial q^\mu} + \frac{\partial^{2\backprime}q^\alpha}{\partial q^\beta\partial q^\mu}y^\beta\right)dq^\mu$$

$$+ \left(\frac{\partial^{\backprime}q^\alpha}{\partial q^\beta}\right)dy^\beta\ ,$$

so that

$${}^{\backprime}\omega = {}^{\backprime}\omega^o\, d^{\backprime}t + {}^{\backprime}\omega^1_\alpha\, d^{\backprime}q^\alpha + {}^{\backprime}\omega^2_\alpha\, d^{\backprime}y^\alpha$$

gives

$$(6.19)\qquad \Gamma^*({}^{\backprime}\omega) = \left({}^{\backprime}\omega_o + {}^{\backprime}\omega^1_\alpha \frac{\partial^{\backprime}q^\alpha}{\partial t} + {}^{\backprime}\omega^2_\alpha\left(\frac{\partial^{\backprime}q^\alpha}{\partial t^2} + y^\mu\frac{\partial^{2\backprime}q^\alpha}{\partial t\partial q^\mu}\right)\right)dt$$

$$+ \left({}^{\backprime}\omega^1_\beta \frac{\partial^{\backprime}q^\beta}{\partial q^\alpha} + {}^{\backprime}\omega^2_\beta\left(\frac{\partial^{\backprime}q^\beta}{\partial q^\alpha\partial t} + y^\mu\frac{\partial^{2\backprime}q^\beta}{\partial q^\alpha\partial q^\mu}\right)\right)dq^\alpha$$

$$+ \frac{\partial^{\backprime}q^\beta}{\partial q^\alpha}{}^{\backprime}\omega^2_\beta\, dy^\alpha\ .$$

This reduces to

$$(6.20)\qquad \Gamma^*({}^{\backprime}\omega) = {}^{\backprime}\omega_o dt + \left({}^{\backprime}\omega^1_\beta \frac{\partial^{\backprime}q^\beta}{\partial q^\alpha} + {}^{\backprime}\omega^2_\beta y^\mu \frac{\partial^{2\backprime}q^\beta}{\partial q^\alpha\partial q^\mu}\right)dq^\alpha$$

$${}^{\backprime}\omega^2_\beta \frac{\partial^{\backprime}q^\beta}{\partial q^\alpha}dy$$

if Γ *is a factor map.*

We note that

$$(6.21)\qquad V\rfloor\omega = {}^{\backprime}V\rfloor{}^{\backprime}\omega = (\Gamma_* V)\rfloor(\overset{-1}{\Gamma^*}\omega)\ ,$$

which follows directly from the above results after a straightforward but messy calculation. We also see that

Each admissible map of K *maps vertical vector fields on* K *onto vertical vector fields on* ${}^{\backprime}K$ *with*

$$(6.22)\qquad {}^{\backprime}\Psi = \Gamma\circ\Psi : J \to {}^{\backprime}K \,|\, {}^{\backprime}t = t;\ {}^{\backprime}q^{\alpha} = {}^{\backprime}\psi^{\alpha}(t) = \gamma^{\alpha}(t;\psi^{\beta}(t));$$

$$ {}^{\backprime}y^{\alpha} = D{}^{\backprime}\psi^{\alpha}(t) = D\gamma^{\alpha}(t;\psi^{\alpha}(t)) \ ,$$

and hence

$$(6.23)\qquad D{}^{\backprime}\Psi : J \to T_{{}^{\backprime}\psi}({}^{\backprime}K) \,|\, \{1; D{}^{\backprime}\psi^{\alpha}(t); D^{2}{}^{\backprime}\psi^{\alpha}(t) \quad .$$

Each admissible map of K *maps the rectilinear field* $\mathcal{t}(t;\ q^{\beta};\ y^{\beta}) = \partial/\partial t + y^{\beta}\ \partial/\partial q^{\beta}$ *of* K *onto the rectilinear field* ${}^{\backprime}\mathcal{t}({}^{\backprime}t;\ {}^{\backprime}q^{\beta};\ {}^{\backprime}y^{\beta}) = \partial/\partial{}^{\backprime}t + {}^{\backprime}y^{\beta}\ \partial/\partial{}^{\backprime}q^{\beta}$ *of* ${}^{\backprime}K$ *.*

Thus, admissible maps relate Kinematic Spaces in such a manner that the rectilinear field of a Kinematic Space is mapped in a one-to-one manner with the Kinematic Space onto the image Kinematic Space. Stated another way, the association of a Kinematic Space with its rectilinear field is an invariant association under admissible maps.

Now that we know how the various quantities behave under mappings $\Gamma\colon K \to {}^{\backprime}K$, we can determine their behavior under point transformations in K and under transformations of coordinate covers. If Id denotes the identity map from K to ${}^{\backprime}K$, then composition of Id^{-1} with Γ yields a mapping

$$(6.24)\qquad \Gamma_{p} = Id^{-1}\circ\Gamma : K \to K$$

that represents a point transformation acting on K . Thus,

$$(6.25)\qquad \Gamma_{p} : K \to K \,|\, \Gamma_{p}(t) = {}^{\backprime}t = t;\ \Gamma_{p}(q^{\alpha}) = {}^{\backprime}q^{\alpha} = \gamma^{\alpha}(t;\ q^{\beta}) \ ;$$

$$\Gamma_{p}(y^{\alpha}) = {}^{\backprime}y^{\alpha} = Z(\gamma^{\alpha})$$

where $({}^{\backprime}t; {}^{\backprime}q^{\alpha}; {}^{\backprime}y^{\alpha})$ are the coordinates of the point in K that is the image of the point $(t;\ q^{\alpha};\ y^{\alpha})$ under the point transformation Γ_p . Point transformations of K can be used to generate coordinate transformations by the process of "dragging along". This process is simplicity itself, for it assigns new coordinates to the old point by setting them equal numerically to the old coordinates of the new point. Accordingly, if $(t'; q^{\alpha'}; y^{\alpha'})$ denote the new coordinates of the old point (that had coordinates $(t; q^{\alpha}; y^{\alpha})$) and if the old point is mapped onto the new point with coordinates $(\grave{t}; \grave{q}^{\alpha}; \grave{y}^{\alpha})$ by the point transformation Γ_p , then

$$(6.26) \qquad t' = \grave{t} , \quad q^{\alpha'} = \grave{q}^{\alpha} , \quad y^{\alpha'} = \grave{y}^{\alpha}$$

defines the *coordinate transformation dragged along by* $\overset{-1}{\Gamma_p}$.

Because Γ_p is generated by the composition of two mappings $\overset{-1}{\mathrm{Id}}$ and Γ , $\Gamma_p = \overset{-1}{\mathrm{Id}}\circ\Gamma$, the mapping properties of $T(K)$ and $\Lambda(K)$ can be used directly in order to determine the effects of point transformations and coordinate transformations on $T(K)$ and $\Lambda(K)$. These are accomplished by allowing Γ_{p*} to act on $T(K)$ and Γ^*_p to act on $\Lambda(K)$ and then use (6.26) if we are interested in coordinate transformations rather than point transformations. There is one very important point that must be noted in this context, however: *we generate neither all possible point transformations nor all possible coordinate transformations of* K *by means of all possible maps* Γ *from Kinematic Space* K *to Kinematic Space* $\grave{K}$. This is easily seen from the fact that general maps $\Gamma: K \to \grave{K}$ are such that they do not preserve the property that K is the group extension domain of E and $\grave{K}$ is the group extension domain of $\grave{E}$, while admissible maps preserve the property that K is the group extension domain of $\mathbb{R}\times C$ and $\grave{K}$ is the group extension domain of $\mathbb{R}\times\grave{C}$. In particular, we see that the set of point transformations and coordinate transformations generated by admissible maps Γ constitute a preferred set that preserves the underlying kinematic relations that obtain from the existence of an inertial frame of reference.

A note of caution must be sounded on the other side as well, so that the reader is not led to believe that we are not permitted to use coordinate transformations or point transformations that do not arise from mappings of K to $`K$. To the contrary, the analysis of the structure of exterior forms on $K$ demands that we use point transformations that do not arise from maps of $K$ to $`K$. For instance, the homotopy operator[†] H that gives the unique decomposition

$$(6.27) \qquad \omega = d(H\omega) + H(d\omega)$$

of any exterior form ω on K, is generated by the point transformations

$$(6.28) \qquad \begin{aligned} `t &= \lambda t + (1-\lambda)t_o , \quad `q^\alpha = \lambda q^\alpha + (1-\lambda)q^\alpha_o , \\ `y^\alpha &= \lambda y^\alpha + (1-\lambda)y^\alpha_o , \quad 0 \leq \lambda \leq 1 \end{aligned}$$

and these can not be generated from any $\Gamma: K \to `K$ since $`y^\alpha \neq Z(`q^\alpha)/Z(`t)$. This does not mean that we can not use the result (6.27); to the contrary, the decomposition (6.27) will figure heavily in the analysis of forces and constraints given in later chapters. The fact that the transformations (6.28) do not arise from any $\Gamma: K \to `K$ can be used, however, to conclude that the action of the operator H can not be realized by any concatenations of kinematically admissible processes unless the exterior forms ω and $d\omega$ are forms on E. The latter result follows upon noting that (6.28) and

$$(6.29) \qquad \begin{aligned} `t &= \lambda t + (1-\lambda)t_o , \quad `q^\alpha = \lambda q^\alpha + (1-\lambda)q^\alpha_o \\ `y^\alpha &= Z(`q^\alpha)/Z(`t) = y^\alpha \end{aligned}$$

agree on E and (6.29) arise from a map of $K \to `K$.

[†]See Appendix, Section 10.

7. HORIZONTAL VECTOR FIELDS AND VARIATIONS

A particularly important class of vector fields on K consists of those designated as *Horizontal Vector Fields*. These vector fields have the generic form

$$(7.1) \qquad U = u_1^\alpha(t;q^\beta)\frac{\partial}{\partial q^\alpha} + Z(u_1^\alpha(t;q^\beta))\frac{\partial}{\partial y^\alpha} .$$

Thus, horizontal vector fields are characterized by the condition that $u^o(t;q^\beta) = 0$; i.e., the coefficient of the basis vector $\partial/\partial t$ is identically zero for horizontal vector fields. We denote the collection of all horizontal vector fields on K by $H(K)$. The collection $H(K)$ is clearly a vector subspace of $T(K)$, and is, in fact, a module over the associative algebra of C^∞-functions on E with respect to the operations $(+,\uparrow)$, as is easily established. We note in passing that if $u_1^\alpha(t;q^\beta) = u^\alpha(q^\beta)$ is satisfied, then the resulting U is the complete lift to K of a vector field $u^\alpha(q^\beta)\partial/\partial q^\alpha$ on C .

The analysis of horizontal vector fields starts with the equations for its orbits:

$$(7.2) \qquad \frac{dt}{d\sigma} = 0 , \quad \frac{dq^\alpha}{d\sigma} = u_1^\alpha(t;q^\beta) ,$$

$$(7.3) \qquad \frac{dy^\alpha}{dt} = \frac{\partial u_1^\alpha(t;q^\beta)}{\partial t} + y^\gamma \frac{\partial u_1^\alpha(t;q^\beta)}{\partial q^\gamma} .$$

It is evident geometrically that the composition of a vertical vector field generated by $\Phi: J \to K|\Phi(t) = t$ with a horizontal vector field has the effect of imbedding Φ in a 1-parameter family $\{\Phi_\sigma\}$ of maps of J to K such that $\Phi_\sigma(t) = t$. We see this analytically by considering the solution of the orbit equations (7.2), (7.3) of a horizontal vector field U subject to the initial data

$$t(0) = t_o , \quad q^\alpha(0) = q_o^\alpha , \quad y^\alpha(0) = y_o^\alpha .$$

This gives, to first order terms in σ ,

$$(7.4)\qquad t(\sigma) = t_o\ ,\quad q^\alpha(\sigma) = q^\alpha_o + \sigma\, u^\alpha_1(t_o;\, q^\beta_o) + o(\sigma)\ ,$$

$$y^\alpha(\sigma) = y^\alpha_o + \sigma\left[\frac{\partial u^\alpha_1(t_o;q^\beta_o)}{\partial t_o} + y^\gamma_o\,\frac{\partial u^\alpha_1(t_o;q^\beta_o)}{\partial q^\gamma_o}\right] + o(\sigma)\ .$$

Composition of Φ with U, where

$$\Phi\colon J \to K\,|\,\{id;\ \phi^\alpha(\cdot);\ D\phi^\alpha(\cdot)\}\ ,$$

is then achieved by assigning the 1-parameter family of initial data by means of Φ ; that is

$$(7.5)\qquad t_o = t\ ,\quad q^\alpha_o = \phi^\alpha(t)\ ,\quad y^\alpha_o = D\phi^\alpha(t)\ ,\quad t\varepsilon J\ .$$

When (7.5) are substituted into (7.4), we obtain the 1-parameter family of maps

(7.6)a

$$\Phi_\sigma\colon J \to K\,|\,\Phi_\sigma(t) = t\ ,\quad \Phi_\alpha(q^\alpha) = \phi^\alpha(t) + \sigma\, u^\alpha_1(t;\phi^\beta(t)) + o(t)\ ,$$

$$\Phi_\sigma(y^\alpha) = D\phi^\alpha(t) + \sigma\left[\frac{\partial u^\alpha_1(t;\phi^\beta(t))}{\partial t} + D\phi^\gamma(t)\,\frac{\partial u^\alpha_1(t;\phi^\alpha(t))}{\partial\phi^\gamma(t)}\right] + o(t)\ ,$$

and $\Phi = \Phi_o$ is imbedded in the family $\{\Phi_\sigma\}$.

Let us now specialize the above results to the case where the horizontal vector field is generated by $u^\alpha_1(t;q^\beta) = u^\alpha(t)$. Under these circumstances we obtain the exact solutions $t(\sigma) = t_o$, $q^\alpha(\sigma) = q^\alpha_o + \sigma\, u^\alpha(t_o)$, $y^\alpha(\sigma) = y^\alpha_o + \sigma\, du^\alpha(t_o)/dt_o$, and the composition of Φ with this exact solution yields

$$(7.6)\text{b}\qquad \Phi_\sigma\colon J \to K\,|\,\Phi_\sigma(t) = t\ ,\quad \Phi_\sigma(q^\alpha) = \phi^\alpha(t) + \sigma\, u^\alpha(t)\ ,$$

$$\Phi_\sigma(y^\alpha) = D\phi^\alpha(t) + \sigma\, Du^\alpha(t)\ .$$

It is customary to write $u^\alpha(t) = \delta\phi^\alpha(t)$, where the functions $\delta\phi^\alpha(t)$ can be changed at will be appropriate changes of the functions $u^\alpha(t)$. The functions $\delta\phi^\alpha(t)$ are referred to as the *variations* of the functions $\phi^{\alpha}(t)$, and correspond exactly with the classic definition of the variation, as shown by (7.6)b:

$$\Phi_{\sigma}: J \to K | \Phi_{\sigma}(t) = t \ , \quad \Phi_{\sigma}(q^{\alpha}) = \phi^{\alpha}(t) + \sigma\, \delta\phi^{\alpha}(t) \ ,$$

$$\Phi_{\sigma}(y^{\alpha}) = D\phi^{\alpha}(t) + \sigma\, D\delta\phi^{\alpha}\,(t) \ .$$

Accordingly, since the classic variation process would yield $\Phi_{\sigma}(y^{\alpha}) = D\phi^{\alpha}(t) + \sigma\, \delta(D\phi^{\alpha}(t))$, we see that *variation and time differentiation commute* in this case.

We now go back to the general case of arbitrary horizontal vector fields. A comparison of (7.6)a and (7.6)b show that we obtain agreement to within second order terms in σ through defining a *general variation* by

$$(7.7) \qquad u_1^{\alpha}(t;\, \phi^{\beta}(t)) = \delta\phi^{\alpha}(t) \ .$$

We can accordingly rewrite (7.6) in this new notation as

$$(7.8) \qquad \Phi_{\sigma}: J \to K | \Phi_{\sigma}(t) = t \ , \quad \Phi_{\sigma}(q^{\alpha}) = \phi^{\alpha}(t) + \sigma\, \delta\phi^{\alpha}(t) + o(\sigma) \ ,$$

$$\Phi_{\sigma}(y^{\alpha}) = D\phi^{\alpha}(t) + \sigma\, D\delta\phi^{\alpha}(t) + o(\sigma) \ ,$$

on noting that

$$\frac{\partial u_1^{\alpha}(t;\phi^{\beta}(t))}{\partial t} + D\phi^{\gamma}(t)\, \frac{\partial u_1^{\alpha}(t;\phi^{\beta}(t))}{\partial \phi^{\gamma}(t)} = Du_1^{\alpha}(t;\phi^{\beta}(t)) = D\delta\phi^{\alpha}(t) \ .$$

Accordingly, if we define $\delta D\phi^{\alpha}(t)$ by $\Phi_{\sigma}(y^{\alpha}) = D\phi^{\alpha}(t) + \sigma\, \delta D\phi^{\alpha}(t) + o(\sigma)$, we again obtain $D\delta\phi^{\alpha}(t) = \delta D\phi^{\alpha}(t)$; that is, *variation and time differentiation commute* in the general case.

Later, we shall need to choose $\delta\phi^{\alpha}(t)$ so that $\delta\phi^{\alpha}(a) = \delta\phi^{\alpha}(b) = 0$, for given numbers (a, b) , in order to obtain

(7.9)

$$\Phi_{\sigma}(q^{\alpha}(a)) = \phi^{\alpha}(a) + o(\sigma) = a^{\alpha} \ , \quad \Phi_{\sigma}(q^{\alpha}(b)) = \phi^{\alpha}(b) + o(\sigma) = b^{\alpha}$$

for given systems of numbers $\{a^{\alpha}\}$ and $\{b^{\alpha}\}$. However,

$$\delta\phi^\alpha(a) = u_1^\alpha(a;\phi^\beta(a)) = u_1^\alpha(a;a^\beta) ,$$

$$\delta\phi^\alpha(b) = u_1^\alpha(b;\phi^\beta(b)) = u_1^\alpha(b;b^\beta) ,$$

and hence the desired results can be achieved by the requirements

$$(7.10) \qquad u_1^\alpha(a;a^\beta) = u_1^\alpha(b;b^\beta) = 0 .$$

This has the further advantage of replacing the approximate results (7.9) by the exact results

$$(7.11) \qquad \Phi_\sigma(q^\alpha(a)) = \phi^\alpha(a) = a^\alpha , \quad \Phi_\sigma(q^\alpha(b)) = \phi^\alpha(b) = b^\alpha$$

since $dq^\alpha/d\sigma = u_1^\alpha(t;q^\beta)$, (7.10) and the initial data $q^\alpha(0) = a^\alpha$ yield $q^\alpha(\sigma) = a^\alpha$.

It is convenient at this point to introduce a notational convention that simplifies the writing of subsequent equations. We have prepared for this by using script letters for quantities that are defined on Kinematic Space; $v_1^\alpha(t;q^\beta;y^\beta)$, $u_1^\alpha(t;q^\beta;y^\beta)$, etc. When such quantities are composed with $\Phi: J \to K$ or acted upon by Φ^* , the result will be denoted by the same symbol written in Latin type. Thus, for instance

$$(7.12) \qquad \Phi^* L(t;q^\beta;y^\beta) = L\circ\Phi = L(t;\, \phi^\beta(t);\, D\phi^\beta(t)) = L ,$$

$$\Phi^* v_o = V_o , \quad \Phi^* v_1^\alpha = V_1^\alpha , \quad \Phi^* v_2^\alpha = V_2^\alpha , \quad \text{etc.}$$

In this context, we note the following for future reference: for any $\Phi: J \to K|\Phi(t) = t$, $\Phi(q^\alpha) = \phi^\alpha(t)$, we have

$$\Phi^* dt = dt , \qquad \Phi^* t = t ,$$

$$\Phi^* dq^\alpha = dt\, D\phi^\alpha , \qquad \Phi^* q^\alpha = \phi^\alpha(t) ,$$

$$\Phi^* dy^\alpha = dt\, D^2\phi^\alpha , \qquad \Phi^* y^\alpha = D\phi^\alpha ,$$

and, for any horizontal vector field U on K , we have

$$\Phi^* u_1^\alpha = \delta\phi^\alpha(t) \ , \qquad \Phi^* u_2^\alpha = D\delta\phi^\alpha(t) = \delta D\phi^\alpha(t) \ ,$$

while

$$(7.13) \qquad \Phi^* d(\cdot) = dt \, D\Phi^*(\cdot) \ .$$

This last statement can be used as an alternative definition of the operator $D = d/dt$.

The reader accustomed to the classic process of variation may be bothered by the more general variation process considered here, wherein $\delta\phi^\alpha(t) = u_1^\alpha(t; \phi^\beta(t))$, rather than $\delta\phi^\alpha(t) = u^\alpha(t)$. Clearly, the classic variation $\delta\phi^\alpha(t) = u^\alpha(t)$ is a special case of the general process $\delta\phi^\alpha(t) = u_1^\alpha(t; \phi^\beta(t))$ that is achieved by the choice $u_1^\alpha(t; q^\beta) = u^\alpha(t)$. In addition, variation and time differentiation commute for both variation processes and we can achieve the exact result $\delta\phi^\alpha(a) = \delta\phi^\alpha(b) = 0$ for both processes. Thus, the classic results can be obtained directly from our analysis by imposing the restriction $u_1^\alpha(t; q^\beta) = u^\alpha(t)$. However, the more general variation process also allows us to consider cases in which $u_1^\alpha(t; q^\beta) = w^\alpha(q^\beta)$, in which case we obtain $\delta\phi^\alpha(t) = w^\alpha(\phi^\beta(t))$. This is particularly useful, for the orbit equations become $dt/d\sigma = 0$, $dq^\alpha/d\sigma = w^\alpha(q^\beta)$, $dy^\alpha/d\sigma = y^\gamma \partial w^\alpha(q^\beta)/\partial q^\gamma$; that is, *they describe the lift to* K *of a flow on configuration space that is specified by* $dq^\alpha/d\sigma = w^\alpha(q^\beta)$. The more general variation process considered here allows us to examine the intrinsic structure associated with flows on configuration space within the variational context. This turns out to be particularly useful in the study of quadrature theory. General variations also have particularly nice mapping properties that are not shared by classic variations, as we shall see in Section 10.

8. VARIATIONS AND CONSTRAINTS

Let U be a horizontal vector field on K and let $\Phi: J \to K | \Phi(t) = t$ be a given map of the kind we have been considering.

Equations (7.7) and (7.8) thus define a map $\Phi_\sigma(U): J\times\mathbb{R} \to K$ which, to within first order terms in σ , is given by

$$(8.1)\quad \Phi_\sigma(U): J\times\mathbb{R} \to K \,|\, \Phi_\sigma(t) = t \ , \quad \Phi_\sigma(q^\alpha) = \phi^\alpha(t) + \sigma\delta\phi^\alpha(t) + o(\sigma) \ ,$$

$$\Phi_\sigma(y^\alpha) = D\phi^\alpha(t) + \sigma D\delta\phi^\alpha(t) + o(\sigma) \ ,$$

where

$$(8.2)\qquad \Phi^* u_1^\alpha = u_1^\alpha(t;\ \phi^\beta(t)) = \delta\phi^\sigma(t) \ .$$

Suppose that we are given a differential form

$$(8.3)\quad \omega(t;q^\beta;y^\beta) = a(t;q^\beta;y^\beta)dt + b_\alpha(t;\ q^\beta;y^\beta)dq^\alpha + h_\alpha(t;q^\beta;y^\beta)dy^\alpha$$

on K . Since $\Phi = \Phi_o$ and $\Phi^* U = \{0;\ \delta\phi^\alpha;\ D\delta\phi^\alpha\}$,

$$\Phi_\sigma^* d\tau = d\tau \ , \quad \Phi_\sigma^* dq^\alpha = D\phi^\alpha dt + \delta\phi^\alpha d\sigma + \sigma D\delta\phi^\alpha dt + o(\sigma) \ ,$$

$$\Phi_\sigma^* dy^\alpha = D^2\phi^\alpha dt + D\delta\phi^\alpha d\sigma + \sigma D^2\phi^\alpha dt + o(\sigma) \ ,$$

we have

$$\begin{aligned}(8.4)\quad \Phi_\sigma^*\omega &= (a\circ\Phi_\sigma)dt + (b_\alpha\circ\Phi_\sigma)(D\phi^\alpha dt + \delta\phi^\alpha d\sigma + \sigma D\delta\phi^\alpha dt + o(\sigma)) \\ &\quad + (h_\alpha\circ\Phi_\sigma)(D^2\phi^\alpha dt + D\delta\phi^\alpha d\sigma + \sigma D^2\delta\phi^\alpha dt + o(\sigma)) \\ &= \Phi^*\omega + (b_\alpha\delta\phi^\alpha + h_\alpha D\delta\phi^\alpha)d\sigma + \sigma\Phi^*\pounds_U\omega + o(\sigma)\end{aligned}$$

where $b_\alpha = b_\alpha\circ\Phi$, $h_\alpha = h_\alpha\circ\Phi$. Accordingly, if $\Phi = \Phi_o$ is required to satisfy a constraint of the form

$$(8.5)\qquad 0 = \Phi^*\omega = (a + b_\alpha D\phi^\alpha + h_\alpha D^2\phi^\alpha)dt \ ,$$

then we must require that Φ_σ satisfy the same constraint to lowest order in σ if the variations are to preserve the constraint. Thus, (8.4) gives the requirement

$$(8.6)\qquad 0 = \Phi_\sigma^*\omega = \Phi^*\omega + (b_\alpha\delta\phi^\alpha + h_\alpha D\delta\phi^\alpha)d\sigma + \sigma\,\Phi^*\pounds_U\omega + o(\sigma) \ .$$

The fact that U is a horizontal vector field $(u_o = 0)$ and (8.3) show that

$$(8.7)\qquad U\rfloor\omega = b_\alpha u_1^\alpha + h_\alpha u_2^\alpha$$

$$= b_\alpha(t;q^\beta;y^\beta)u_1^\alpha(t;q^\beta) + h_\alpha(t;q^\beta;y^\beta)Z(u_1^\alpha(t;q^\beta)) ,$$

and hence the condition (8.6) can be written as

$$(8.8)\qquad 0 = \Phi^*\omega + \Phi^*(U\rfloor\omega)d\sigma + \sigma\,\Phi^*\pounds_u\omega + o(\sigma) .$$

Since $\Phi^*\omega = 0$ by (8.5), we have established the following result.

If $\Phi: J \to K|\Phi(t) = t$ *satisfies a constraint of the form*

$$(8.9)\qquad \Phi^*\omega = 0 ,$$

where $\omega(t;q^\beta;y^\beta)$ *is a given differential form on* K *, then the variations* $\delta\phi^\alpha = \Phi^*u_1^\alpha(t;\, q^\beta)$ *generated by a horizontal vector field* U *preserve the constraints to lowest order terms in* σ *if and only if*

$$(8.10)\qquad \Phi^*(U\rfloor\omega) = 0 .$$

Careful note should be made of the fact that, although we have constraints of the general form

$$(8.11)\qquad 0 = \Phi^*\omega = a(t;\,\phi^\beta(t);\,D\phi^\beta(t)) + b_\alpha(t;\,\phi^\beta(t);$$

$$D\phi^\beta(t))D\phi^\alpha(t) + h_\alpha(t;\,\phi^\beta(t);\,D\phi^\beta(t))D^2\phi^\alpha(t) ,$$

the corresponding constraints on the variations are

$$(8.12)\qquad 0 = b_\alpha(t;\,\phi^\beta(t);\,D\phi^\beta(t))\delta\phi^\alpha(t) + h_\alpha(t;\,\phi^\beta(t);$$

$$D\phi^\beta(t))D\delta\phi^\alpha(t) .$$

Thus, constraints that are nonlinear in the velocities and acceleration dependent yield constraints on the variations that are *linear functions of* $\{\delta\phi^\alpha\}$, $\{D\delta\phi^\alpha\}$ *and independent of* $\{D^2\delta\phi^\alpha\}$. This result will allow us, in subsequent chapters, to handle significantly more general systems of constraints than have been reported in the literature.

Clearly, this is an important point, so we shall develop the same result from an alternative point of view. Since $\Phi_\sigma(U)$ is obtained from Φ by composition with the trajectories of the horizontal vector field U, and $\Phi = \Phi_0(U)$, a computation to first order terms in σ and use of (8.1) show that

$$(8.13)\qquad \Phi_\sigma^*\omega - \Phi_0^*\omega = \Phi_\sigma^*\omega - \Phi^*\omega = \Phi^*\pounds_{\sigma U}\omega + o(\sigma) .$$

Hence

$$(8.14)\qquad \Phi_\sigma^*\omega = \Phi^*\omega + \Phi^*\left(\sigma\pounds_U\omega + d\sigma \wedge (U \rfloor \omega)\right) + o(\sigma) .$$

Accordingly, the conditions $\Phi^*\omega = 0$, $\Phi_\sigma^*\omega = 0(\sigma)$ give

$$(8.15)\qquad 0 = \Phi^*(U \rfloor \omega)d\sigma + \sigma\, \Phi^*\pounds_U\omega + o(\sigma) ,$$

and we obtain $0 = \Phi^*(U \rfloor \omega)$ to lowest terms in σ. It should be carefully noted that we do not obtain $\Phi_\sigma^*\omega = 0$, but rather

$$(8.16)\qquad \Phi_\sigma^*\omega = \Phi^*\sigma\pounds_U\omega + o(\sigma) = 0(\sigma)$$

under the conditions $\Phi^*\omega = 0$, $\Phi^*(U \rfloor \omega) = 0$. Thus, the varied curve Φ_σ in K does not satisfy the constraint $\Phi_\sigma^*\omega = 0$ except for $\sigma=0$; that is, except on the original curve $\Phi = \Phi_0$. Be this as it may, those readers who are familiar with the classical treatment of dynamical systems with constraints will see that the conditions $\Phi^*\omega = 0$, $\Phi_\sigma^*\omega = 0(\sigma)$ are exactly what are used classically.† From the analytic point of view, there is no choice in the matter due to the incommensurability of $d\sigma$ and σ. Now $\Phi^*(U \rfloor \omega) = 0$ gives the constraint on the variation generated by the horizontal vector field U, while $\Phi_\alpha^*\omega$ gives the variation of the constraint. The above result thus shows that "constraints on

†Pars (Ref. 17).

variations" and "variations of constraints" are essentially distinct constructs.

9. VARIATIONAL PROCESSES

The easiest way of contrasting the results obtained in the last section is to compare them with those that obtain from the more standard variational processes. Let $L(t;\, q^{\alpha};\, y^{\beta})$ be a given function that is defined on all of K and taking its values in $\mathbb{R}$. If $\Phi\colon J \to K \mid \Phi(t) = t$ is a given mapping of J into K, then L gives rise to a function $L(t;\phi^{\alpha}(t);D\phi^{\alpha}(t))$ through the relation

$$(9.1) \qquad \Phi^{*}L = L\,.$$

If U is a horizontal vector field, we define the *variation operator* δ that is generated by U by means of

$$(9.2) \qquad \delta L = \frac{\partial L}{\partial \phi^{\alpha}}\,\delta\phi^{\alpha} + \frac{\partial L}{\partial D\phi^{\alpha}}\,D\delta\phi^{\alpha}\,, \qquad \delta\phi^{\alpha} = \Phi^{*}u_1^{\alpha}\,.$$

A trivial calculation shows that

$$(9.3) \qquad \delta L = \lim_{\sigma\to o}\left[\frac{L(t;\phi^{\beta}+\sigma\delta\phi^{\beta}+o(\sigma);D\phi^{\beta}+\sigma D\delta\phi^{\beta}+o(\sigma))-L(t;\phi^{\beta};D\phi^{\beta})}{\sigma}\right]$$

$$= \lim_{\sigma\to o}\left(\frac{\Phi_{\sigma}^{*}-\Phi_{o}^{*}}{\sigma}\right)L\,.$$

Since $\delta\phi^{\alpha}(t) = u_1^{\alpha}(t;\, \phi^{\beta}(t)) = \Phi^{*}u_1^{\alpha}(t;\, q^{\beta})$ and

$$U = u_1^{\alpha}\,\partial/\partial q^{\alpha} + u_2^{\alpha}\,\partial/\partial y^{\alpha}\,,$$

we have

$$U \rfloor dL = u_1^{\alpha}\frac{\partial L}{\partial q^{\alpha}} + u_2^{\alpha}\frac{\partial L}{\partial y^{\alpha}}\,,$$

so that

$$(9.4) \qquad \delta L = \Phi^{*}(U \rfloor dL)\,.$$

However, $\pounds_U L = U \rfloor dL + d(U \rfloor L) = U \rfloor dL$ and we see that

$$(9.5) \qquad \delta L = \Phi^*(U \rfloor dL) = \Phi^*(\pounds_U L) .$$

Thus, δL, as a function on $J \subset \mathbb{R}$, is obtained from either $U \rfloor dL$ or $\pounds_U L$ on K by the mapping Φ^* from $\Lambda(K)$ to $\Lambda(J)$.

These results can be taken further by noting that any horizontal vector field U has $u_o = 0$. Thus, we see that

$$(9.6) \qquad \delta(Ldt) = \Phi^*(U \rfloor d(Ldt)) = \Phi^*(\pounds_U(Ldt)) ,$$

since $\pounds_U(Ldt) = U \rfloor d(Ldt) + d(U \rfloor Ldt)$ and $U \rfloor Ldt = LU \rfloor dt = Lu_o = 0$. Accordingly, *the variation of* Ldt *is obtained from the action of* Φ^* *on the differential form* $\pounds_U L(t; q^\beta; y^\beta)dt$ on K.

If $\omega(t; q^\beta; y^\beta) \varepsilon \Lambda(K)$ is an arbitrary exterior form on K, then $\pounds_U \omega$ is well defined. However, $\mathbb{R}$ is 1-dimensional and hence any Φ^* that comes from a $\Phi: J \to K$ annihilates all elements of $\Lambda(K)$ except those belonging to $\Lambda^o(K)$ and to $\Lambda^1(K)$; that is Φ^* annihilates all but functions on K and differential forms on K. We can accordingly define the operator δ on all of $\Lambda(K)$ by

$$(9.7) \qquad \delta(\Phi^* \omega) = \Phi^*(\pounds_U \omega) .$$

This allows us to go back and reexamine the constraint question from the standpoint of variational processes. We know that

$$(9.8) \qquad \omega(\sigma) = \omega(0) + \sigma \pounds_U \omega + o(\sigma) ,$$

and hence

$$(9.9) \qquad \Phi^* \omega(\sigma) = \Phi^* \omega(0) + \sigma \Phi^*(\pounds_U \omega(0)) + o(\sigma) .$$

On the other hand, we have shown that

$$(9.10) \qquad \Phi^*_\sigma \omega(0) = \Phi^* \omega(0) + \Phi^*(U \rfloor \omega(0))d\sigma + \sigma \Phi^*(\pounds_U \omega(0)) + o(\sigma) .$$

A comparison of (9.9) and (9.10) under the constraint $\Phi^*\omega(0) = 0$ makes the distinction between "constraints on variations" and "variations of constraints" much clearer. In effect, variations of constraints obtain for constant values of the parameter σ, while constraints of variations account for the change $d\sigma$ that generates $\omega(d\sigma)$ from $\omega(0)$.

The next thing we note is that (A§8)

$$d\pounds_{u}\omega = \pounds_{u}d\omega , \tag{9.11}$$

and hence

$$\Phi^*(d\pounds_{u}\omega) = \Phi^*(\pounds_{u}d\omega) . \tag{9.12}$$

Thus, since $\Phi^*d\omega = D(\Phi^*\omega)dt$, (9.12) says that variation and time differentiation commute:

$$D\delta(\cdot) = \delta D(\cdot) .$$

This leads directly to the basic result

$$\int_a^b \delta(Ldt) = \delta\int_a^b Ldt = \delta\int_a^b \Phi^*(Ldt) = \delta\int_{\Phi} Ldt = \int_{\Phi}\pounds_{u}(Ldt) . \tag{9.13}$$

We now turn to the question of the explicit evaluation of $\delta(Ldt)$. It follows immediately from the definition of δ and $\Phi^*u^1_{\alpha} = \delta\phi^{\alpha}(t)$, $\Phi^*u^{\alpha}_2 = D\delta\phi^{\alpha}(t)$, that

$$\begin{aligned}\delta(Ldt) &= \Phi^*u \rfloor d(Ldt) = \Phi^*\left(\frac{\partial L}{\partial q^{\alpha}}u^{\alpha}_1 dt + \frac{\partial L}{\partial y^{\alpha}}u^{\alpha}_2 dt\right) \\ &= \left(\frac{\partial L}{\partial \phi^{\alpha}}\delta\phi^{\alpha} + \frac{\partial L}{\partial D\phi^{\alpha}}D\delta\phi^{\alpha}\right)dt = \left(\frac{\partial L}{\partial \phi^{\alpha}} - D\frac{\partial L}{\partial D\phi^{\alpha}}\right)\delta\phi^{\alpha}dt \\ &\quad + D\left(\frac{\partial L}{\partial D\phi^{\alpha}}\delta\phi^{\alpha}\right)dt .\end{aligned} \tag{9.14}$$

Thus, if we define the *Euler-Lagrange derivative* of L with respect to ϕ^{α} by

$$(9.15)\qquad \{E|L\}_{\phi^\alpha} = D\left(\frac{\partial L}{\partial D\phi^\alpha}\right) - \frac{\partial L}{\partial \phi^\alpha} ,$$

we have

$$(9.16)\qquad \delta(Ldt) = -\{E|L\}_{\phi^\alpha}\,\delta\phi^\alpha\, dt + D\left(\frac{\partial L}{\partial D\phi^\alpha}\,\delta\phi^\alpha\right) dt .$$

It then follows immediately from (9.15) that

$$(9.17)\qquad \left\{E\Big|\frac{dg(t;\phi^\beta)}{dt}\right\}_{\phi^\alpha} = 0$$

for all functions $\phi^\alpha(t)$ and hence

$$(9.18)\qquad \delta\left(\frac{dg(t;\phi^\beta)}{dt}\,dt\right) = \delta(dg(t;\phi^\beta)) = D\left(\frac{\partial g}{\partial \phi^\alpha}\delta\phi^\alpha\right)dt .$$

Lastly, we note that

$$(9.19)\qquad \{E|aL_1 + bL_2\}_{\phi^\alpha} = a\{E|L_1\}_{\phi^\alpha} + b\{E|L_2\}_{\phi^\alpha}$$

provided a and b are constants, and

$$(9.20)\qquad \{E|L\}_{\phi^\alpha}\, dt = \Phi^*\left(d\left(\frac{\partial \mathcal{L}}{\partial y^\alpha}\right) - \frac{\partial \mathcal{L}}{\partial q^\alpha}\, dt\right) .$$

Thus, the Euler-Lagrange derivative has a fundamental meaning as the Φ^* map of the differential form $d\left(\frac{\partial \mathcal{L}}{\partial y^\alpha}\right) - \frac{\partial \mathcal{L}}{\partial q^\alpha}\, dt$ on K .

10. MAPPING PROPERTIES OF VARIATIONS

The results obtained in the previous Section for variation processes were based upon Kinematic Space being referred to the given global coordinate cover $(t;\, q^\beta;\, y^\beta)$ that obtains from the underlying assumption of the existence of an inertial frame of reference. It is therefore incumbent upon us to examine what happens to general variation processes under admissible mappings $\Gamma\colon K \to {}^{\backprime}K$.

Let $\Gamma\colon K \to {}^{\backprime}K | \Gamma(t) = t$ be an admissible map (preserve t). If

$$(10.1)\qquad U = u_1^\alpha(t;\ q^\beta)\partial/\partial q^\alpha + Z(u_1^\alpha(t;\ q^\beta))\partial/\partial y^\alpha$$

is an arbitrary horizontal vector field on K, the results established in Section 6 give us

$$(10.2)\qquad \Gamma_* U = \grave{u}_o\ \partial/\partial\grave{t} + \grave{u}_1^\alpha\ \partial/\partial\grave{q}^\alpha + \grave{Z}(\grave{u}_1^\alpha - \grave{y}^\alpha \grave{u}_o)\partial/\partial y^\alpha ,$$

where

$$\grave{u}_o(\grave{t};\ \grave{q}^\beta) = u_o \circ \Gamma^{-1} = 0$$

$$\grave{u}_1^\alpha(\grave{t};\ \grave{q}^\beta) = \left(\frac{\partial\grave{q}^\alpha}{\partial t} u_o + \frac{\partial\grave{q}^\alpha}{\partial q^\beta} u_1^\beta\right)\circ \Gamma^{-1} = \left(\frac{\partial\grave{q}^\alpha}{\partial q^\beta} u_1^\beta\right)\circ \Gamma^{-1}$$

$$\grave{u}_2^\alpha(\grave{t};\ \grave{q}^\beta;\ \grave{y}^\beta) = \grave{Z}(\grave{u}_1^\alpha - \grave{y}^\alpha \grave{u}_o) = \grave{Z}(\grave{u}_1^\alpha)$$

$$= \grave{Z}\left[\left(\frac{\partial\grave{q}^\alpha}{\partial q^\beta} u_1^\beta\right)\circ \Gamma^{-1}\right] = \left[Z\left(\frac{\partial\grave{q}^\alpha}{\partial q^\beta} u_1^\beta\right)\right]\circ \Gamma^{-1} .$$

We therefore have the following result.

The image of a horizontal vector field under an admissible map $\Gamma: K \to \grave{K} | \grave{t} = t$, $\grave{q}^\alpha = \gamma^\alpha(t;\ q^\beta)$, $\grave{y}^\alpha = Z(\gamma^\alpha)$ *is given by*

$$(10.3)\qquad \grave{u}_o = 0 ,$$

$$(10.4)\qquad \grave{u}_1^\alpha = \left(\frac{\partial\grave{q}^\alpha}{\partial q^\beta} u_1^\beta(t;\ q^\mu)\right)\circ \Gamma^{-1} ,$$

$$(10.5)\qquad \grave{u}_2^\alpha = \grave{Z}(\grave{u}_1^\alpha) = \left[Z\left(\frac{\partial\grave{q}^\alpha}{\partial q^\beta} u_1^\beta(t;\ q^\mu)\right)\right]\circ \Gamma^{-1} .$$

Admissibility of Γ *thus preserves the condition* $u_o = 0$, *and maps horizontal vector fields onto horizontal vector fields.* We note in passing that a horizontal vector field U for which $u_1^\alpha(t;\ q^\beta) = u^\alpha(t)$ is mapped by an admissible Γ onto a horizontal vector field $\grave{U}$ for which $\grave{u}_1^\alpha(t;\ q^\beta) \neq \grave{u}^\alpha(t)$ in general. Thus, in particular, *an admissible* Γ *does not map a classic variation* $\delta\phi^\alpha(t) = u^\alpha(t)$ *onto a classic variation,* in general. It is perhaps for this reason above all others that we have chosen to replace the classic variation process by the more general variation process introduced in Section 7.

Consider an arbitrary admissible map

$$\text{(10.6)} \qquad \Gamma\colon K \to \grave{K} \,|\, \grave{t} = t \ , \qquad \grave{q}^{\alpha} = \gamma^{\alpha}(t;\, q^{\beta}) \ ,$$

$$\grave{y}^{\alpha} \;=\; Z(\gamma^{\alpha}) \;=\; \left(\frac{\partial}{\partial t} + y^{\beta}\frac{\partial}{\partial q^{\beta}}\right)\gamma^{\alpha}(t;\, q^{\mu}) \ .$$

Any function $L(t;\, q^{\beta};\, y^{\beta})$ defined on K induces the function

$$\text{(10.7)} \qquad \grave{L}(\grave{t};\, \grave{q}^{\beta};\, \grave{y}^{\beta}) \;=\; \overset{-1}{\Gamma}{}^{*}L \;=\; L\circ\overset{-1}{\Gamma} \ ,$$

on $\grave{K}$ and any map $\Phi\colon J \to K|\Phi(t) = t$ induces the map

$$\text{(10.8)} \qquad \grave{\Phi} \;=\; \Gamma\circ\Phi\colon J \to \grave{K}|\grave{\Phi}(\grave{t}) \;=\; \grave{t} \ .$$

We thus have

$$\text{(10.9)} \qquad \grave{\Phi}^{*} \;=\; \Phi^{*}\circ\Gamma^{*} \ .$$

Now, $\delta(Ldt) = \Phi^{*}\{U\rfloor(dL\wedge dt)\}$, and $U\rfloor(dL\wedge dt) = \Gamma^{*}(\grave{U}\rfloor(d\grave{L}\wedge d\grave{t}))$, where $\grave{U} = \Gamma_{*}U$ is given above. A simple combination of these two results gives

$$\delta(Ldt) \;=\; \Phi^{*}\Big[\Gamma^{*}(\grave{U}\rfloor(d\grave{L}\wedge d\grave{t}))\Big] \;=\; \Phi^{*}\circ\Gamma^{*}(\grave{U}\rfloor(d\grave{L}\wedge d\grave{t}))$$

$$\;=\; \grave{\Phi}^{*}(\grave{U}\rfloor(d\grave{L}\wedge d\grave{t})) \;=\; \delta(\grave{L}d\grave{t}) \ ,$$

and we conclude the following:

The variation $\delta(Ldt)$ *is invariant under admissible maps* $\Gamma\colon K \to \grave{K}$; *and we have*

$$\text{(10.10)} \quad \delta(Ldt) \;=\; \Phi^{*}(U\rfloor dL\wedge dt) \;=\; \grave{\Phi}^{*}(\grave{U}\rfloor d\grave{L}\wedge d\grave{t}) \;=\; \delta(\grave{L}d\grave{t})$$

where

$$\text{(10.11)} \qquad \grave{L} = \overset{-1}{\Gamma}{}^{*}L = L\circ\overset{-1}{\Gamma} \ , \quad \grave{\Phi} = \Gamma\circ\Phi \ , \quad \grave{\Phi}^{*} = \Phi^{*}\circ\Gamma^{*} \ ,$$

and $`U = \Gamma_* U$ *is set out by the results given in* (10.3-10.5).

Although the above proof is an almost "1-liner" compared with the classical method of proof of this important result, we still need to obtain the explicit transformation properties of the Euler-Lagrange derivative for future reference. We first note that

$$d`L\wedge d`t \;=\; d`L\wedge dt \;=\; \frac{\partial`L}{\partial`q^\alpha}\, d`q^\alpha\wedge dt + \frac{\partial`L}{\partial`y^\alpha}\, d`y^\alpha\wedge dt \;.$$

Thus, since $`u_o = 0$, we obtain (see (10.4) and (10.5))

$$`U \lrcorner (d`L\wedge d`t) \;=\; \left[\frac{\partial`L}{\partial`q^\alpha}\frac{\partial`q^\alpha}{\partial q^\beta}\, u_1^\beta + \frac{\partial`L}{\partial`y^\alpha}\, Z\!\left(\frac{\partial`q^\alpha}{\partial q^\beta}\, u_1^\beta\right)\right] dt$$

when use is made of (10.3)-(10.5). When (9.16) is used in conjunction with (10.10), we have

$$\delta(Ldt) \;=\; \Phi^*(U\lrcorner(dL\wedge dt)) \;=\; \left[-\{E|L\}_{\phi^\beta}\delta\phi^\beta + D\!\left(\frac{\partial L}{\partial D\phi^\alpha}\,\delta\phi^\alpha\right)\right]dt$$

$$= \;`\Phi^*(`U\lrcorner(d`L\wedge d`t))$$

$$= \;`\Phi^*\left[\frac{\partial`L}{\partial`q^\alpha}\frac{\partial`q^\alpha}{\partial q^\beta}\, u_1^\beta + \frac{\partial`L}{\partial`y^\alpha}\, Z\!\left(\frac{\partial`q^\alpha}{\partial q^\beta}\, u_1^\beta\right)\right]dt$$

$$= \;`\Phi^*\left[\frac{\partial`L}{\partial`q^\alpha}\right]\Phi^*\left[\frac{\partial`q^\alpha}{\partial q^\beta}\, u_1^\beta\right]dt + `\Phi^*\left[\frac{\partial`L}{\partial`y^\alpha}\right]\Phi^*\left[Z\!\left(\frac{\partial`q^\alpha}{\partial q^\beta}\, u_1^\beta\right)\right]dt$$

$$= \;\left[\frac{\partial`L}{\partial`\phi^\alpha}\frac{\partial`q^\alpha(t;\phi)}{\partial\phi^\beta}\,\delta\phi^\beta + \frac{\partial`L}{\partial D`\phi^\alpha}\, D\!\left(\frac{\partial`q^\alpha(t;\phi)}{\partial\phi^\beta}\,\delta\phi^\beta\right)\right]dt$$

$$= \;\left[-\{E|`L\}_{`\phi^\alpha}\frac{\partial`q^\alpha(t;\phi)}{\partial\phi^\beta}\,\delta\phi^\beta + D\!\left(\frac{\partial`L}{\partial D`\phi^\alpha}\frac{\partial`q^\alpha(t;\phi)}{\partial\phi^\beta}\,\delta\phi^\beta\right)\right]dt\;,$$

where

$$(10.12)\qquad \{E|`L\}_{`\phi^\alpha} \;=\; D\!\left(\frac{\partial`L}{\partial D`\phi^\alpha}\right) - \frac{\partial`L}{\partial`\phi^\alpha}\;.$$

We have thus established the following result.

The transformation properties of $\delta(Ldt)$ *under admissible maps is given by*

$$(10.13)\qquad \delta(Ldt) = \left[-\{E|L\}_{\phi^\beta}\delta\phi^\beta + D\left(\frac{\partial L}{\partial D\phi^\alpha}\,\delta\phi^\alpha\right)\right]dt$$

$$= \left[-\{E|\grave{}L\}_{\grave{}\phi^\alpha}\frac{\partial\grave{}q^\alpha}{\partial q^\beta}\,\delta\phi^\beta + D\left(\frac{\partial\grave{}L}{\partial D\grave{}\phi^\alpha}\,\frac{\partial\grave{}q^\alpha}{\partial q^\beta}\,\delta\phi^\beta\right)\right]dt\ .$$

11. EXTENDED VARIATIONS

A more general variation process will be needed in subsequent Chapters. These variations come about through the replacement of horizontal vector fields by arbitrary elements

$$(11.1)\qquad V = v_o\frac{\partial}{\partial t} + v_1^\alpha\frac{\partial}{\partial q^\alpha} + Z(v_1^\alpha - y^\alpha v_o)\frac{\partial}{\partial y^\alpha}$$

of $T(K)$. Composition of Φ with V , rather than with a horizontal vector field U , gives the family of maps

$$(11.2)\qquad \hat{\Phi}_\sigma(V): J\times\mathbb{R} \to K\Big|\Big\{t + \sigma V_o(t,\phi) + o(\sigma);\ \phi^\alpha + \sigma V_1^\alpha(t,\phi) + o(\sigma)\ ;$$

$$D\phi^\alpha + \sigma\left(\frac{dV_1^\alpha(t,\phi)}{dt} - D\phi^\alpha\,\frac{dV_o(t,\phi)}{dt}\right) + o(\sigma)\Big\}\ ,$$

where $V_o(t,\phi) = \Phi^*v_o(t;\ q^\beta)$, $V_1^\alpha(t,\phi) = \Phi^*v_1^\alpha(t;\ q^\beta)$. If we define the *extended variation* of L by

$$(11.3)\qquad \Delta(V)(L) = \lim_{\sigma\to 0}\left(\frac{\hat{\Phi}_\sigma(V) - \hat{\Phi}_o(V)}{\sigma}\right)(L)$$

$$= \frac{\partial L}{\partial t}V_o + \frac{\partial L}{\partial\phi^\alpha}V_1^\alpha + \frac{\partial L}{\partial D\phi^\alpha}(DV_1^\alpha - DV_oD\phi^\alpha)\ ,$$

we obtain

$$(11.4)\qquad \Delta(V)(L) = \Phi^*(V\rfloor dL) = \Phi^*(\pounds_V L)\ .$$

However, $\pounds_V(Ldt) = (\pounds_V L)dt + L(\pounds_V dt) = (\pounds_V L)dt + Ld(V\rfloor dt) = (\pounds_V L)dt + Ldv_o$. We thus define the *extended variation* of (Ldt) by

$$(11.5)\qquad \Delta(V)(Ldt) = \Phi^*(\pounds_V(Ldt))$$

and obtain

$$\Delta(V)(L\,dt) = [\Delta(V)(L)]dt + L\,DV_o dt \ . \tag{11.6}$$

If we now follow exactly the same steps as given in the previous Section, we see that

$$\begin{aligned}\Delta(V)(L\,dt) &= \Bigg[-\{E|L\}_{\phi^\alpha}(V_1^\alpha - V_o D\phi^\alpha) \\ &\quad + D\left(\frac{\partial L}{\partial D\phi^\alpha} V_1^\alpha - \left(\frac{\partial L}{\partial D\phi^\alpha} D\phi^\alpha - L\right)V_o\right)\Bigg]dt \\ &= \Bigg[\frac{\partial L}{\partial t} V_o + \frac{\partial L}{\partial \phi^\alpha} V_1^\alpha + \frac{\partial L}{\partial D\phi^\alpha} DV_1^\alpha \\ &\quad - \left(\frac{\partial L}{\partial D\phi^\alpha} D\phi^\alpha - L\right)DV_o\Bigg]dt \ .\end{aligned} \tag{11.7}$$

CHAPTER II

NEWTONIAN MOTIONS IN KINEMATIC SPACE

12. THE MASS TENSOR AND NEWTONIAN MOTIONS

Let C denote 3N-dimensional Configuration Space, E denote (3N+1)-dimensional Event Space, and K denote (6N+1)-dimensional Kinematic Space. For the time being, we refer K to the global inertial coordinate cover $(t;\ q^\beta;\ y^\beta)$ postulated at the beginning. Accordingly, E is referred to the inertial coordinate cover $(t;\ q^\beta)$ and C is referred to the inertial coordinate cover (q^β).

We assume that ordinary 3-dimensional physical space E_3 is populated with N particles of given masses $m_1,\ m_2,\ \ldots,\ m_N$. The numbers m_i, $i=1,\ldots,N$ can be used to define a symmetric, second order covariant tensor field M on C whose components relative to the inertial coordinate cover (q^β) are given by

$$((m_{\alpha\beta})) = \mathrm{diagonal}(m_1, m_1, m_1, m_2, \ldots, m_{N-1}, m_N, m_N, m_N) \tag{12.1}$$

The tensor M is referred to as the *mass tensor*. It gives rise to a map M of $T(C)\times T(C)$ into $\mathbb{R}$ that is given by

$$M: T(C)\times T(C) \to \mathbb{R} \,|\, T(y^\alpha) = \frac{1}{2}\, y^\alpha\, m_{\alpha\beta}\, y^\beta . \tag{12.2}$$

(Recall that $\{y^\alpha\}$ can be considered as an element of $T(C)$ since factor maps of K arise from the lifts of maps of C and $\{y^\alpha\}$ transforms as a tangent vector in C under factor maps.) The map M can also be considered as defining a map of $T(C)$ into $\Lambda^1(C)$.

Let

$$\Phi: J \to K \,|\, \{\mathrm{id};\ \phi^\alpha(\cdot);\ D\phi^\alpha(\cdot)\} \tag{12.3}$$

be a given map of $J \subset \mathbb{R}$ into K. Φ defines a curve in K whose tangent vector field is given by

$$(12.4)\qquad D\Phi\colon J \to T_{\Phi}(K)|\{1;\ D\phi^{\alpha}(\cdot);\ D^2\phi^{\alpha}(\cdot)\}\ .$$

The mass tensor M can now be used to define the components of the covariant *momentum flux tensor* $\{m_{\alpha\beta}D^2\phi^{\beta}\}$ that is associated with the map $D\Phi$.

Let W be a 1-form on K such that

$$(12.5)\qquad W \wedge dq^1 \wedge dq^2 \wedge \ldots \wedge dq^{3N} = 0\ .$$

It then follows that

$$(12.6)\qquad W = f_{\alpha}(t;\ q^{\beta};\ y^{\beta})dq^{\alpha}\ .$$

We refer to W as a *work* 1-form on K since

$$(12.7)\qquad \Phi^* W = f_{\alpha}(t;\ \phi^{\beta}(t);\ D\phi^{\beta}(t))\ D\phi^{\alpha}(t)\ dt$$

can be interpreted as the differential element of work that is done on the system by the forces

$$(12.8)\qquad f_{\alpha} = \Phi^* f_{\alpha} = f_{\alpha}(t;\ \phi^{\beta}(t);\ D\phi^{\beta}(t))$$

in the course of traversing the curve in K that is defined by Φ.

The above notation allows us to give more exact definition of a Newtonian Motion.

A map $\Phi\colon[a\le t\le b] \to K|\Phi(t) = t;\ \Phi(q^{\alpha}) = \phi^{\alpha}(t);\ \Phi(y^{\alpha}) = D\phi^{\alpha}(t)$ *is a* *Newtonian Motion* *for the forces* $\{\Phi^* f_{\alpha}\}$ *if and only if*

$$(12.9)\qquad m_{\alpha\beta}D^2\phi^{\beta}(t) = f_{\alpha}(t;\ \phi^{\beta}(t);\ D\phi^{\beta}(t))$$

is satisfied for all t *in the interval* $a\le t\le b$.

The conditions given in order that Φ be a Newtonian Motion for $\{\Phi^* f_{\alpha}\}$ are very inconvenient, for they involve components of tensors rather than scalars or scalar-valued differential forms. The next two Sections are devoted to obtaining equivalent formulations of a more usable form.

13. THE FIRST REFORMULATION

We have already seen that the forces $\{f_\alpha = \Phi^* \mathfrak{f}_\alpha\}$ arise naturally from the 1-form

$$(13.1)\qquad \mathcal{W} = \mathfrak{f}_\alpha(t;\ q^\beta;\ y^\beta)\ dq^\alpha$$

on K. Let

$$(13.2)\qquad V(t;\ q^\beta;\ y^\beta) = v_o \frac{\partial}{\partial t} + v_1^\alpha \frac{\partial}{\partial q^\alpha} + Z(v_1^\alpha - y^\alpha v_o)\frac{\partial}{\partial y^\alpha}$$

be an arbitrary element of $T(K)$, then

$$(13.3)\qquad V \rfloor \mathcal{W} = v_1^\alpha(t;\ q^\beta)\mathfrak{f}_\alpha(t;\ q^\beta;\ y^\beta)$$

is an element of $\Lambda^o(K)$ and

$$(13.4)\qquad \Phi^*(V \rfloor \mathcal{W}) = \Phi^*\left(v_1^\alpha(t;\ q^\beta)\right)\mathfrak{f}_\alpha\left(t;\ \phi^\beta(t);\ D\phi^\beta(t)\right) .$$

Let us define the 1-form A on K by

$$(13.5)\qquad A = y^\alpha\ m_{\alpha\beta}\ dy^\beta ,$$

where $m_{\alpha\beta}$ are the components of the mass tensor. It is then a trivial matter to verify that (12.2) and (13.5) yield

$$(13.6)\qquad A = dT$$

and hence A is an exact 1-form on K. Proceeding as above, innermultiplication with V gives

$$(13.7)\qquad V \rfloor A = y^\alpha m_{\alpha\beta} v_2^\beta = y^\alpha m_{\alpha\beta} Z(v_1^\beta - y^\beta v_o)$$

and hence

$$(13.8)\qquad \Phi^*(V \rfloor A) = D\phi^\alpha m_{\alpha\beta}\{DV_1^\beta - D\phi^\beta DV_o\}$$

where

$$(13.9)\qquad V_1^\beta(t) = \Phi^* v_1^\beta = v_1^\beta(t;\phi^\alpha(t)) \;,\quad V_o(t) = v_o(t;\phi^\alpha(t)) \;.$$

Up to this point V has been an arbitrary element of $T(K)$. If we now restrict V to be a horizontal vector field

$$(13.10)\qquad U = u_1^\alpha(t;\, q^\beta)\frac{\partial}{\partial q^\alpha} + Z(u_1^\alpha)\frac{\partial}{\partial y^\alpha} \;,$$

that is $v_o=0$, then $V_o=0$, $\Phi^* u_1^\alpha = \delta\phi^\alpha(t) = V_1^\alpha$, and $\Phi^* u_2^\alpha = D\delta\phi^\alpha(t) = DV_1^\beta - D\phi^\beta DV_o = DV_1^\beta$. Accordingly, (13.9) becomes

$$(13.11)\qquad \Phi^*(U\lrcorner A) = D\phi^\alpha m_{\alpha\beta} D\delta\phi^\beta = D(\delta\phi^\beta m_{\alpha\beta} D\phi^\beta) - \delta\phi^\beta m_{\alpha\beta} D^2\phi^\alpha \;,$$

and we obtain

$$(13.12)\qquad (m_{\alpha\beta}D^2\phi^\beta - f_\alpha(t;\phi^\beta;D\phi^\beta))\delta\phi^\alpha$$
$$= D(\delta\phi^\beta m_{\alpha\beta} D\phi^\alpha) - \Phi^*(U\lrcorner(A+W)) \;.$$

If $\Phi:[a\le t\le b]\to K|\Phi(t)=t$ *is a Newtonian motion for the forces* $\{f_\alpha(t;\phi^\alpha;D\phi^\alpha)\}$ *that define the work 1-form* W, *then*

$$(13.13)\qquad \Phi^*(U\lrcorner(A+W)) = D(\delta\phi^\alpha m_{\alpha\beta} D\phi^\beta) \;,\quad A=dT$$

for all horizontal vector fields U *that give rise to the variations* $\{\delta\phi^\alpha\}$ *by means of* $\Phi^* u_1^\alpha = \delta\phi^\alpha$.

In order to obtain the implication the other way and to eliminate the term $D(\delta\phi^\alpha m_{\alpha\beta} D\phi^\beta)$, we place further restrictions on the horizontal vector fields U and on the map Φ so that we may use the fundamental lemma of the calculus of variations. We therefore assume the map Φ gives a curve in K that connects the point with coordiantes (a^α) in C at time a with the point (b^α) in C at time b:

$$(13.14)\qquad \phi^\alpha(a) = a^\alpha \;,\quad \phi^\alpha(b) = b^\alpha \;.$$

It is then natural to require the variations to preserve these endpoints; that is,

$$u_1^\alpha(a;a^\beta) = u_1^\alpha(b;b^\beta) = 0 \ , \quad \delta\phi^\alpha(a) = \delta\phi^\alpha(b) = 0 \ . \tag{13.15}$$

Let us now integrate both sides of (13.13) with respect to t over the closed interval $[a,b]$. This gives

$$\int_a^b \Phi^*(U \rfloor (A+W))dt = 0 \tag{13.16}$$

for all horizontal vector fields satisfying (13.15) if Φ is a Newtonian motion that satisfies the boundary data (13.14). Conversely, if (13.16) holds for all horizontal vector fields such that $u_1^\alpha(a;a^\beta) = u_1^\alpha(b;b^\beta) = 0$, the fundamental lemma of the calculus of variations shows that Φ gives a Newtonian motion that connects the point (a^β) in C at time $t=a$ with the point (b^β) in C at time $t=b$ †.

A map $\Phi:[a\le t\le b] \to K|\Phi(t)=t$ *is a Newtonian motion that connects the point* $(a^\beta)\varepsilon C$ *at* $t=a$ *with the point* $(b^\beta)\varepsilon C$ *at* $t=b$, *for the forces* $\{f_\alpha(t;\phi^\beta;D\phi^\beta)\}$ *that define the work* 1-*form* W , *if and only if*

$$\int_a^b \Phi^*(U \rfloor (A+W))dt = 0 \ , \quad A=dT \tag{13.17}$$

holds for all horizontal vector fields U *such that*

$$u_1^\alpha(a;\ a^\beta) = u_2^\alpha(b;\ b^\beta) = 0 \ . \tag{13.18}$$

14. THE SECOND REFORMULATION - LAGRANGE'S EQUATIONS

We now obtain a second reformulation whereby the properties of variations established in Section 9 can be used directly. Since

$$A = dT \ , \quad T = \frac{1}{2}\, y^\alpha m_{\alpha\beta} y^\beta \ , \tag{14.1}$$

---

† Simply take $u_1^\alpha(t;a^\beta) = u^\alpha(t)$ to obtain the classic variation process.

it follows that

$$(14.2)\qquad \mathcal{U}\rfloor A \;=\; \mathcal{U}\rfloor dT \;=\; \pounds_{\mathcal{U}}T - d(\mathcal{U}\rfloor T) \;=\; \pounds_{\mathcal{U}}T\ .$$

Accordingly, (9.5) yields

$$(14.3)\qquad \delta\mathrm{T} \;=\; \Phi^*(\pounds_{\mathcal{U}}T) \;=\; \Phi^*(\mathcal{U}\rfloor A)\ ,$$

where

$$(14.4)\qquad \mathrm{T} \;=\; \Phi^*T \;=\; \frac{1}{2}\,D\phi^{\alpha}m_{\alpha\beta}D\phi^{\beta}$$

is the *Kinetic Energy* associated with the map Φ. Further, since $\mathcal{U}$ is horizontal

$$(14.5)\qquad \delta(\mathrm{T}dt) \;=\; \Phi^*\pounds_{\mathcal{U}}(\mathrm{T}dt) \;=\; \Phi^*((\pounds_{\mathcal{U}}T)dt+T\pounds_{\mathcal{U}}dt)$$

$$=\; \Phi^*((\pounds_{\mathcal{U}}T)dt) \;=\; \Phi^*(\pounds_{\mathcal{U}}T)dt \;=\; \Phi^*(\mathcal{U}\rfloor A)dt\ ,$$

and

$$(14.6)\qquad \Phi^*(\mathcal{U}\rfloor(\mathcal{W}\wedge dt)) \;=\; \Phi^*((\mathcal{U}\rfloor\mathcal{W})dt-\mathcal{W}(\mathcal{U}\rfloor dt))$$

$$=\; \Phi^*((\mathcal{U}\rfloor\mathcal{W})dt) \;=\; \Phi^*(\mathcal{U}\rfloor\mathcal{W})dt\ .$$

The condition given by (13.17) *is equivalent to the condition*

$$(14.7)\qquad \int_a^b \Phi^*\left\{\pounds_{\mathcal{U}}(\mathrm{T}dt) + \mathcal{U}\rfloor(\mathcal{W}\wedge dt)\right\} \;=\; 0\ .$$

Thus, satisfaction of (14.7), *for all horizontal vector fields* $\mathcal{U}$ *such that* $u_1^{\alpha}(a;a^{\beta}) = u_1^{\alpha}(b;b^{\beta}) = 0$, *is necessary and sufficient that* Φ *define a Newtonian motion that connects the point* (a^{α}) *at* $t=a$ *with the point* (b^{α}) *at time* $t=b$ *in* C *for the forces associated with* $\mathcal{W}$.

If we set $L = T$ $(\mathrm{L}=\mathrm{T})$ in (9.6) and use (9.16) and (13.1), it follows immediately that (14.7) yields

$$(14.8) \qquad \int_a^b \left[-\{E|T\}_{\phi^\alpha} + f_\alpha(t;\phi^\alpha;D\phi^\alpha)\right]\delta\phi^\alpha dt \;=\; 0 \;,$$

where $\{E|T\}_{\phi^\alpha} = D\left(\frac{\partial T}{\partial D\phi^\alpha}\right) - \frac{\partial T}{\partial \phi^\alpha}$ is the Euler-Lagrange derivative of T that is defined by (9.15). The fundamental lemma of the calculus of variations now shows that (14.8) can hold for all $\delta\phi^\alpha(t)$ such that $\delta\phi^\alpha(a) = \delta\phi^\alpha(b) = 0$ if and only if

$$(14.9) \qquad \{E|T\}_{\phi^\alpha} \;=\; f_\alpha(t;\phi^\alpha;D\phi^\alpha)$$

holds for all t in the interval $[a,b]$. The equations (14.9) are known as *Lagrange's Equations*. These equations are very useful in solving specific problems even though they appear to be in non-invariant form. This is, however, an appearance only, for Lagranges equations arise from the scalar-valued statement (13.17) and hence have an invariant meaning as we show in the next Section.

15. TRANSFORMATION PROPERTIES OF LAGRANGE'S EQUATIONS

Most of the work has already been done, for we have found how $\delta(Tdt) = \Phi^*(\pounds_{U}(Tdt))$ behaves under admissible maps. Let $\Gamma\colon K \to {}^{\backprime}K$ be an admissible map (i.e., a time preserving map). Since Γ_* maps any horizontal vector field U on K into a horizontal vector field ${}^{\backprime}U$ on ${}^{\backprime}K$ with ${}^{\backprime}u_0 = 0$, and $\Gamma(t) = {}^{\backprime}t = t$, it follows that $U \rfloor (W \wedge dt) = \Gamma^*[{}^{\backprime}U \rfloor ({}^{\backprime}W \wedge d{}^{\backprime}t)] = ({}^{\backprime}U \rfloor {}^{\backprime}W)dt = {}^{\backprime}u_1^\beta\, {}^{\backprime}\!f_\beta = {}^{\backprime}u_1^\beta (f_\alpha \circ \overset{-1}{\Gamma}) \frac{\partial q^\alpha}{\partial {}^{\backprime}q^\beta}\, dt$ even though

$$(15.1) \qquad {}^{\backprime}W \;=\; \overset{-1}{\Gamma}{}^* W \;=\; \overset{-1}{\Gamma}{}^*\left(f_\alpha(t;\, q^\beta;\, y^\beta)\, dq^\alpha\right)$$

$$= \; (f_\alpha \circ \overset{-1}{\Gamma}) \frac{\partial q^\alpha}{\partial {}^{\backprime}t}\, dt + (f_\alpha \circ \overset{-1}{\Gamma}) \frac{\partial q^\alpha}{\partial {}^{\backprime}q^\beta}\, d{}^{\backprime}q^\beta \;.$$

Thus, in particular, we have from (10.13) that

$$(15.2)\qquad \int_a^b \Phi^*\left(\pounds_{\mathcal{U}}(T\,dt) + \mathcal{U}\lrcorner(\mathcal{W}\wedge dt)\right)$$

$$= \int_a^b\left[-\{E|L\}_{\phi^\beta}+f_\beta\right]\delta\phi^\beta dt + \int_a^b D\left(\frac{\partial L}{\partial D\phi^\alpha}\,\delta\phi^\alpha\right)dt$$

$$= \int_a^b \left[-\{E|{}^\backprime L\}_{{}^\backprime\phi^\alpha} + {}^\backprime f_\alpha\right]\frac{\partial\, {}^\backprime q^\alpha}{\partial q^\beta}\,\delta\phi^\beta dt + \int_a^b D\left(\frac{\partial\, {}^\backprime L}{\partial D\,{}^\backprime\phi^\alpha}\,\frac{\partial\, {}^\backprime q^\alpha}{\partial q^\beta}\,\delta\phi^\beta\right)dt$$

$$= \int_a^b {}^\backprime\Phi^*\left(\pounds_{{}^\backprime\mathcal{U}}({}^\backprime T\,d\,{}^\backprime t) + {}^\backprime\mathcal{U}\lrcorner({}^\backprime\mathcal{W}\wedge d\,{}^\backprime t)\right) .$$

Observing that $\delta\phi^\alpha(a) = \delta\phi^\alpha(b)$, the third equality in (15.2) and the fundamental lemma of the calculus of variations gives us the following result.

Lagrange's equations transform covariantly under admissible maps Γ *of Kinematic space. In particular, we have*

$$(15.3)\qquad \left[\{E|{}^\backprime T\}_{{}^\backprime\phi^\alpha} - {}^\backprime f_\alpha\right]\Phi^*\left(\frac{\partial\,{}^\backprime q^\alpha}{\partial q^\beta}\right) = \{E|T\}_{\phi^\beta} - f_\beta ,$$

where ${}^\backprime T = T\circ\overset{-1}{\Gamma}$, ${}^\backprime\Phi = \Gamma\circ\Phi$

$$(15.4)\qquad {}^\backprime\mathcal{W} = (\mathit{f}_\alpha\circ\overset{-1}{\Gamma})\frac{\partial q^\alpha}{\partial\,{}^\backprime t}\,d\,{}^\backprime t + (\mathit{f}_\alpha\circ\overset{-1}{\Gamma})\frac{\partial q^\beta}{\partial\,{}^\backprime q^\alpha}\,d\,{}^\backprime q^\alpha ,$$

and

$$(15.5)\qquad {}^\backprime f_\alpha = {}^\backprime\Phi^*\left[(\mathit{f}_\beta\circ\overset{-1}{\Gamma})\frac{\partial q^\beta}{\partial\,{}^\backprime q^\alpha}\right].$$

It is of interest to note that although ${}^\backprime\mathcal{W}$ acquires a term involving $d\,{}^\backprime t$ if Γ is not a factor map (if $\frac{\partial q^\beta({}^\backprime t;{}^\backprime q^\beta)}{\partial\,{}^\backprime t} \neq 0$) , it is only the terms in ${}^\backprime\mathcal{W}$ that involve $d\,{}^\backprime q^\alpha$ that appear in the transformed Lagrange's equations. We also note that the above result generalizes what is reported in the classic literature wherein only factor maps are considered. The inclusion of admissible rather than just factor maps may be paraphrased as follows: *Lagrange's equations retain their covariant transformation proper-*

ties under maps to noninertial frames that preserve time. In this context, we note that (6.7) yields

$$(15.6) \qquad y^{\alpha} = \frac{\partial q^{\alpha}}{\partial \grave{}q^{\beta}}\left(\grave{}y^{\beta} - \frac{\partial \grave{}q^{\beta}}{\partial t}\right)$$

for admissible maps Γ. Thus, since M is a second order covariant tensor on C, the scalar nature of T gives

$$(15.7) \qquad T = \frac{1}{2} y^{\alpha} m_{\alpha\beta} y^{\beta}$$

$$= \frac{1}{2}\left(\grave{}y^{\rho} - \frac{\partial \grave{}q^{\rho}}{\partial t}\right)\frac{\partial q^{\alpha}}{\partial \grave{}q^{\rho}} m_{\alpha\beta} \frac{\partial q^{\beta}}{\partial \grave{}q^{\eta}}\left(\grave{}y^{\eta} - \frac{\partial \grave{}q^{\eta}}{\partial t}\right)$$

$$= \frac{1}{2}\left(\grave{}y^{\rho} - \frac{\partial \grave{}q^{\rho}}{\partial t}\right)\grave{}m_{\rho\eta}\left(\grave{}y^{\eta} - \frac{\partial \grave{}q^{\eta}}{\partial t}\right) = \grave{}T .$$

We thus see that the current formulation correctly accounts for how kinetic energy changes when we transform to new q's by means of time-dependent transformation equations (to noninertial reference frames if the t-dependence is other than linear and additive). This result (15.7) is also what would obtain if we calculate the mapping of

$$\left(\frac{ds}{dt}\right)^2 = \frac{d\phi^{\alpha}}{dt} m_{\alpha\beta} \frac{d\phi^{\beta}}{dt} = 2\Phi^*(T)$$

under $\grave{}q^{\alpha} = \gamma^{\alpha}(t;\, q^{\beta})$ and then compose with Φ:

$$\left(\frac{ds}{dt}\right)^2 = \left(\frac{d\grave{}\phi^{\rho}}{dt} - \frac{\partial \gamma^{\rho}}{\partial t}\right)\grave{}m_{\rho\eta}\left(\frac{d\grave{}\phi^{\eta}}{dt} - \frac{\partial \gamma^{\eta}}{\partial t}\right) = 2\grave{}\Phi^*(\grave{}T) .$$

CHAPTER III

ANALYSIS OF FORCES

The analysis given in the last Chapter shows that the forces enter into our formulation of mechanics through the work 1-form

$$\mathcal{W} = \mathfrak{f}_\alpha(t;\ q^\beta;\ y^\beta)\ dq^\alpha$$

that occurs in the fundamental equation

$$0 = \int_a^b \Phi^*\{\pounds_{\mathcal{U}}(Tdt) + \mathcal{U}\rfloor(\mathcal{W}\wedge dt)\}\ .$$

This chapter is devoted to the analysis and classification of possible systems of forces through the use of the results of the calculus of exterior forms. The results of this calculus are summarized in the Appendix and referenced in the text where appropriate by specific citation of the numerical statements of the appendix. We use the designation (A.§) for such citations.

16. CONSERVATIVE FORCES

We first note that the 1-form of work has the basic structure

$$\mathcal{W} = \mathfrak{f}_\alpha(t;\ q^\beta;\ y^\beta)dq^\alpha\ , \tag{16.1}$$

and this is subsumed by the condition

$$0 = \mathcal{W}\wedge dq^1\wedge dq^2\wedge\ldots\wedge dq^{3N}\ . \tag{16.2}$$

Clearly, the simplest possible situation that can arise is that in which the 1-form $\mathcal{W}$ is closed.

A 1-form $\mathcal{W}$ on K characterizes a system of *Conservative Forces* if and only if

$$(16.3)\qquad \mathcal{W}\wedge dq^1\wedge dq^2\wedge\ldots\wedge dq^{3N} = 0\ ,$$

$$(16.4)\qquad d\mathcal{W} = 0\ ,$$

in which case we say that $\mathcal{W}$ is *conservative*.

Since a conservative $\mathcal{W}$ is closed, it is exact (A§10), and hence there exists a scalar valued function $\hat{V}(t;\ q^\beta;\ y^\beta)$ on K such that

$$(16.5)\qquad \mathcal{W} = -\,d\hat{V} = -\frac{\partial\hat{V}}{\partial t}dt - \frac{\partial\hat{V}}{\partial q^\alpha}dq^\alpha - \frac{\partial\hat{V}}{\partial y^\alpha}dy^\alpha\ .$$

(The minus sign in (16.5) comes about from the history of the subject and is used here in order to simplify comparison with classic treatments.) Substitution of (16.5) into the remaining condition (16.3) gives

$$(16.6)\qquad 0 = \frac{\partial\hat{V}}{\partial t}dt\wedge dq^1\wedge\ldots\wedge dq^{3N} + \frac{\partial\hat{V}}{\partial y^\alpha}dy^\alpha\wedge dq^1\wedge\ldots\wedge dq^{3N}\ .$$

The 3N+1 forms $dt\wedge dq^1\wedge\ldots\wedge dq^{3N}$, $dy^\alpha\wedge dq^1\wedge\ldots\wedge dq^{3N}$, $\alpha=1,\ldots,3N$, being linearly independent (3N+1)-forms shows that the condition (16.6) can be satisfied if and only if

$$(16.7)\qquad \frac{\partial\hat{V}(t;\ q^\beta;\ y^\beta)}{\partial t} = 0\ ,\qquad \frac{\partial\hat{V}(t;\ q^\beta;\ y^\beta)}{\partial y^\beta} = 0\ .$$

Thus $\hat{V} = \hat{V}(q^\beta)$ and we have established the following result.

If $\mathcal{W}$ *is a conservative work 1-form on* K *then there exists a scalar valued function* $\hat{V}(q^\beta)$ *on* C *, called the potential of* $\mathcal{W}$ *, such that*

$$(16.8)\qquad \mathcal{W} = -\,d\hat{V}(q^\beta)\ .$$

When we write out (16.8), we have

$$(16.9)\qquad \mathcal{W} = -\frac{\partial\hat{V}}{\partial q^\alpha}dq^\alpha\ ,$$

and a comparison with (16.1) yields the classic result that the components of a conservative system of forces are the components of a gradient vector on configuration space:

$$(16.10) \qquad \mathcal{f}_\alpha = -\partial\hat{V}(q^\beta)/\partial q^\alpha .$$

Although it seems trivial from the above method of derivation, it should be noted that we have proved rather than assumed that the potential function for a conservative system is independent of time. The more customary method of treating conservative forces is based upon the assumption that the work 1-form is closed as a 1-form on configuration space. Since t does not occur explicitly in the structure of configuration space, an additional assumption has to be made to the effect that the potential is time independent. It is thus clear that $\hat{V} = \hat{V}(q^\beta)$ comes about because we have formulated the problem in Kinematic Space K so that we start with the possibility $\hat{V} = \hat{V}(t;\ q^\beta;\ y^\beta)$ and obtain the explicit conditions (16.7) from the defining equations (16.3), (16.4). It is also clear from (16.8) that $W = W(q^\beta) = \mathcal{f}_\alpha(q^\gamma)dq^\alpha$ is conservative if and only if

$$(16.11) \qquad \frac{\partial \mathcal{f}_\alpha}{\partial q^\beta} = \frac{\partial \mathcal{f}_\beta}{\partial q^\alpha} ,$$

which is the classical result if W assumed, *ad initio*, to be defined on Configuration space.

If W is conservative, then

$$(16.12) \qquad W \wedge dt = -\, d\hat{V}(q^\beta) \wedge dt = -\, d(\hat{V}(q^\beta)dt) ,$$

and hence

$$(16.13) \qquad U \rfloor (W \wedge dt) = -\, U \rfloor d(\hat{V}(q^\beta)dt) = -\, \pounds_U(\hat{V}(q^\beta)dt)$$

for any horizontal vector field U on K. This shows that

$$\int_a^b \Phi^*\{\pounds_U(Tdt) + U \rfloor (W \wedge dt)\} = \int_a^b \Phi^*\{\pounds_U(T-\hat{V}) \wedge dt\} ,$$

and use of the calculations given in Section 9 yields the following results.

A map $\Phi:[a\le t\le b] \to K|\Phi(t)=t,\ \phi(q^\alpha)=\phi^\alpha(t),\ \phi^\alpha(a)=a^\alpha,\ \phi^\alpha(b)=b^\alpha$ *gives a Newtonian motion for a conservative work 1-form*

$$(16.14)\qquad W = -\,d\hat{V}(q^\beta)$$

with potential $\hat{V}(q^\beta)$ *if and only if*

$$(16.15)\qquad 0 = \int_a^b \Phi^*\{\pounds_U(T-\hat{V})\,dt\}$$

holds for all horizontal vector fields U *such that*

$$(16.16)\qquad u_1^\alpha(a;a^\beta) = u_1^\alpha(b;b^\beta) = 0\ .$$

Thus, Φ *must satisfy*

$$(16.17)\qquad 0 = \{E|L\}_{\phi^\alpha}\ ,\quad \alpha=1,\dots,3N\ ,$$

where

$$(16.18)\qquad L = \Phi^*L\ ,\quad L = T-\hat{V}(q^\beta)\ .$$

The function $L(t;\ q^\beta;\ y^\beta) = T(y^\beta) - \hat{V}(q^\beta)$ is called the *Lagrangian* for the given conservative system, $\hat{V}(\phi^\beta) = \Phi^*\hat{V}(q^\beta)$ is called the *potential energy* associated with the conservative forces $F_\alpha(\phi^\beta) = \Phi^*f_\alpha(q^\beta) = -\partial\hat{V}(\phi^\beta)/\partial\phi^\alpha$, and (16.17) are *Lagrange's Equations* for the given conservative system.

The reason why forces of the kind considered above are called conservative is that the equations of motion, (16.17), admit an integral of the energy. In order to see this, we first write out the equations (16.17) by use of equations (9.15) that define the Euler-Lagrange operator $\{E|\cdot\}_{\phi^\alpha}$:

$$(16.19)\qquad 0 = D\left(\frac{\partial L}{\partial D\phi^\alpha}\right) - \frac{\partial L}{\partial\phi^\alpha}\ ,\quad \alpha=1,\dots,3N\ .$$

If we multiply (16.19) by $D\phi^\alpha$ and sum on α , we obtain

$$(16.20)\qquad 0 = D\left(\frac{\partial L}{\partial D\phi^\alpha}\right)D\phi^\alpha - \frac{\partial L}{\partial \phi^\alpha} D\phi^\alpha$$

$$= D\left(\frac{\partial L}{\partial D\phi^\alpha} D\phi^\alpha\right) - \frac{\partial L}{\partial D\phi^\alpha} D^2\phi^\alpha - \frac{\partial L}{\partial \phi^\alpha} D\phi^\alpha + \frac{\partial L}{\partial t} - \frac{\partial L}{\partial t}$$

$$= D\left(\frac{\partial L}{\partial D\phi^\alpha} D\phi^\alpha - L\right) + \frac{\partial L}{\partial t} .$$

However, $T = \Phi^* T = \frac{1}{2} D\phi^\alpha m_{\alpha\beta} D\phi^\beta$ is homogeneous of degree 2 in $\{D\phi^\alpha\}$ and independent of t explicitly, while $\hat{V} = \Phi^*\hat{V}(q^\beta) = \hat{V}(\phi^\beta)$ is also independent of t. Thus $\frac{\partial L}{\partial t} = \frac{\partial}{\partial t}(T-\hat{V}) = 0$, $\frac{\partial L}{\partial D\phi^\alpha} D\phi^\alpha - L = 2T-(T-\hat{V}) = T+\hat{V}$ and (16.20) yields $D(T+\hat{V}) = 0$. We thus have the energy quadrature

$$(16.21)\qquad T + \hat{V} = E_o = \text{constant.}$$

The constant E_o, being the sum of the kinetic energy and the potential energy is referred to as the *total energy*. The following fact is noted for future reference.

Any functions $\{\phi^\alpha(t)\}$ *that satisfy Lagrange's equations*

$$(16.22)\qquad 0 = \{E|L\}_{\phi^\alpha} ,\quad \alpha=1,\ldots,3N ,$$

for a given Lagrangian function $L(t;\, q^\beta;\, y^\beta)$ *are such that they satisfy the differential identity*

$$(16.23)\qquad D\left(\frac{\partial L}{\partial D\phi^\alpha} D\phi^\alpha - L\right) = -\frac{\partial L}{\partial t} ,$$

and

$$(16.24)\qquad E(t;\, \phi^\beta;\, D\phi^\beta) = \frac{\partial L}{\partial D\phi^\alpha} D\phi^\alpha - L$$

is the total energy of the system that obtains from the total energy function

$$(16.25)\qquad E(t;\, q^\beta;\, y^\beta) = \frac{\partial L}{\partial y^\alpha} y^\alpha - L$$

on Kinematic Space, K.

Another result can be obtained in the coordinate system $(t;\ q^{\beta};\ y^{\beta})$ that is very useful in problems in celestial mechanics. If we multiply the equations (16.19) by ϕ^{α} and sum on α, we obtain

$$(16.26)\qquad 0 = \phi^{\alpha} D\left(\frac{\partial L}{\partial D\phi^{\alpha}}\right) - \phi^{\alpha}\frac{\partial L}{\partial \phi^{\alpha}}$$

$$= D\left(\phi^{\alpha}\frac{\partial L}{\partial D\phi^{\alpha}}\right) - \frac{\partial L}{\partial D\phi^{\alpha}} D\phi^{\alpha} - \frac{\partial L}{\partial \phi^{\alpha}}\phi^{\alpha} .$$

Since $L = T - \hat{V} = \frac{1}{2} D\phi^{\alpha} m_{\alpha\beta} D\phi^{\beta} - \hat{V}(\phi^{\beta})$, we have

$$(16.27)\qquad D\left(\phi^{\alpha}\frac{\partial L}{\partial D\phi^{\alpha}}\right) = D(\phi^{\alpha} m_{\alpha\beta} D\phi^{\beta})$$

$$= D^{2}(\tfrac{1}{2}\phi^{\alpha} m_{\alpha\beta}\phi^{\beta}) ,$$

while

$$(16.28)\qquad \frac{\partial L}{\partial \phi^{\alpha}}\phi^{\alpha} + \frac{\partial L}{\partial D\phi^{\alpha}} D\phi^{\alpha} = -\frac{\partial \hat{V}}{\partial \phi^{\alpha}}\phi^{\alpha} + D\phi^{\alpha}\frac{\partial T}{\partial D\phi^{\alpha}}$$

$$= -\frac{\partial \hat{V}}{\partial \phi^{\alpha}}\phi^{\alpha} + 2T .$$

Now

$$(16.29)\qquad I(\phi^{\beta}) = \frac{1}{2}\phi^{\alpha} m_{\alpha\beta}\phi^{\beta}$$

is the *moment of inertia* of the system and thus (16.26)-(16.29) yield the result that

$$(16.30)\qquad D^{2} I(\phi^{\beta}) = 2T - \phi^{\alpha}\frac{\partial \hat{V}}{\partial \phi^{\alpha}} .$$

The energy integral, (16.21) can then be used to obtain

$$(16.31)\qquad D^{2} I(\phi^{\beta}) = 2(E_{o} - \hat{V}) - \phi^{\alpha}\frac{\partial \hat{V}}{\partial \phi^{\alpha}} .$$

Finally, if $\hat{V}$ is a homogeneous function of the ϕ's of degree k, (16.31) and (16.21) yield

$$(16.32)\qquad D^2I(\phi^\beta) = 2E_o - (k+2)\hat{V} = (2+k)T - kE_o ,$$

and $D^2I(\phi^\beta) = 2E_o$ if $k=-2$.

17. VARIATIONAL FORCES

The next simplest case is that in which the exterior form $\mathcal{W}\wedge dt$ is closed.

A 1-form $\mathcal{W}$ on K characterizes a system of *Variational Forces* if and only if

$$(17.1)\qquad \mathcal{W}\wedge dq^1\wedge\ldots\wedge dq^{3N} \doteq 0 ,$$

$$(17.2)\qquad d(\mathcal{W}\wedge dt) = 0 ,$$

in which case $\mathcal{W}$ is referred to as *variational.*

In this instance, it is easier to use (17.1) first, rather than (17.2), to infer the existence of functions $\{f_\alpha(t;\ q^\beta;\ y^\beta)\}$ such that

$$(17.3)\qquad \mathcal{W} = f_\alpha(t;\ q^\beta;\ y^\beta)dq^\alpha .$$

When this is substituted into (17.2), we obtain

$$(17.4)\qquad 0 = df_\alpha\wedge dq^\alpha\wedge dt = \frac{\partial f_\alpha}{\partial q^\beta}dq^\beta\wedge dq^\alpha\wedge dt + \frac{\partial f_\alpha}{\partial y^\beta}dy^\beta\wedge dq^\alpha\wedge dt .$$

Thus, since $dq^\beta\wedge dq^\alpha\wedge dt$, $\beta<\alpha$, and $dy^\beta\wedge dq^\alpha\wedge dt$ are linearly independent, (17.4) can be satisfied if and only if

$$(17.5)\qquad \frac{\partial f_\alpha}{\partial q^\beta} = \frac{\partial f_\beta}{\partial q^\alpha} ,\quad \frac{\partial f_\alpha}{\partial y^\beta} = 0 .$$

These equations in turn can be satisfied if and only if there exists a scalar valued function $h(t;\ q^\beta)$ such that

$$(17.6)\qquad f_\alpha = -\frac{\partial h_\alpha(t;q^\beta)}{\partial q^\alpha}\ .$$

A 1-*form* W *on* K *characterizes a system of variational forces and* W *is variational if and only if there exists a function* $h(t;\, q^\beta)$ *on* E *such that*

$$(17.7)\qquad W = -\frac{\partial h(t;q^\beta)}{\partial q^\alpha}\, dq^\alpha\ ,$$

$$(17.8)\qquad f_\alpha = -\frac{\partial h(t;q^\beta)}{\partial q^\alpha}\ .$$

Variational forces thus have a potential, but the potential is time dependent.

We now obtain the governing equations for variational systems and some of their consequences. It follows directly from (17.7) that

$$(17.9)\qquad W\wedge dt = -\frac{\partial h(t;q^\beta)}{\partial q^\alpha}\, dq^\alpha\wedge dt = -\, dh(t;\, q^\beta)\wedge dt\ ,$$

$$= -\, d(h(t;\, q^\beta)dt)$$

and hence

$$(17.10)\qquad U\rfloor(W\wedge dt) = -\, U\rfloor d(h(t;\, q^\beta)dt)$$

$$= -\, \pounds_U(h(t;\, q^\beta)dt)$$

for any horizontal field U on K . This shows that

$$\int_a^b \Phi^*\{\pounds_U(Tdt) + U\rfloor(W\wedge dt)\}$$

$$= \int_a^b \Phi^*\{\pounds_U(T-h(t;\, q^\beta))dt\}\ ,$$

and hence we obtain the following result.

A map $\Phi:[a\le t\le b]\to K|\Phi(t)=t,\ \ \phi^\alpha(a)=a^\alpha,\ \ \phi^\alpha(b)=b^\alpha$ *gives a Newtonian motion for a variational work form*

$$(17.11)\qquad W = -\frac{\partial h(t;q^\beta)}{\partial q^\alpha}\,dq^\alpha$$

if and only if

$$(17.12)\qquad 0 = \{E|L\}_{\phi^\alpha}\ ,\quad \alpha=1,\dots,3N\ ,$$

where

$$(17.13)\qquad L = \Phi^* L\ ,\quad L = T - h(t;\ q^\beta)\ .$$

The total energy for the system follows directly from (16.24) and (17.13), so that

$$(17.14)\qquad E(t;\ \phi^\alpha;\ D\phi^\alpha) = T + h(t;\ \phi^\alpha)$$

where $h(t;\phi^\alpha) = \Phi^* h(t;q^\alpha)$, while the identity (16.23) yields

$$(17.15)\qquad D(T+h(t;\phi^\alpha)) = \frac{\partial h(t;\phi^\alpha)}{\partial t}\ .$$

Accordingly, variational systems do not possess an energy quadrature unless $\partial h(t;\phi^\alpha)/\partial t = 0$, in which case they are conservative systems. A variational system with $\partial h(t;\phi^\alpha)/\partial t \neq 0$ is a *non-conservative* system.

18. THE CANONICAL REPRESENTATIONS OF DARBOUX

In order to make headway in those cases where the work 1-form is neither conservative nor variational we have to rely on a collection of general results from the calculus of exterior forms that yield standard or canonical representations of 1-forms in terms of their differential structure. This section is devoted to an analysis based on the representation theorem credited to Darboux. The Darboux theorem and its associated computational algorithms are given in Section 9 of the Appendix to which the reader is referred for the details.

Let $W = W(t;\ q^\beta;\ y^\beta)$ be a given work 1-form on K, so that W satisfies

$$(18.1)\qquad W \wedge dq^1 \wedge \ldots \wedge dq^{3N} = 0 .$$

The 1-form W can be used to generate a sequence of exterior forms of increasing degree by

$$(18.2)\qquad I_1 = W , \quad I_2 = dW ;$$

$$I_{2k+1} = I_1 \wedge I_{2k} , \quad I_{2k+2} = dI_{2k+1} = I_2 \wedge I_{2k} .$$

Let $k=k(W)$, $0 \leq k \leq 6N+1 = \dim K$, be the integer (determined by W) such that $I_{k+m} \equiv 0$ for all integers $m>0$ and $I_k \not\equiv 0$ at at least one point of K. The integer $k(W)$ is called the *class* of the 1-form W. The *rank* of W is the largest even integer $2\rho(W) \leq k(W)$. The *index* of W is defined by $\varepsilon(W) = k(W) - 2\rho(W)$ so that $\varepsilon(W)=0$ if $k(W)$ is an even integer, while $\varepsilon(W)=1$ if $k(W)$ is an odd integer. A point P of K is said to be a *regular* point of W if $I_{k(W)}(P) \neq 0$. If $I_{k(W)}(P) = 0$, then P is said to be a *critical* point of W. The Darboux theorem states that there exist $\rho(W)$ scalar valued functions $a_i(t;\ q^\beta;\ y^\beta)$, $i=1,\ldots,\rho$, $\rho(W)$ scalar valued functions $b^i(t;\ q^\beta;\ y^\beta)$, $i=1,\ldots,\rho$, and a scalar valued function $b^{\rho+1}(t;\ q^\beta;\ y^\beta)$ if $\varepsilon(W) \neq 0$ such that[†]

$$(18.3)\qquad W = -\sum_{i=1}^{\rho} a_i(t;\ q^\beta;\ y^\beta)\, db^i(t;\ q^\beta;\ y^\beta) - \varepsilon\, db^{\rho+1}(t;\ q^\beta;\ y^\beta) ,$$

and the functions $\{a_i\}$, $\{b^i\}$, $\{b^{\rho+1}\}$ are functionally independent at each regular point of W. In addition, the functions $\{a_i, b^i\}$ can be computed from the sequence $I_1, I_2, \ldots, I_{k(W)}$ by a

[†]The minus signs that occur in (18.3) are there to preserve historical continuity and to conform with common usage.

systematic use of gradient transformations and similarity transformations as shown in the Appendix.

We now have to cut this general result down to fit work 1-forms that are required to satisfy (18.1). Since the functions $\{a_i, b^i\}$ are functionally independent, a substitution of (18.3) into (18.1) yields

$$0 = db^i(t;\ q^\beta;\ y^\beta) \wedge dq^1 \wedge dq^2 \wedge \ldots \wedge dq^{3N}, \quad i = \begin{cases} 1,\ldots,\rho \text{ for } \varepsilon=0 \\ 1,\ldots,\rho+1 \text{ for } \varepsilon=1 \end{cases}$$

and hence we obtain the results

$$(18.4) \qquad b^i = b^i(q^\beta), \quad i = \begin{cases} 1,\ldots,\rho(W)\leq 3N & \text{for } \varepsilon(W)=0 \\ 1,\ldots,\rho(W)+1\leq 3N & \text{for } \varepsilon(W)=1 . \end{cases}$$

This establishes the following basic result.

Any work 1-form W *on* K *, of class* $k(W)$ *, rank* $\rho(W)$ *and index* $\varepsilon(W)$ *, can be represented at regular points of* K *in terms of* $k(W)$ *functionally independent functions* $a_i(t;\ q^\beta;\ y^\beta)$, $i=1,\ldots,\rho(W)$, $b^i(q^\beta)$, $i=1,\ldots,\rho(W)+\varepsilon(W)$ *in the form*

$$(18.5) \qquad W = -\sum_{i=1}^{\rho(W)} a_i(t;\ q^\beta;\ y^\beta)\,db^i(q^\beta) - \varepsilon(W)\,db^{\rho(W)+1}(q^\beta) ,$$

and $\rho(W)+\varepsilon(W) \leq 3N$.

Please note that W can have any class between 0 and 6N+1 even though the b's depend only upon the q's , for the class is determined not only by the db's , but also by the da's . If P is not a regular point of W , then P is a critical point of W and some of the b's and a's become functionally dependent at such points; but we still have (18.5). The reason why such points are called critical points can be made clear from the observation that a conservative W is a work 1-form with $k(W)=\varepsilon(W)=1$, $\rho(W)=0$, $W=-db^1(q^\beta)$, for $I_1=W\not\equiv 0$, $I_2=dW\equiv 0$. Thus, a point P where $I_1=W=0$ is a point where $db^1(q^\beta)=0$; a critical point of the potential function $b^1(q^\beta)=\hat{V}(q^\beta)$. Critical points for $k(W)>1$ can be shown to have similar classical interpretations.

The Darboux representation, (18.5), can be combined with

$$W = f_\alpha(t;\, q^\beta;\, y^\beta)\,dq^\alpha$$

to obtain the representation

$$(18.6)\qquad f_\alpha = \sum_{i=1}^{\rho(W)} a_i(t;\, q^\beta;\, y^\beta)\frac{\partial b^i(q^\beta)}{\partial q^\alpha} + \varepsilon(W)\frac{\partial b^{\rho(W)+1}(q^\beta)}{\partial q^\alpha}$$

for the forces associated with W. Accordingly, we see that any t or y^α dependence of the forces comes about multiplicatively rather than differentially in this representation. The various cases can be distinguished, however, by dW, for (18.5) shows that

$$dW \wedge dq^1 \wedge \ldots \wedge dq^{3N} = 0 \leftrightarrow a_i = a_i(q^\beta)\ ,$$

$$(18.7)\qquad dW \wedge dt \wedge dq^1 \wedge \ldots \wedge dq^{3N} = 0 \leftrightarrow a_i = a_i(t;\, q^\beta)\ ,$$

$$dW \wedge dq^1 \wedge \ldots \wedge dq^{3N} \wedge dy^1 \wedge \ldots \wedge dy^{3N} = 0 \leftrightarrow a_i = a_i(a^\beta;\, y^\beta)\ .$$

We call any work 1-form W with class $k(W)>1$ a *nonconservative* work form, and the forces associated with such work 1-forms *nonconservative* forces.

It is essential to note that the functions $\{a_i, b^i\}$ that enter into the representation (18.5) are not uniquely determined by W. In addition, the form of the right-hand side of (18.5) is not unique if $k(W)>1$. This follows by simply noting that

$$a_1 db^1 = d(a_1 b^1) - b^1 da_1\ .$$

Thus, in particular, (18.5) can be written in the equivalent form

$$(18.8)\qquad W = -\,d\left\{\varepsilon(W)b^{\rho+1}(q^\beta) + \sum_{i=1}^{\rho(W)} a_i(t;\, q^\beta;\, y^\beta)b^i(q^\beta)\right\}$$

$$+ \sum_{i=1}^{\rho(W)} b^i(q^\beta)\,da_i(t;\, q^\beta;\, y^\beta)\ ,$$

with a different partitioning of the total differentials. However, the representation of the forces are given by (18.6) remains invariant under such changes in the representation of $\mathcal{W}$.

19. MOTIONS UNDER NONCONSERVATIVE FORCES

Now that we have specific representations for the forces, it is a straightforward matter to obtain the equations of motion. We first note that (18.5) yields

$$(19.1)\qquad \Phi^*(\mathcal{U}\rfloor(\mathcal{W}\wedge dt)) = -\varepsilon(\mathcal{W})\Phi^*(\mathcal{U}\rfloor db^{\rho(\mathcal{W})+1}\wedge dt) - \sum_{i=1}^{\rho(\mathcal{W})}\Phi^*(a_i)\Phi^*(\mathcal{U}\rfloor db^i\wedge dt)$$
$$= -\Phi^*\pounds_{\mathcal{U}}(b^{\rho(\mathcal{W})+1}dt) - \sum_{i=1}^{\rho(\mathcal{W})}\Phi^*(a_i\pounds_{\mathcal{U}}(b^i dt)) .$$

The fundamental condition

$$0 = \int_a^b \Phi^*(\pounds_{\mathcal{U}}(Tdt) + \mathcal{U}\rfloor(\mathcal{W}\wedge dt))$$

thus becomes

$$(19.2)\qquad 0 = \int_a^b \Phi^*\left\{\pounds_{\mathcal{U}}((T-\varepsilon(\mathcal{W})b^{\rho(\mathcal{W})+1})dt)) - \sum_{i=1}^{\rho(\mathcal{W})} a_i\pounds_{\mathcal{U}}(b^i dt)\right\} .$$

If we use the identity $\pounds_{\mathcal{U}}(ab) = a\pounds_{\mathcal{U}}b + b\pounds_{\mathcal{U}}a$, we can obtain the equivalent fundamental condition

$$(19.3)\qquad 0 = \int_a^b \Phi^*\left\{\pounds_{\mathcal{U}}\left((T-\varepsilon(\mathcal{W})b^{\rho(\mathcal{W})+1} - \sum_{i=1}^{\rho(\mathcal{W})} a_i b^i)dt\right) + \sum_{i=1}^{\rho(\mathcal{W})} b^i\pounds_{\mathcal{U}}(a_i dt)\right\} ,$$

that corresponds with the representation (18.8). Since (19.2) and (19.3) are equivalent, we base our analysis on (19.2).

We know from the results established in Chapter II that (19.2) will be satisfied for all horizontal vector fields $\mathcal{U}$ with $u_\alpha^1(a;a^\beta) = u_\alpha^1(b;b^\beta) = 0$ if and only if

$$(19.4)\qquad \{E|L\}_{\phi^\alpha} = -\sum_{i=1}^{\rho(W)} a_i(t;\ \phi^\beta;\ D\phi^\beta)\frac{\partial b^i(\phi^\beta)}{\partial\phi^\alpha}\ ,\quad \alpha=1,\ldots,3N\ ,$$

where the Lagrangian for the system is

$$(19.5)\qquad L = T - \varepsilon(W)b^{\rho(W)+1}(q^\beta)$$

and

$$(19.6)\qquad \mathrm{L} = \Phi^* L = \mathrm{T} - \varepsilon(W)b^{\rho(W)+1}(\phi^\beta)\ .$$

Thus, (19.4) are Lagrange's equations for nonconservative systems and we have established the following result.

A map $\Phi\colon [a\le t\le b] \to K|\Phi(t) = t\ ,\ \phi^\alpha(a) = a^\alpha\ ,\ \phi^\alpha(b) = b^\alpha$ *is a Newtonian motion for a work* 1-*form*

$$(19.7)\qquad W = -\sum_{i=1}^{\rho(W)} a_i(t;\ q^\beta;\ y^\beta)db^i(q^\beta) - \varepsilon(W)db^{\rho(W)+1}(q^\beta)$$

of rank $2\rho(W)$ *and index* $\varepsilon(W)$ *if and only if*

$$(19.8)\qquad \{E|L\}_{\phi^\alpha} = -\sum_{i=1}^{\rho(W)} a_i(t;\ \phi^\beta;\ D\phi^\beta)\frac{\partial b^i(q^\beta)}{\partial q^\alpha}\ ,\quad i=1,\ldots,3N\ ,$$

where

$$(19.9)\qquad \mathrm{L} = \Phi^* L\ ,\quad L = T - \varepsilon(W)b^{\rho(W)+1}(q^\beta)\ .$$

If we multiply (19.8) by $D\phi^\alpha$ and sum on α, calculations similar to those made in the previous two Sections show that

$$(19.10)\qquad D(\mathrm{T}+\varepsilon(W)b^{\rho(W)+1}(\phi^\beta)) = -\sum_{i=1}^{\rho(W)} a_i(t;\ \phi^\beta;\ D\phi^\beta)Db^i(\phi^\beta)\ ,$$

and hence the total energy for the system is given by

$$(19.11)\qquad E(t;\ \phi^\beta;\ D\phi^\beta) = \mathrm{T} + \varepsilon(W)b^{\rho(W)+1}(\phi^\beta)\ .$$

The term "nonconservative" for systems with $k(W)>1$ is thus legitimate since (19.10) shows that such systems do not, in general, admit an energy quadrature.

Of the various kinds of nonconservative systems that can

arise, the most important are those that reflect the fact that interactions in the real world are governed by the second law of thermodynamics. We therefore distinguish two classes of nonconservative systems:

A nonconservative system is said to be *dissipative* if and only if

$$DE(t;\ \phi^{\beta};\ D\phi^{\beta}) \ \leq \ 0\ . \tag{19.12}$$

A nonconservative system is said to be *strictly dissipative* if and only if

$$DE(t;\ \phi^{\beta};\ D\phi^{\beta}) \ < \ 0 \quad \text{for} \quad T>0\ . \tag{19.13}$$

A combination of (19.10) and (19.11) show that *the nonconservative systems considered in this section are dissipative if and only if*

$$\sum_{i=1}^{\rho(W)} a_i(t;\ \phi^{\beta};\ D\phi^{\beta}) \frac{\partial b^i(\phi^{\beta})}{\partial \phi^{\alpha}} D\phi^{\alpha} \ \geq \ 0 \tag{19.14}$$

and are strictly dissipative if and only if

$$\sum_{i=1}^{\rho(W)} a_i(t;\ \phi^{\beta};\ D\phi^{\beta}) \frac{\partial b^i(\phi^{\beta})}{\partial \phi^{\alpha}} D\phi^{\alpha} \ > \ 0 \quad \text{for} \quad T>0\ . \tag{19.15}$$

Suppose that we have a strictly dissipative system for which $E(t;\ \phi^{\alpha};\ D\phi^{\alpha})$ is bounded from below. Since $E = T+\varepsilon(W)b^{\rho(W)+1}(\phi^{\beta})$ and $T\geq 0$, this will always be the case when $\varepsilon(W) = 0$, and will be satisfied for $\varepsilon(W) = 1$ when $b^{\rho(W)+1}(\phi^{\beta})$ is bounded from below as a function on configuration space, C. If the equations of motion, (19.8) have solutions for arbitrarily large values of t, an integration of the inequality

$$DE(t;\ \phi^{\beta};\ D\phi^{\beta}) \ < \ 0 \quad \text{for} \quad T>0 \tag{19.16}$$

leads to a contradiction to the assumption that $E(t;\ \phi^{\beta};\ D\phi^{\beta})$ is bounded from below unless

$$(19.17)\qquad \lim_{t\to\infty} DE(t;\ \phi^\beta(t);\ D\phi^\beta(t)) = 0 = -\lim_{t\to\infty}\sum_{i=1}^{\rho} a_i Db^i .$$

If this condition is met, then we also have

$$\lim_{t\to\infty} T(\phi^\alpha(t);\ D\phi^\alpha(t)) = \lim_{t\to\infty}\left(\frac{1}{2}\, D\phi^\alpha m_{\alpha\beta} D\phi^\beta\right) = 0 ,$$

because $DE<0$ for $T>0$. Now, $\det(m_{\alpha\beta}) \neq 0$ and hence we see that

$$(19.18)\qquad \lim_{t\to\infty} D\phi^\alpha(t) = 0 .$$

Further, since $-\sum_{i=1}^{\rho} a_i Db^i$ is the power developed by the nonconservative part of the forces, we see that *the "large time limit" of the "nonconservative power" is equal to zero for strictly dissipative systems.*

20. THE DISSIPATIVE AND NONDISSIPATIVE PARTS OF $\mathcal{W}$

We have just seen that the nonconservative aspects of a 1-form

$$(20.1)\qquad \mathcal{W} = -\sum_{i=1}^{\rho(\mathcal{W})} a_i(t;\ q^\beta;\ y^\beta)\, db^i(q^\beta) - \varepsilon(\mathcal{W})\, db^{\rho(\mathcal{W})+1}(q^\beta) ,$$

are contained in the *nonpotential* part

$$(20.2)\qquad \bar{\mathcal{W}} = \mathcal{W} + \varepsilon(\mathcal{W})\, db^{\rho(\mathcal{W})+1} = -\sum_{i=1}^{\rho(\mathcal{W})} a_i(t;\ q^\beta;\ y^\beta)\, db^i(q^\beta)$$

of $\mathcal{W}$. Further analysis of $\bar{\mathcal{W}}$ would therefore be useful in obtaining a fuller understanding of nonconservative 1-forms; in particular, we need to clarify whether or not "velocity dependent forces" always preclude a quadrature of the energy equation

$$(20.3)\qquad D(T + \varepsilon(\mathcal{W})\, b^{\rho(\mathcal{W})+1}) = -\sum_{i=1}^{\rho(\mathcal{W})} a_i(t;\ \phi^\beta;\ D\phi^\beta)\, Db^i(\phi^\beta) .$$

Define the scalar-valued function $R(t;\ q^\beta;\ y^\beta)$ by

$$(20.4)\qquad R(t;\ q^\beta;\ y^\beta) = \sum_{i=1}^{\rho(\mathcal{W})} \frac{\partial b^i(q^\beta)}{\partial q^\alpha}\, y^\alpha \int_0^1 a_i(t;\ q^\beta;\ \lambda y^\beta)\, d\lambda .$$

It is then a straightforward matter to show that

$$(20.5)\qquad \sum_{i=1}^{\rho(W)} a_i \frac{\partial b^i}{\partial q^\alpha} = \frac{\partial R}{\partial y^\alpha} + N_\alpha ,$$

where

$$(20.6)\qquad N_\alpha(t;\, q^\beta;\, y^\beta) = \int_0^1 y^\gamma \left\{ \frac{\partial b^i(q^\mu)}{\partial q^\alpha} \frac{\partial a_i(t;q^\beta;\lambda y^\beta)}{\partial(\lambda y^\gamma)} - \frac{\partial b^i(q^\mu)}{\partial q^\gamma} \frac{\partial a_i(t;q^\mu;\lambda y^\mu)}{\partial(\lambda y^\alpha)} \right\} \lambda d\lambda .$$

Thus, R and N_α have the following properties:

$$(20.7)\qquad R(t;\, q^\beta;\, 0) = 0 , \quad N_\alpha(t;\, q^\beta;\, 0) = 0 ,$$

$$(20.8)\qquad y^\alpha N_\alpha(t;\, q^\beta;\, y^\beta) = 0 .$$

Now (20.2), (20.4) show that

$$(20.9)\qquad \bar{W} = -\left(\frac{\partial R}{\partial y^\alpha} + N_\alpha\right) dq^\alpha ,$$

and hence (20.1) and (20.2) yield

$$(20.10)\qquad W = -\left[\varepsilon(W) \frac{\partial b^{\rho(W)+1}}{\partial q^\alpha} + \frac{\partial R}{\partial y^\alpha} + N_\alpha\right] dq^\alpha .$$

The forces associated with W thus become

$$(20.11)\qquad \mathcal{f}_\alpha = -\left[\varepsilon(W) \frac{\partial b^{\rho(W)+1}}{\partial q^\alpha} + \frac{\partial R}{\partial y^\alpha} + N_\alpha\right] .$$

It is easily seen from combining (19.8) and (20.4) that the equations of motion become

$$(20.12)\qquad \{E|L\}_{\phi^\alpha} = -\frac{\partial R(t;\phi^\beta;D\phi^\beta)}{\partial D\phi^\alpha} - N_\alpha(t;\, \phi^\beta;\, D\phi^\beta) ,$$

where

$$(20.13)\qquad L = \Phi^*(T - \varepsilon(W) b^{\rho(W)+1}) , \quad R = \Phi^* R, \quad N_\alpha = \Phi^* N_\alpha .$$

On the other hand, since (20.8) gives

(20.14) $$0 = \Phi^*(y^\alpha N_\alpha) = N_\alpha(t;\ \phi^\beta;\ D\phi^\beta) D\phi^\alpha ,$$

the energy equation, (20.3) becomes

(20.15) $$D(T+\varepsilon(W) b^{\rho(W)+1}) = -\frac{\partial R(t;\phi^\beta;D\phi^\beta)}{\partial D\phi^\alpha} D\phi^\alpha ;$$

the terms $\{N_\alpha(t;\ \phi^\beta;\ D\phi^\beta)\}$ *do not contribute to the energy equation of a nonconservative system.* We also note that *a nonconservative system with* $\partial R/\partial D\phi^\alpha = 0$ *admits an energy quadrature* $T+\varepsilon(W)b^{\rho(W)+1}$ = constant *even though* $\{E|T-\varepsilon(W)b^{\rho(W)+1}\}_{\phi^\alpha} \neq 0$ *and the forces* $\{f_\alpha\}$ *depend on the velocity variables* $\{D\phi^\beta\}$.

If $N_\alpha(t;\ q^\beta;\ y^\beta) = 0$, we obtain

(20.16) $$\{E|L\}_{\phi^\alpha} = -\frac{\partial R(t;\phi^\beta;D\phi^\beta)}{\partial D\phi^\alpha} ,$$

so that we can identify $R = \Phi^* R$ with the *Rayleigh dissipation function* of classic analytical dynamics. It is then natural, in view of (20.15), to refer to $\{N_\alpha(t;\ \phi^\beta;\ D\phi^\beta)\}$ as the *nondissipative part* of the nonconservative forces $\{f_\alpha(t;\ \phi^\beta;\ D\phi^\beta)\}$ = $\Phi^*\{f_\alpha(t;\ q^\beta;\ y^\beta)\}$.

The decomposition (20.5) is particularly useful in connection with characterizing dissipative and strictly dissipative systems, for the definitions of such systems and (20.15) show that *the system is dissipative if and only if*

(20.17)a $$\frac{\partial R(t;\phi^\beta;D\phi^\beta)}{\partial D\phi^\alpha} D\phi^\alpha \geq 0 ,$$

and is strictly dissipative if and only if

(20.17)b $$\frac{\partial R(t;\phi^\beta;D\phi^\beta)}{\partial D\phi^\alpha} D\phi^\alpha > 0 \quad \text{for} \quad T>0 .$$

Accordingly, dissipative systems obtain in those cases where the Rayleigh dissipation function can be represented by

(20.18) $$R(t;\ q^\beta;\ y^\beta) = \int_0^1 r(t;\ q^\beta;\ \lambda y^\beta)\frac{d\lambda}{\lambda}$$

with

$$(20.19)\qquad r(t;\, q^\beta;\, y^\beta) \geq 0\,, \quad r(t;\, q^\beta;\, 0) = 0\,,$$

while strictly dissipative systems obtain in those cases where (20.18) holds with

$$(20.20)\qquad r(t;\, q^\beta;\, y^\beta) > 0 \text{ for } T>0\,, \quad r(t;\, q^\beta;\, 0) = 0\,.$$

Suppose that we have a nonconservative system where $f_\alpha = f_\alpha(q^\beta)$, in which case $\frac{\partial f_\alpha}{\partial q^\beta} \neq \frac{\partial f_\beta}{\partial q^\alpha}$ for at least one point in C. This gives

$$(20.21)\qquad -W = \sum_{i=1}^{\rho} a_i(q^\beta)db^i(q^\beta) + \varepsilon(W)db^i(q^\beta)$$

and (20.4) - (20.6) yield

$$(20.22)\qquad -R(t;\, q^\beta;\, y^\beta) = \sum_{i=1}^{\rho(W)} y^\alpha \frac{\partial b^i(q^\beta)}{\partial q^\alpha} a_i(q^\beta)\,,$$

$$N_\alpha(t;\, q^\beta;\, y^\beta) = 0\,.$$

It then follows that

$$(20.23)\qquad \frac{\partial R}{\partial D\phi^\alpha} D\phi^\alpha = -\sum_{i=1}^{\rho(W)} a_i(\phi^\beta) \frac{\partial b^i(\phi^\beta)}{\partial q^\alpha} D\phi^\alpha$$

can be made negative for any $\{\phi^\alpha(t)\}$ by an appropriate assignment of $\{D\phi^\alpha(t)\}$. Thus, *a nonconservative system of forces with C as domain can not be dissipative for all motions of the system; it is always possible to increase the total energy of the system in a neighborhood of the initial data by an appropriate choice of the initial data.*

We consider the example of forces that are linear in the velocities since these are important in applications:

$$(20.24)\qquad f_\alpha = -\frac{\partial \hat{V}(q^\mu)}{\partial q^\alpha} - A_{\alpha\beta}(q^\mu)y^\beta\,.$$

In this instance, (20.4) and (20.6) yield

$$(20.25)\qquad R = \tfrac{1}{4}(A_{\alpha\beta}(q^{\mu}) + A_{\beta\alpha}(q^{\mu}))y^{\alpha}y^{\beta} ,$$

$$(20.26)\qquad N_{\alpha} = \tfrac{1}{2}(A_{\alpha\beta}(q^{\mu}) - A_{\beta\alpha}(q^{\mu}))y^{\beta} ,$$

and the energy equation yields

$$(20.27)\qquad D(T+\hat{V}) = - (A_{\alpha\beta}(\phi^{\mu}) + A_{\beta\alpha}(\phi^{\mu}))D\phi^{\alpha}D\phi^{\beta} .$$

Such systems thus possess an energy quadrature

$$(20.28)\qquad T + \hat{V} = \text{constant}$$

provided

$$(20.29)\qquad A_{\alpha\beta}(\phi^{\mu}) = - A_{\beta\alpha}(\phi^{\mu}) ,$$

in which case the equations of motion become

$$(20.30)\qquad \{E|T-\hat{V}\}_{\phi^{\alpha}} = - \tfrac{1}{2}(A_{\alpha\beta}(\phi^{\mu}) - A_{\beta\alpha}(\phi^{\mu}))D\phi^{\beta}$$

$$= - A_{\alpha\beta}(\phi^{\mu})D\phi^{\beta} .$$

Finally, since (20.29) holds, we can find a 1-form $A = A(q^{\mu}) = A_{\alpha}(q^{\mu})dq^{\alpha}$ such that $dA = A_{\beta\alpha}dq^{\beta} \wedge dq^{\alpha}$ and (20.30) become

$$(20.31)\qquad \{E|T-\hat{V}\}_{\phi^{\alpha}} = - \frac{1}{2}\left(\frac{\partial A_{\beta}}{\partial\phi^{\alpha}} - \frac{\partial A_{\alpha}}{\partial\phi^{\beta}}\right)D\phi^{\beta} .$$

21. THE UNIQUE POTENTIAL REPRESENTATION

The Darboux representation, although very useful, has a glaring weakness, namely, the nonuniqueness of the functions $\{a_i(t;\, q^{\beta};\, y^{\beta}),\ b^i(q^{\beta})\}$. The assignment of the total energy by means of $E = T+\varepsilon(W)b^{\rho(W)+1}(\phi^{\beta})$ may thus be spurious. This nonuniqueness can be eliminated altogether if we use the representation of 1-forms that comes about by use of the identity $dH(\cdot) + Hd(\cdot) = id(\cdot)$. Since this decomposition theorem and its consequences are given a full treatment in Section 10 of the Appendix, we simply extract the salient results here and apply them to work 1-forms.

Let $\{t_o, q_o^\alpha, y_o^\alpha\}$ be a given point of K. The decomposition theorem shows that there exists, for each given $\mathcal{W}\varepsilon\Lambda^1(K)$, one and only one scalar valued function $p(t; q^\beta; y^\beta)$ and one and only one 1-form

(21.1) $$\mathcal{Q}(t; q^\beta; y^\beta) = \mathcal{Q}^o(t; q^\beta; y^\beta)dt + \mathcal{Q}_\alpha^1(t; q^\beta; y^\beta)dq^\alpha + \mathcal{Q}_\alpha^2(t; q^\beta; y^\beta)dy^\beta$$

such that

(21.2) $$p(t_o; q_o^\beta; y_o^\beta) = 0, \quad \mathcal{Q}(t_o; q_o^\beta; y_o^\beta) = 0,$$

(21.3) $$\mathcal{Q}^o(t; q^\beta; y^\beta)(t-t_o) + \mathcal{Q}_\alpha^1(t; q^\beta; y^\beta)(q^\alpha-q_o^\alpha) + \mathcal{Q}_\alpha^2(t; q^\beta; y^\beta)(y^\beta-y_o^\beta) = 0,$$

(21.4) $$\mathcal{W}(t; q^\beta; y^\beta) = -dp(t; q^\beta; y^\beta) - \mathcal{Q}(t; q^\beta; y^\beta).$$

The quantities $p(t; q^\beta; y^\beta)$, $\mathcal{Q}(t; q^\beta; y^\beta)$ are determined in terms of $\mathcal{W}$ by (21.4) and

(21.5) $$p(t; q^\beta; y^\beta) = -\int_o^1 \{(t-t_o)\tilde{\omega}^o(\lambda) + (q^\alpha-q_o^\alpha)\tilde{\omega}^1(\lambda) + (y^\alpha-y_o^\alpha)\tilde{\omega}_\alpha^2(\lambda)\}d\lambda,$$

where

(21.6) $$\mathcal{W}(t; q^\beta; y^\beta) = \omega^o(t; q^\beta; y^\beta)dt + \omega_\alpha^1(t; q^\beta; y^\beta)dq^\alpha + \omega_\alpha^2(t; q^\beta; y^\beta)dy^\alpha,$$

and

$$(21.7) \qquad \tilde{W}(\lambda) = W(\lambda t+(1-\lambda)t_o;\ \lambda q^\alpha+(1-\lambda)q_o^\alpha;\ \lambda y^\alpha+(1-\lambda)y_o^\alpha) .$$

The function $p(t;\ q^\beta;\ y^\beta)$ may thus be interpreted, legitimately, as a potential energy since (21.5) shows that p is the negative of the generalized work done in the linear process $\lambda P+(1-\lambda)P_o$ that connects the points P and P_o on K. In terms of the homotopy operator H introduced in Section 10 of the Appendix, (21.5) becomes $p = -H(W)$. In this context, we have $Q = -H(dW)$, and (21.4) becomes $W = dH(W) + H(dW)$. We also note that (21.4) implies

$$(21.8) \qquad dW = - dQ$$

and that the minus signs in (21.4) have been introduced in order to preserve the interpretation of $p(t;\ q^\beta;\ y^\beta)$ as a generalized potential function for the forces.

We now cut down this general result so that it applies to work 1-forms that satisfy

$$(21.9) \qquad W \wedge dq^1 \wedge \ldots \wedge dq^{3N} = 0 .$$

Substitution of (21.4) into (21.9) and using (21.1), we obtain

$$0 = \left(\frac{\partial p}{\partial t} + Q^o\right)dt \wedge dq^1 \wedge \ldots \wedge dq^{3N}$$

$$+ \left(\frac{\partial p}{\partial y^\alpha} + Q_\alpha^2\right)dy^\alpha \wedge dq^1 \wedge \ldots \wedge dq^{3N} .$$

The linear independence of $dt \wedge dq^1 \wedge \ldots \wedge dq^{3N}$, $dy^\alpha \wedge dq^1 \wedge \ldots \wedge dq^{3N}$, $\alpha=1,\ldots,3N$ thus shows that (21.9) can be satisfied if and only if

$$(21.10) \qquad \frac{\partial p}{\partial t} = - Q^o , \qquad \frac{\partial p}{\partial y^\alpha} = - Q_\alpha^2 .$$

On the other hand, if we start with

$$W = f_\alpha(t;\ q^\beta;\ y^\beta)dq^\alpha ,$$

then (21.4) and (21.5) will yield quantities $p(t;\ q^\beta;\ y^\beta)$ and $\mathcal{Q}(t;\ q^\beta;\ y^\beta)$ that satisfy the conditions (21.10). We accordingly have the equivalent expressions

$$(21.11)\qquad \mathcal{W} = \mathscr{f}_\alpha dq^\alpha = -\left(\frac{\partial p}{\partial q^\alpha} + \mathcal{Q}^1_\alpha\right)dq^\alpha$$

$$= -\ dp(t;\ q^\beta;\ y^\beta) - \mathcal{Q}(t;\ q^\beta;\ y^\beta)\ ,$$

with

$$(21.12)\qquad \frac{\partial p}{\partial t} = -\ \mathcal{Q}^o\ ,\qquad \frac{\partial p}{\partial y^\alpha} = -\ \mathcal{Q}^2_\alpha\ ,$$

$$p(t;\ q^\beta;\ y^\beta) = -\int_o^1 (q^\alpha - q^\alpha_o)\mathscr{f}_\alpha(\lambda t + (1-\lambda)t_o;$$

$$\lambda q^\beta + (1-\lambda)q^\beta_o;\ \lambda y^\beta + (1-\lambda)y^\beta_o)d\lambda\ .$$

It now remains to obtain the equations of motion. If $\mathcal{U}$ is any horizontal vector field, (21.11) yields

$$(21.13)\qquad \mathcal{U}\rfloor(\mathcal{W}\wedge dt) = -\ \mathcal{U}\rfloor d(pdt) - (\mathcal{U}\rfloor\mathcal{Q})dt$$

$$= -\ \pounds_{\mathcal{U}}(pdt) - \mathcal{U}\rfloor\mathcal{Q}dt\ .$$

The fundamental integral condition,

$$0 = \int_a^b \Phi^*\{\pounds_{\mathcal{U}}(Tdt) + \mathcal{U}\rfloor(\mathcal{W}\wedge dt)\}$$

thus yields

$$(21.14)\qquad 0 = \int_a^b \Phi^*\{\pounds_{\mathcal{U}}(T-p)dt - \mathcal{U}\rfloor\mathcal{Q}dt\}\ .$$

Now,

$$\int_a^b \Phi^*\pounds_{\mathcal{U}}(T-p)dt = -\int_a^b \{E|L\}_{\phi^\alpha}\delta\phi^\alpha dt\ ,$$

where

$$L = \Phi^*(T-p) = T - p(t;\ \phi^\beta;\ D\phi^\beta) .$$

Also

$$\int_a^b - \Phi^* \mathcal{W} \rfloor \mathcal{Q} dt = - \int_a^b \{Q^1_\alpha(t;\ \phi^\alpha;\ D\phi^\alpha)\delta\phi^\alpha + Q^2_\alpha(t;\ \phi^\alpha;\ D\phi^\alpha) D\delta\phi^\alpha\} dt$$

$$= \int_a^b \{DQ^2_\alpha(t;\ \phi^\alpha;\ D\phi^\alpha) - Q^1_\alpha(t;\ \phi^\alpha;\ D\phi^\alpha)\}\delta\phi^\alpha dt$$

since $\delta\phi^\alpha(a) = \delta\phi^\alpha(b) = 0$. When these are substituted into (21.4) we obtain the following result.

A map $\Phi: [a \le t \le b] \to K | \Phi(t) = t$, $\phi^\alpha(a) = a^\alpha$, $\phi^\alpha(b) = b^\alpha$ *is a Newtonian motion for a work 1-form*

(21.15) $$\mathcal{W} = - dp(t;\ q^\beta;\ y^\beta) - \mathcal{Q}(t;\ q^\beta;\ y^\beta)$$

(21.16) $$\frac{\partial p}{\partial t} = - \mathcal{Q}^o , \quad \frac{\partial p}{\partial y^\alpha} = - \mathcal{Q}^2_\alpha$$

if and only if

(21.17) $$\{E|L\}_{\phi^\alpha} = DQ^2_\alpha(t;\ \phi^\beta;\ D\phi^\beta) - Q^1_\alpha(t;\ \phi^\beta;\ D\phi^\beta) ,$$

where

$$L = \Phi^* L , \quad L = T - p(t;\ q^\beta;\ y^\beta) ,$$

is the Lagrangian function and

(21.18) $$\Phi^* \mathcal{Q} = (Q^o + Q^1_\alpha D\phi^\alpha + Q^2_\alpha D^2\phi^\alpha) dt .$$

The total energy of the system is easily seen to be given by

(21.19) $$E(t;\ \phi^\alpha;\ D\phi^\alpha) = \frac{\partial L}{\partial D\phi^\alpha} D\phi^\alpha - L$$

$$= T + p - \frac{\partial p}{\partial D\phi^\alpha} D\phi^\alpha ,$$

and satisfies

(21.20) $$D(E) = \frac{\partial p}{\partial t} + (DQ^2_\alpha - Q^1_\alpha)D\phi^\alpha$$

$$= - Q^o - Q^1_\alpha D\phi^\alpha - Q^2_\alpha D^2\phi^\alpha + D(Q^2_\alpha D\phi^\alpha) ,$$

where we have used $\frac{\partial p}{\partial t} = -Q^o$. If we use (21.18) to eliminate E in (21.19), we obtain

$$D\left(T + p - \frac{\partial p}{\partial D\phi^\alpha} D\phi^\alpha - Q^2_\alpha D\phi^\alpha\right) = \Phi^*(D\Phi \rfloor \mathcal{Q}) .$$

However, $\frac{\partial p}{\partial D\phi^\alpha} = -Q^2_\alpha$, and so we see that

$$D(T+p) = \Phi^*(D\Phi \rfloor \mathcal{Q}) .$$

An example that often arises in practice is that in which

(21.21) $$\mathcal{W} = \mathcal{t} \rfloor dA(q^\beta) ,$$

where $A(q^\beta) = A_\alpha(q^\beta)dq^\alpha$ is a 1-form on C and $\mathcal{t}(t;\, q^\beta;\, y^\beta) = \partial/\partial t + y^\beta\, \partial/\partial q^\beta$ is the rectilinear field on K . Since only $dA(q^\beta)$ enters into $\mathcal{W}$, we can assume, without loss of generality that $A(q^\beta)$ belongs to the class of antiexact 1-forms on C (see Section 10 of Appendix):

(21.22) $$A = HdA , \quad HA = 0$$

where H is the homotopy operator defined in (A.§10). Thus, since $p(t;\, q^\beta;\, y^\beta) = -H\mathcal{W}$, the choise $t_o = 0$, $y^\alpha_o = 0$, $q^\alpha_o = 0$ gives

(21.23) $$p(t;\, q^\beta;\, y^\beta) = - \int_o^1 \mathcal{r} \rfloor (\mathcal{t} \rfloor dA)(\lambda t;\, \lambda q^\beta;\, \lambda y^\beta)\frac{d\lambda}{\lambda}$$

$$= \int_o^1 \mathcal{t}(\lambda t;\, \lambda q^\beta;\, \lambda y^\beta) \rfloor (\mathcal{r} \rfloor dA)(\lambda t;\, \lambda q^\beta;\, \lambda y^\beta)\frac{d\lambda}{\lambda}$$

where

$$\mathcal{r} = (t-t_o)\frac{\partial}{\partial t} + (q^\alpha - q^\alpha_o)\frac{\partial}{\partial q^\alpha} + (y^\alpha - y^\alpha_o)\frac{\partial}{\partial y^\alpha}$$

$$= t\,\frac{\partial}{\partial t} + q^\alpha\,\frac{\partial}{\partial q^\alpha} + y^\alpha\,\frac{\partial}{\partial y^\alpha}\ .$$

Since $\mathcal{r}\rfloor dA$ has no dt component and $\mathcal{t}$ is linear in $\{y^\alpha\}$, (21.23) yields

(21.24) $$p = \mathcal{t}\rfloor \int_0^1 (\mathcal{r}\rfloor dA)(\lambda t;\ \lambda q^\beta;\ \lambda y^\beta)d\lambda = \mathcal{t}\rfloor H(dA)$$

$$= \mathcal{t}\rfloor A = y^\alpha A_\alpha(q^\beta)$$

because $HdA = A$. We then have

(21.25) $$\mathcal{Q}^1_\alpha = -\,y^\beta\,\frac{\partial A_\alpha}{\partial q^\beta}\ ,\quad \mathcal{Q}^2_\alpha = -\,\frac{\partial p}{\partial y^\alpha} = -\,A_\alpha\ ,\quad \mathcal{Q}^o = 0$$

from $W = -dp - \mathcal{Q}$ and (21.16). Thus

(21.26) $$p = A_\alpha(\phi^\beta)D\phi^\alpha\ ,\quad Q^o = 0\ ,\quad Q^1_\alpha = -\,\frac{\partial A_\alpha}{\partial \phi^\beta}\,D\phi^\beta\ ,\quad Q^2_\alpha = -A_\alpha$$

and hence

(21.27) $$DQ^2_\alpha - Q^1_\alpha = -\,\frac{\partial A_\alpha}{\partial \phi^\beta}\,D\phi^\beta + \frac{\partial A_\alpha}{\partial \phi^\beta}\,D\phi^\beta = 0\ .$$

The equations of motion, (21.17), become

(21.28) $$\{E|T-p\}_{\phi^\alpha} = \{E|T - A_\mu(\phi^\beta)D\phi^\mu\}_{\phi^\alpha} = 0\ ,$$

while (21.19), (21.20) and (21.23) give

(21.29) $$E(t;\ \phi^\alpha;\ D\phi^\alpha) = T\ ,\quad D(T) = 0\ .$$

The motion thus has an energy quadrature T=constant - the motion preserves kinetic energy.

It is of interest to inquire into what happens when $A(q^\beta)$ is replaced by a 1-form $B(t;\ q^\beta)$ on E such that $B = HdB$:

(21.30) $$B(t;\ q^\beta) = B^o(t;\ q^\beta)dt + B^1_\alpha(t;\ q^\beta)dq^\alpha\ .$$

We can not use $\mathcal{W} = t \lrcorner d\mathcal{B}$ since

$$(21.31) \qquad d\mathcal{B} = \left(\frac{\partial \mathcal{B}^1_\alpha}{\partial t} - \frac{\partial \mathcal{B}^o}{\partial q^\alpha}\right) dt \wedge dq^\alpha + \frac{\partial \mathcal{B}^1_\alpha}{\partial q^\beta} dq^\beta \wedge dq^\alpha$$

gives

$$(21.32) \qquad t \lrcorner d\mathcal{B} = \left(\frac{\partial \mathcal{B}^1_\alpha}{\partial t} - \frac{\partial \mathcal{B}^o}{\partial q^\alpha} + \left(\frac{\partial \mathcal{B}^1_\alpha}{\partial q^\beta} - \frac{\partial \mathcal{B}^1_\beta}{\partial q^\alpha}\right) y^\beta\right) dq^\alpha$$

$$- \left(\frac{\partial \mathcal{B}^1_\alpha}{\partial t} - \frac{\partial \mathcal{B}^o}{\partial q^\alpha}\right) y^\alpha dt \ ,$$

and $\mathcal{W}$ would not satisfy $\mathcal{W} \wedge dq^1 \wedge \ldots \wedge dq^{3N} = 0$. However,

$$\left(\frac{\partial \mathcal{B}^1_\alpha}{\partial t} - \frac{\partial \mathcal{B}^o}{\partial q^\alpha}\right) y^\alpha = \frac{\partial}{\partial t} (t \lrcorner \mathcal{B}) - \frac{\partial \mathcal{B}^o}{\partial t} - \frac{\partial \mathcal{B}^o}{\partial q^\alpha} y^\alpha$$

$$= \frac{\partial}{\partial t} (t \lrcorner \mathcal{B}) - t[\mathcal{B}^o]$$

where $t[\mathcal{B}^o] = \left(\frac{\partial}{\partial t} + y^\beta \frac{\partial}{\partial q^\beta}\right) (\mathcal{B}^o)$. We must therefore consider

$$(21.33) \qquad \mathcal{W} = t \lrcorner d\mathcal{B} + \left(t[\mathcal{B}^o] - \frac{\partial}{\partial t} (t \lrcorner \mathcal{B})\right) dt$$

$$= \left(\frac{\partial \mathcal{B}^1_\alpha}{\partial t} - \frac{\partial \mathcal{B}^o}{\partial q^\alpha} + \left(\frac{\partial \mathcal{B}^1_\alpha}{\partial q^\beta} - \frac{\partial \mathcal{B}^1_\beta}{\partial q^\alpha}\right) y^\beta\right) dq^\alpha \ .$$

CHAPTER IV

ANALYSIS OF CONSTRAINTS

22. THE CLASSIC THEORY AND ITS EXTENSION

The classic theory of constraints is formulated in Event Space, E by giving a collection of $R<3N+1$ linearly independent differential forms:

$$\omega^i(t;\, q^\beta) = \gamma^i(t;\, q^\beta)dt + h^i_\alpha(t;\, q^\beta)dq^\alpha , \quad i=1,\ldots,R ,$$

$$\omega^1 \wedge \omega^2 \wedge \ldots \wedge \omega^R \neq 0 .$$

If $\Phi\colon J \to E|\phi(t)=t$, $\phi(q^\alpha) = \phi^\alpha(t)$ is a map of J into E , Φ is said to satisfy the given constraints $\omega^i=0$ if and only if

$$\Phi^*\omega^i = 0 , \quad i=1,\ldots,R .$$

When these constraint equations are written out in terms of the functions γ^i , h^i_α that specify the ω's , we obtain

$$\gamma^i(t;\, \phi^\alpha) + h^i_\alpha(t;\, \phi^\alpha)D\phi^\alpha = 0 , \quad i=1,\ldots,R .$$

An inspection of these equations shows that *the classic theory only allows one to impose constraints that are linear in the velocity variables,* $\{D\phi^\alpha(t)\}$ *and independent of the acceleration variables* $\{D^2\phi^\alpha(t)\}$. This is a very severe limitation.

Our formulation of mechanics uses Kinematic Space rather than Event Space as the underlying arena. Since Kinematic Space obtains from Event Space by the process of group extension, it appears reasonable to expect that the formulation of constraint problems in Kinematic Space would allow us to eliminate the severe restrictions that are demanded by the classic theory. This is indeed the case, as we shall now show.

We assume that we have R given and independent differential forms $\omega^i(t;\, q^\beta;\, y^\beta)$ on K :

$$(22.1)\qquad \omega^i(t;\, q^\beta;\, y^\beta) = \omega^{0i}(t;\, q^\beta;\, y^\beta)dt + \omega^{1i}_\alpha(t;\, q^\beta;\, y^\beta)dq^\alpha + \omega^{2i}_\alpha(t;\, q^\beta;\, y^\beta)dy^\alpha\ ,\quad i=1,\ldots,R\ ,$$

$$(22.2)\qquad \omega^1\wedge\omega^2\wedge\ldots\wedge\omega^R \neq 0\ .$$

If $\Phi\colon J\to K|\Phi(t)=t$, $\Phi(q^\alpha)=\phi^\alpha(t)$, $\Phi(y^\alpha)=D\phi^\alpha(t)$ is a map of J to K , then Φ is said to satisfy the given system of *extended constraints* $\omega^i=0$, $i=1,\ldots,R$ if and only if

$$(22.3)\qquad \Phi^*\omega^i = 0\ ,\quad i=1,\ldots,R\ .$$

When (22.1) is used to write out these extended constraints explicitly, we obtain

$$(22.4)\qquad 0 = \omega^{0i}(t;\, \phi^\beta;\, D\phi^\beta) + \omega^{1i}_\alpha(t;\, \phi^\beta,\, D\phi^\beta)D\phi^\alpha + \omega^{2i}_\alpha(t;\, \phi^\beta;\, D\phi^\beta)D^2\phi^\alpha\ ,\quad i=1,\ldots,R\ .$$

Extended constraints thus allow for quite general dependence on the velocity variables $\{D\phi^\alpha\}$; the only restriction is that we have no more than a linear dependence on the acceleration variables $\{D^2\phi^\alpha\}$. In particular, if we have $\omega^{2i}_\alpha=0$, (22.4) becomes

$$0 = \omega^{0i}(t;\, \phi^\beta;\, D\phi^\beta) + \omega^{1i}_\alpha(t;\, \phi^\beta;\, D\phi^\beta)D\phi^\alpha\ ,\quad i=1,\ldots,R\ ,$$

which constitutes a nontrivial extension of the classic theory even in this particular case.

A simple, but useful example of extended constraints is provided by the following problem. Suppose that we have a conservative system for which

$$T = T_1(y^1,\ldots,y^r) + T_2(y^{r+1},\ldots,y^{3N}) ,$$

$$\hat{V} = \hat{V}_1(q^1,\ldots,q^r) + \hat{V}_2(q^1,q^2,\ldots,q^{3N}) :$$

that is, the system contains a subsystem with kinetic energy function $T_1(y^1,\ldots,y^r)$ and potential energy $\hat{V}_1(q^1,\ldots,q^r)$. It is then meaningful to inquire as to what additional forces must be brought into play in order that the full system shall admit a quadrature that represents the total energy of the given subsystem:

$$(22.5) \qquad \Phi^*(T_1+\hat{V}_1) = \text{constant} .$$

If we form the total energy function $E_1 = T_1+\hat{V}_1$ of the subsystem and view (22.5) as a constraint function, then this constraint function is equivalent to the differential constraint.

$$(22.6) \qquad \Phi^*(dE_1) = \Phi^*(dT_1 + d\hat{V}_1) = 0 ;$$

that is, we obtain the constraint 1-form

$$(22.7) \qquad \omega^1 = dT_1 + d\hat{V}_1 .$$

Since T_1 is quadratic in the variables $y^1,\ldots,y^r$, the constraint represented by (22.7) is an extended constraint. Such considerations provide a basis for investigating the partition of energy in a dynamical system comprised of subsystems by calculating the forces necessary to achieve specific partitions.

In another vein, suppose that we need to determine what forces must be brought into play so that a given conservative system with kinetic energy function T and potential energy function $\hat{V}$ will be such that it tends to a state of rest for large time. Accordingly, if we set

$$(22.8) \qquad \Phi^* dT = \Phi^* T f(q^1,\ldots,q^{3N})dt , \quad f(q^1,\ldots,q^{3N}) < 0 ,$$

we obtain the constraint form

(22.9) $$\omega^1 = dT - Tf(q^\beta)dt$$

whose satisfaction will force Φ^*T to be monotone decreasing along the orbit of Φ .

23. NEWTONIAN MOTIONS OF CONSERVATIVE SYSTEMS WITH EXTENDED CONSTRAINTS

We limit our considerations in this Chapter, in the interests of conceptual simplicity, to conservative systems with extended constraints. Since the analysis is primarily concerned with the constraints, this is no undue restriction. The general case will be taken up in subsequent chapters.

The dynamics of conservative systems is completely specified by giving the kinetic energy function $T = \frac{1}{2} y^\alpha m_{\alpha\beta} y^\beta$ and the potential energy function $\hat{V}(q^\beta)$; that is, we know the t-independent Lagrangian function

(23.1) $$L(q^\beta;\ y^\beta) = T - \hat{V}\ .$$

Newtonian motions $\Phi\colon [a\underline{<}t\underline{<}b] \to K|\Phi(t)=t$, $\phi^\alpha(a)=a^\alpha$, $\phi^\alpha(b)=b^\alpha$ are then obtained through satisfaction of

(23.2) $$0 = \int_a^b \Phi^*(\pounds_{U} L dt)$$

for all horizontal vector fields U such that $u^1_\alpha(a;a^\beta) = u^1_\alpha(b;b^\beta) = 0$; that is

(23.3) $$\{E|L\}_{\phi^\alpha} = 0\ ,\quad \alpha=1,\ldots,3N\ .$$

The problem that now confronts us is that of imposing a given system of constraints, $\omega^i(t;\ q^\beta;\ y^\beta) = 0$, on the motion Φ . Since Φ connects the points (a^α) and (b^α) in C , the first thing we have to require is that there is at least one Φ that satisfies the constraint conditions

(23.4) $$\Phi^*\omega^i = 0\ ,\quad i=1,\ldots,R$$

and the conditions $\phi^\alpha(a)=a^\alpha$, $\phi^\alpha(b)=b^\alpha$. This is a nontrivial problem, even if it is usually ignored in the classic treatment of contraints. The situation that arises most frequently is that the point with coordinates (b^α) in C has to be chosen by "integration" of the constraint equations (23.4) starting from the point with coordinates (a^α) in C . Let us assume, however, that the points (a^α) and (b^α) have been chosen in C so that they are consistent with the constraints (23.3).

It is clear that only in rare circumstances will the dynamical equations $\{E|L\}_{\phi^\alpha} = 0$ have solutions that give rise to mappings Φ satisfying all of the constraint equations $\Phi^*\omega^i = 0$. In general, forces in addition to $-\partial\hat{V}/\partial q^\alpha$ will have to come into play in order to force the system to satisfy the given constraints. Further, the components of the horizontal fields U , that occur in (23.2), can no longer be chosen subject only to the conditions $u_1^\alpha(a;a^\beta) = u_1^\alpha(b;b^\beta) = 0$. Satisfaction of the constraints implies relations that must also be satisfied by the horizontal vector fields U if the variations generated by U are to be consistent with the constraints. These relations have been obtained in Section 8:

$$(23.5) \qquad \Phi^*(U \lrcorner \omega^i) = 0 , \qquad i=1,\dots,R .$$

As is well known, calculation of the forces needed to maintain the constraints and relaxation of the R conditions (23.5) can both be accomplished by introducing R new variables λ_i , $i=1,\dots,R$, known as *Lagrange multipliers* into the problem.

A map Φ: $[a\le t\le b] \to K|\Phi(t)=t$, $\Phi(q^\alpha)=\phi^\alpha(t)$, $\phi^\alpha(a)=a^\alpha$, $\phi^\alpha(b)=b^\alpha$ *is a Newtonian motion for the given Lagrangian* $L(q^\beta;\, y^\beta) = T-\hat{V}(q^\beta)$ *subject to given independent constraints* $\omega^i=0$, $i=1,\dots,R$, *if and only if* Φ *and the variables* $\lambda_1,\dots,\lambda_R$ *are such that* (1)

$$(23.6) \qquad \Phi^*\omega^i = 0 , \quad i=1,\dots,R ,$$

and (2)

$$(23.7) \qquad 0 = \int_a^b \Phi^*\{\pounds_{\mathcal{U}}(L\,dt) + \lambda_i \mathcal{U} \rfloor (\omega^i \wedge dt)\}$$

holds for all horizontal vector fields $\mathcal{U}$ *that satisfy* $u_1^\alpha(a;a^\beta) = u_1^\alpha(b;b^\beta) = 0$.

24. FORCES THAT MAINTAIN GIVEN CONSTRAINTS AND RELATED TOPICS

We have seen previously that

$$(24.1) \qquad \int_a^b \Phi^*\pounds_{\mathcal{U}}(L\,dt) = -\int_a^b \{E|L\}_{\phi^\alpha}\delta\phi^\alpha dt ,$$

where

$$(24.2) \qquad \delta\phi^\alpha(t) = u_1^\alpha(t;\ \phi^\beta(t)) .$$

We need therefore only analyze the second set of terms in (23.7). Since $\mathcal{U}$ is horizontal and

$$(24.3) \qquad \omega^i = w^{0i}dt + w_\alpha^{1i}dq^\alpha + w_\alpha^{2i}dy^\alpha ,$$

we have

$$(24.4) \qquad \mathcal{U}\rfloor(\omega^i \wedge dt) = (w_\alpha^{1i}u_1^\alpha + w_\alpha^{2i}u_2^\alpha)dt = (w_\alpha^{1i}u_1^\alpha + w_\alpha^{2i}Zu_1^\alpha)dt .$$

Accordingly

$$(24.5) \qquad \Phi^*\mathcal{U}\rfloor(\omega^i \wedge dt) = (W_\alpha^{1i}\delta\phi^\alpha + W_\alpha^{2i}D\delta\phi^\alpha)dt$$

$$= (W_\alpha^{1i} - DW_\alpha^{2i})\delta\phi^\alpha dt + D(W_\alpha^{2i}\delta\phi^\alpha)dt ,$$

where

$$(24.6) \qquad W_\alpha^{1i} = \Phi^* w_\alpha^{1i} = w_\alpha^{1i}(t;\ \phi^\beta;\ D\phi^\beta) ,$$

$$W_\alpha^{2i} = \Phi^* w_\alpha^{2i} , \qquad W^{0i} = \Phi^* w^{0i} .$$

When (24.1) and (24.6) are substituted into the basic integral (23.7), and we note that $u_1^\alpha(a;a^\beta) = u_1^\alpha(b;b^\beta) = 0$ implies $\delta\phi^\alpha(a) = \delta\phi^\alpha(b) = 0$, we obtain the following results.

A map $\Phi: [a \le t \le b] \to K | \Phi(t)=t$, $\Phi(q^\alpha)=\phi^\alpha(t)$, $\phi^\alpha(a)=a^\alpha$, $\phi^\alpha(b)=b^\alpha$ *is a Newtonian motion for the Lagrangian* L *subject to the independent constraints* $\omega^i=0$, $i=1,\dots,R$ *if and only if* $\{\phi^\alpha(t)\}$ *and the* R *functions* $\lambda_1,\dots,\lambda_R$ *satisfy the* $3N+R$ *equations*

$$\{E|L\}_{\phi^\alpha} = -D(\lambda_i W_\alpha^{2i}) + \lambda_i W_\alpha^{1i}, \quad \alpha=1,\dots,3N, \tag{24.7}$$

$$0 = W^{0i}(t;\phi^\beta;D\phi^\beta) + W_\alpha^{1i}(t;\phi^\beta;D\phi^\beta)D\phi^\alpha + W_\alpha^{2i}(t;\phi^\beta;D\phi^\beta)D^2\phi^\alpha, \quad i=1,\dots,R. \tag{24.8}$$

It is now a trivial matter to recognize the forces that maintain the constraints (24.8); we simply compare (24.7) with $\{E|T\}_{\phi^\alpha} = f_\alpha$ and obtain (recall that $L=T-\hat{V}$)

$$f_\alpha = -\frac{\partial \hat{V}}{\partial \phi^\alpha} - D(\lambda_i W_\alpha^{2i}) + \lambda_i W_\alpha^{1i}.$$

Thus, since $-\partial\hat{V}/\partial\phi^\alpha$ constitute the forces that come from the work 1-form of the conservative system, the forces $\{f_\alpha^c\}$ that maintain the constraints are given by

$$f_\alpha^c = -D(\lambda_i W_\alpha^{2i}) + \lambda_i W_\alpha^{1i}. \tag{24.9}$$

It is of interest to note that the right-hand side of (24.9) would be the negative of the Euler-Lagrange operator on Q if $\lambda_i W_\alpha^{2i} = \partial Q/\partial D\phi^\alpha$, $\lambda_i W_\alpha^{1i} = \partial Q/\partial\phi^\alpha$; of course, this is not the case in general.

In order to obtain the energy equation associated with the equations of motion (24.7), we multiply (24.7) by $D\phi^\alpha$ and sum on α to obtain

$$D(E) = D(T+\hat{V}) = \lambda_i W^{1i}_\alpha D\phi^\alpha - D(\lambda_i W^{2i}_\alpha) D\phi^\alpha$$

$$= + \lambda_i (W^{1i}_\alpha D\phi^\alpha + W^{2i}_\alpha D^2\phi^\alpha) - D(\lambda_i W^{2i}_\alpha D\phi^\alpha) .$$

An elimination of the terms $W^{1i}_\alpha D\phi^\alpha + W^{2i}_\alpha D^2\phi^\alpha$ by means of the constraint equations (24.8) thus gives

(24.10) $$D(E+\lambda_i W^{2i}_\alpha D\phi^\alpha) = - \lambda_i W^{0i}(t;\ \phi^\beta;\ D\phi^\beta) .$$

If the functions W^{0i} all vanish, we obtain an energy quadrature $E+\lambda_i W^{2i}_\alpha D\phi^\alpha$ = constant. Constraints with this property are so important they are given a specific name. A system of constraints $\omega^i=0$, $i=1,\ldots,R$ are said to be *catastatic* if $W^{0i}=0$, $i=1,\ldots,R$. An invariant formulation of these conditions is given by

(24.11) $$\omega^i \wedge dq^1 \wedge \ldots \wedge dq^{3N} \wedge dy^1 \wedge \ldots \wedge dy^{3N} = 0 , \quad i=1,\ldots,R .$$

For catastatic constraints, we have

(24.12) $$0 = \Phi^*\omega^i = (W^{1i}_\alpha D\phi^\alpha + W^{2i}_\alpha D^2\phi^\alpha) dt$$

while (23.5) and $\Phi^* u^1_\alpha = \delta\phi^\alpha$, $\Phi^* u^2_\alpha = D\delta\phi^\alpha$ give

(24.13) $$0 = W^{1i}_\alpha \delta\phi^\alpha + W^{2i}_\alpha D\delta\phi^\alpha .$$

Accordingly, we obtain

$$0 = (\Phi+\sigma\delta\Phi+0(\sigma))^*\omega^i ,$$

and hence the varied motion $\Phi+\sigma\delta\Phi$ satisfies the constraints if Φ satisfies them. This does not happen for noncatastatic systems. Conservative systems with catastatic constraints admit an energy integral

(24.14) $$T + \hat{V} + \lambda_i W^{2i}_\alpha D\phi^\alpha = \text{constant}$$

and are dissipative if and only if

(24.15) $$D(\lambda_i W^{2i}_{\alpha} D\phi^{\alpha}) \geq 0 .$$

They are strictly dissipative if and only if

(24.16) $$D(\lambda_i W^{2i}_{\alpha} D\phi^{\alpha}) > 0 \quad \text{for} \quad T>0 .$$

25. INTEGRABLE CONSTRAINTS - APPLICATION OF THE FROBENIUS THEOREM

There are certain circumstances in which one can affect significant simplification of the constraints. These circumstances are delineated by the Frobenius theorem (A§9) which states the following: *If* ω^i , $i=1,\ldots,R$ *are* R *linearly independent differential forms on* K *such that*

(25.1) $$d\omega^i \wedge \omega^1 \wedge \omega^2 \wedge \ldots \wedge \omega^R = 0 , \quad i=1,\ldots,R ,$$

then there exist a nonsingular R-by-R *matrix* $((A^i_j))$ *of functions on* K *and* R *functionally independent functions* B^i , $i=1,\ldots,R$, *on* K *such that*

(25.2) $$\omega^i = A^i_j \, dB^j .$$

Further, if the conditions (25.1) *are met, there exists an* R-by-R *matrix* $((\Gamma^i_j))$ *of* 1-*forms on* K *such that*

(25.3) $$d\omega^i = \Gamma^i_j \wedge \omega^j ,$$

and the system of forms ω^i *is said to be completely integrable.*

The classic theory of constraints on Event Space distinguishes a preferred class of *Holonomic* constraints; namely those of the form $\omega^i = d\psi^i = 0$. Strange as it may seem, the classic theory does not make use of the Frobenius theorem to replace a system of constraints $0 = \omega^i \neq d\psi^i$ by an equivalent system of holonomic constraints when the ω's satisfy (25.1). Even though our setting if the (6N+1)-dimensional Kinematic Space, rather than the (3N+1)-dimensional Event Space of the classic theory, the

Frobenius theorem, when applicable, leads to significant simplifications. Suppose that our given system of 1-forms ω^i satisfies the conditions (25.1). We then know that there exists a nonsingular matrix $((A^i_j))(t;\, q^\beta;\, y^\beta)$ of functions on K and R functions $B^i(t;\, q^\beta;\, y^\beta)$ on K such that

$$(25.4) \qquad \omega^i = A^i_j \, dB^j .$$

Thus, since $((A^i_j))$ is nonsingular, the given constraints $\omega^i=0$ are satisfied if and only if the constraints

$$(25.5) \qquad dB^j = \frac{\partial B^j}{\partial t}\, dt + \frac{\partial B^j}{\partial q^\alpha}\, dq^\alpha + \frac{\partial B^j}{\partial y^\alpha}\, dy^\alpha = 0 , \quad i=1,\dots,R$$

are satisfied. We can therefore replace the constraints specified by the ω's by the equivalent ones (25.5) throughout the last section. This leads to the following results.

A map $\Phi\colon [a\le t\le b] \to K|\Phi(t) = t$ *is a Newtonian motion for the Lagrangian* L *subject to the* R *linearly independent constraints* $\omega^i=0$, $i=1,\dots,R$ *that satisfy*

$$d\omega^i \wedge \omega^1 \wedge \omega^2 \wedge \dots \wedge \omega^R = 0 , \quad i=1,\dots,R$$

if and only if the functions $\{\phi^\alpha(t)\}$ *and* $\bar\lambda_i$ *are such that*

$$(25.6) \qquad \{E|L\}_{\phi^\alpha} = - D\left(\bar\lambda_j \frac{\partial B^j}{\partial D\phi^\alpha}\right) + \bar\lambda_j \frac{\partial B^j}{\partial \phi^\alpha} ,$$

$$(25.7) \qquad DB^j(t;\, \phi^\alpha;\, D\phi^\alpha) = 0 , \quad j=1,\dots,R ,$$

where $B^j = \Phi^* B^j$ *and the scalar functions* $B^j(t;\, q^\alpha;\, y^\alpha)$ *are related to the* ω's *by*

$$(25.8) \qquad \omega^i = A^i_j(t;\, q^\beta;\, y^\beta)\, dB^j(t;\, q^\beta;\, y^\beta)$$

with $\det(A^i_j) \neq 0$.

It follows immediately from (24.10) that the energy equation is given by

$$(25.9)\qquad D\left(E + \bar{\lambda}_i \frac{\partial B^i}{\partial D\phi^\alpha} D\phi^\alpha\right) = -\bar{\lambda}_i \frac{\partial B^i}{\partial t} .$$

Examination of (25.8) shows that if the ω's are catastatic, then the $d\mathcal{B}$'s are catastatic, in which case we have the energy quadrature

$$(25.10)\qquad T + \hat{V} + \bar{\lambda}_i \frac{\partial B^i}{\partial D\phi^\alpha} D\phi^\alpha = \text{constant}.$$

Clearly, we also have the quadratures

$$(25.11)\qquad B^j(t;\ \phi^\alpha;\ D\phi^\alpha) = \text{constant},$$

as follows from (25.7). Thus problems with completely integrable constraints become equivalent to problems that admit a given system of quadratures $B^i = \text{constant}$.

Since we can replace the ω's by the $d\mathcal{B}$'s in the equations of motion and in the constraints, when the latter are completely integrable, we can do the same thing in the fundamental integral (23.7). This amounts to replacing $\lambda_i \mathcal{U} \lrcorner (\omega^i \wedge dt)$ by

$$\bar{\lambda}_i \mathcal{U} \lrcorner (d\mathcal{B}^i \wedge dt) = \bar{\lambda}_i \mathcal{U} \lrcorner d(\mathcal{B}^i dt) = \bar{\lambda}_i \pounds_{\mathcal{U}} \mathcal{B}^i dt .$$

We accordingly obtain the equivalent fundamental integral

$$(25.12)\qquad 0 = \int_a^b \Phi^*\{\pounds_{\mathcal{U}}(L dt) + \bar{\lambda}_i \pounds_{\mathcal{U}}(\mathcal{B}^i dt)\} .$$

It is also clear that one formulation can be mapped onto the other by $\bar{\lambda}^i = \lambda^j A^i_j$, although it is much easier to go back to the beginning and replace the ω's by the $d\mathcal{B}$'s as we have done above.

Suppose we are given a completely integrable system of 1-forms ω^i on Event Space and a Lagrangian $L = T - \hat{V}$, and that we have used the Frobenius theorem to replace the constraints $\omega^i = 0$ by the equivalent system

$$(25.13)\qquad d\mathcal{B}^i(t;\ q^\beta) = \frac{\partial \mathcal{B}^i}{\partial t} dt + \frac{\partial \mathcal{B}^i}{\partial q^\alpha} dq^\alpha = 0 , \quad i=1,\ldots,R .$$

We then have the governing equations

$$B^i(t;\ \phi^\beta(t)) \quad = \quad \text{constant}, \tag{25.14}$$

$$\{E|L\}_{\phi^\alpha} \quad = \quad \bar{\lambda}_i \frac{\partial B^i}{\partial q^\alpha}\ , \tag{25.15}$$

$$D(T+\hat{V}) \quad = \quad \bar{\lambda}_i \frac{\partial B^i}{\partial t}\ . \tag{25.16}$$

The known mapping properties of $\{E|L\}_{\phi^\alpha}$ and $\dfrac{\partial B^i(t;q^\beta)}{\partial q^\alpha}$ suggests that significant simplification can be achieved by an appropriately chosen admissible map $\Gamma\colon K \to {}^\backprime K$. We know that the R functions $\mathcal{B}^i(t;\ q^\beta)$ are functionally independent since

$$\omega^1 \wedge \omega^2 \wedge \ldots \wedge \omega^R \;=\; \det(A^i_j)\, d\mathcal{B}^1 \wedge d\mathcal{B}^2 \wedge \ldots \wedge d\mathcal{B}^R \;\neq\; 0\ .$$

We assume that the $\{q^\beta\}$ are so ordered that $\mathcal{B}^1, \mathcal{B}^2, \ldots, \mathcal{B}^R, q^{R+1}, \ldots, q^{3N}$ are functionally independent, and construct the admissible map Γ defined by

$$ {}^\backprime t = t\ ;\quad {}^\backprime q^i = \mathcal{B}^i(t;\ q^\beta)\ ,\quad i=1,\ldots,R\ ; \tag{25.17}$$

$$ {}^\backprime q^{\underline{i}} = q^{\underline{i}}\ ,\quad \underline{i} = R+1,\ldots,3N\ .$$

Under these circumstances, we have $\partial\mathcal{B}^i/\partial q^\alpha = \partial{}^\backprime q^i/\partial q^\alpha$, and hence (25.15) yield

$$0 \;=\; \left[\{E|L\}_{\phi^\alpha} - \bar{\lambda}_i \frac{\partial B^i}{\partial q^\alpha}\right]\frac{\partial q^\alpha}{\partial {}^\backprime q^\beta} \;=\; \{E|{}^\backprime L\}_{{}^\backprime\phi^\beta} - \bar{\lambda}_i \frac{\partial {}^\backprime q^i}{\partial {}^\backprime q^\beta}\ .$$

We thus have

$$\bar{\lambda}_i \quad = \quad \{E|{}^\backprime L\}_{{}^\backprime\phi^i}\ ,\quad i = 1,\ldots,R\ , \tag{25.18}$$

$$0 \quad = \quad \{E|{}^\backprime L\}_{{}^\backprime\phi^{\underline{j}}}\ ,\quad \underline{j} = R+1,\ldots,3N\ , \tag{25.19}$$

$${}^\backprime q^i \quad = \quad \text{constant}\ ,\quad i = 1,\ldots,R\ . \tag{25.20}$$

The problem is therefore reduced to solving only the $3N-R$ equations (25.19) on $`K$ since the first $R$ of the $`q$'s are constants and the R equations (25.18) serve to determine the R multipliers $\bar{\lambda}_1,\ldots,\bar{\lambda}_R$.

The above results for completely integrable constraints on E has led mechanicians to define the *degrees of freedom* of a constrained system to be

$$\text{Deg} = 3N - R \tag{25.21}$$

where R is the number of independent constraints. We adopt the same definition for general problems with constraints on K (extended constraint problems). Since $\dim(K) = 6N+1$, the number of degrees of freedom of an extended constraint problem can be negative! This should not come as a great surprise, for *any variational system with* N *particles will always satisfy a system of* $6N$ *extended constraints*; that is, the system can be viewed as one with $-3N$ degrees of freedom in the classical sense. In order to see this, we observe that a general solution of the equations of motion of a variational system with N particles is given by $\phi^\alpha(t) = \rho^\alpha(t;\, c_1,\ldots,c_{6N})$, $\alpha=1,\ldots,3N$, where c_1 through c_{6N} are the $6N$ integration constants of the system. This then gives $D\phi^\alpha(t) = \partial\rho(t;\, c_1,\ldots,c_{6N})/\partial t$, and we have $6N$ equations in the $12N + 1$ variables ϕ^α, $D\phi^\alpha$, t, $c_1,\ldots,c_{6N}$. Since the integration constants are independent, we can use the implicit function theorem to solve for the c's as functions of the variables ϕ^α, $D\phi^\alpha$, t so as to obtain

$$c_\Gamma = Q_\Gamma(t;\, \phi^\alpha;\, D\phi^\alpha) \;, \quad \Gamma = 1,\ldots,6N \;.$$

Thus, if we take the exterior derivative of these equations, we see that

$$0 = \Phi^*\omega_\Gamma \;, \quad \omega_\Gamma = dQ_\Gamma(t;\, q^\beta;\, y^\beta) \;;$$

that is, the system satisfies $6N$ extended constraints. In fact, it is obvious that any quadrature of a dynamical system will yield an extended constraint upon exterior differentiation. Probably the reason that this has not been recognized in the classical theory is that the classical theory does not consider constraints of the form $\Phi^* d\mathcal{Q}(t;\ q^\alpha;\ y^\alpha) = 0$, for there can be no (y^α)-dependence in the classical theory.

We note that the mapping technique used above does not apply to completely integrable extended constraint problems because

$$`q^i = B^i(t;\ q^\alpha;\ y^\alpha)$$

does not define an admissible map.

As an example, consider the case for a single particle of unit mass, $N=1$,

$$L = \tfrac{1}{2}\{(y^1)^2 + (y^2)^2 + (y^3)^2\} - q^3, \tag{25.22}$$

and the constraint forms are

$$\omega^1 = (e^{q^2} - 3e^{4q^2+q^1})dq^1 + (2e^{q^2} - 7e^{4q^2+q^1})dq^2 + (2q^3e^{q^2} + 5e^{4q^2+q^1})dq^3$$

$$\begin{aligned}\omega^2 = {} & \left[(1+t^2)(q^1)^2 + 6e^{q^1}(q^3)^2\right] dq^1 \\ & + \left[2(1+t^2)(q^1)^2 + 14e^{q^1}(q^3)^2\right] dq^2 \\ & + \left[2q^3(1+t^2)(q^1)^2 - 10e^{q^1}(q^3)^2\right] dq^3 .\end{aligned} \tag{25.23}$$

It is a trivial matter to see that $d\omega^1 \neq 0$, $d\omega^2 \neq 0$, while a lengthy calculation shows that $d\omega^1 \wedge \omega^1 \wedge \omega^2 = 0$, $d\omega^2 \wedge \omega^1 \wedge \omega^2 = 0$. The Frobenius theorem can thus be used, and we obtain

$$\begin{Bmatrix}\omega^1 \\ \omega^2\end{Bmatrix} = \begin{pmatrix} e^{q^2} & -e^{4q^2+q^1} \\ (1+t^2)(q^1)^2 & 2e^{q^1}(q^3)^2 \end{pmatrix} \begin{Bmatrix} d(q^1 + 2q^2 + (q^3)^2) \\ d(3q^1 + 7q^2 - 5q^3 \end{Bmatrix} . \tag{25.24}$$

The constraint forms (25.23) can thus be replaced by the equivalent system of holonomic constraint forms

$$(25.25)\qquad \bar{\omega}^1 = d(q^1 + 2q^2 + (q^3)^2)\ , \quad \bar{\omega}^2 = d(3q^1 + 7q^2 - 5q^3)\ .$$

The governing equations for the constrained motion are therefore

$$(25.26)\qquad \begin{cases} D(\phi^1 + 2\phi^2 + (\phi^3)^2) = 0 \\ D(3\phi^1 + 7\phi^2 - 5\phi^3) = 0 \end{cases} ,$$

$$(25.27)\qquad \begin{cases} D^2\phi^1 = \bar{\lambda}_1 + 3\bar{\lambda}_2 \\ D^2\phi^2 = 2\bar{\lambda}_1 + 7\bar{\lambda}_2 \\ D^2\phi^3 + 1 = 2\phi^3\bar{\lambda}_1 - 5\bar{\lambda}_2\ . \end{cases}$$

We have seen that significant simplification can be achieved by using the integrable constraints to construct a factor map. In this case we set

$$(25.28)\qquad {}^{\backprime}q^1 = q^1 + 2q^2 + (q^3)^2\ , \quad {}^{\backprime}q^2 = 3q^1 + 7q^2 - 5q^3\ ,$$
$${}^{\backprime}q^3 = q^3$$

with the inverse

$$(25.29)\qquad q^1 = 7\,{}^{\backprime}q^1 - 2\,{}^{\backprime}q^2 - 10\,{}^{\backprime}q^3 - 7({}^{\backprime}q^3)^2\ ,$$
$$q^2 = -\,3\,{}^{\backprime}q^1 + {}^{\backprime}q^2 + 5\,{}^{\backprime}q^3 + 3({}^{\backprime}q^3)^2\ , \quad q^3 = {}^{\backprime}q^3\ .$$

Under this factor map, the constraints become

$$(25.30)\qquad d\,{}^{\backprime}q^1 = 0\ , \quad d\,{}^{\backprime}q^2 = 0\ ,$$

while $y^\alpha = Z(q^\alpha)$ yields

$$(25.31)\qquad y^1 = 7\,{}^{\backprime}y^1 - 2\,{}^{\backprime}y^2 - (10 + 14\,{}^{\backprime}q^3)\,{}^{\backprime}y^3 ,$$

$$y^2 = -\,3\,{}^{\backprime}y^1 + {}^{\backprime}y^2 + (5 + 6\,{}^{\backprime}q^3)\,{}^{\backprime}y^3 , \quad y^3 = {}^{\backprime}y^3 .$$

Thus, since ${}^{\backprime}L = L(y({}^{\backprime}q, {}^{\backprime}y); q({}^{\backprime}q))$, (25.22) yields

$$(25.32)\qquad {}^{\backprime}L = \frac{1}{2}(7\,{}^{\backprime}y^1 - 2\,{}^{\backprime}y^2 - (10 + 14\,{}^{\backprime}q^3)\,{}^{\backprime}y^3)^2$$

$$+ \frac{1}{2}(-3\,{}^{\backprime}y^1 + {}^{\backprime}y^2 + (5 + 6\,{}^{\backprime}q^3)\,{}^{\backprime}y^3)^2 + \frac{1}{2}({}^{\backprime}y^3)^2 - {}^{\backprime}q^3 .$$

The new equations are therefore

$$(25.33)\qquad {}^{\backprime}q^1 = \text{constant} = c_1 , \quad {}^{\backprime}q^2 = c_2$$

$$(25.34)\qquad \bar{\lambda}_1 = D\Big[(7D\,{}^{\backprime}\phi^1 - 2D\,{}^{\backprime}\phi^2 - (10 + 14\,{}^{\backprime}\phi^3)D\,{}^{\backprime}\phi^3)7$$

$$+ (-3D\,{}^{\backprime}\phi^1 + D\,{}^{\backprime}\phi^2 + (5 + 6\,{}^{\backprime}\phi^3)D\,{}^{\backprime}\phi^3)(-3)\Big]$$

$$= -\,D\Big[(85 + 116\,{}^{\backprime}\phi^3)D\,{}^{\backprime}\phi^3\Big] ,$$

$$(25.35)\qquad \bar{\lambda}_2 = D\Big[(7D\,{}^{\backprime}\phi^1 - 2D\,{}^{\backprime}\phi^2 - (10 + 14\,{}^{\backprime}\phi^3)D\,{}^{\backprime}\phi^3)(-2)$$

$$+ (-3D\,{}^{\backprime}\phi^1 + D\,{}^{\backprime}\phi^2 + (5 + 6\,{}^{\backprime}\phi^3)D\,{}^{\backprime}\phi^3)\Big]$$

$$= D\Big[(25 + 34\,{}^{\backprime}\phi^3)D\,{}^{\backprime}\phi^3\Big] ,$$

$$(25.36)\qquad 0 = D\Big[(7D\,{}^{\backprime}\phi^1 - 2D\,{}^{\backprime}\phi^2 - (10 + 14\,{}^{\backprime}\phi^3)D\,{}^{\backprime}\phi^3)(-10 - 14\,{}^{\backprime}\phi^3)$$

$$+ (-3D\,{}^{\backprime}\phi^1 + D\,{}^{\backprime}\phi^2 + (5 + 6\,{}^{\backprime}\phi^3)D\,{}^{\backprime}\phi^3)(5 + 6\,{}^{\backprime}\phi^3) + D\,{}^{\backprime}\phi^3\Big]$$

$$+ (7D\,{}^{\backprime}\phi^1 - 2D\,{}^{\backprime}\phi^2 - (10 + 14\,{}^{\backprime}\phi^3)D\,{}^{\backprime}\phi^3)(14D\,{}^{\backprime}\phi^3)$$

$$-\,(-3\,{}^{\backprime}\phi^1 + D\,{}^{\backprime}\phi^2 + (5 + 6\,{}^{\backprime}\phi^3)D\,{}^{\backprime}\phi^3)(6D\,{}^{\backprime}\phi^3) + 1$$

$$= D\Big[\{(10 + 14\,{}^{\backprime}\phi^3)^2 + (5 + 6\,{}^{\backprime}\phi^3)^2 + 1\}D\,{}^{\backprime}\phi^3\Big]$$

$$-\,(170 + 232\,{}^{\backprime}\phi^3)D\,{}^{\backprime}\phi^3 + 1 .$$

Equations (25.34) and (25.35) determine the Lagrange multipliers

$\bar{\lambda}_1$, $\bar{\lambda}_2$ while (25.36) is the only surviving equation of motion that represents the motion of the system with $3-2 = 1$ degree of freedom. Since the constraints are catastatic, (25.36) admits an integral of the energy

$$\frac{1}{2}\{(10+14\phi^3)^2 + (5+6\phi^3)^2 + 1\}(D\phi^3)^2 + \phi^3 = E_o \tag{25.37}$$

and so

$$c_3 + t = \pm \int \sqrt{\frac{(10+14\phi^3)^2+(5+6\phi^3)^2+1}{E_o-\phi^3}}\, d\phi^3 .$$

There is one further simplification that can be obtained when the constraints are completely integrable but dependent on the y's. We can start with (25.4) and then specify the constraints in terms of the equations

$$dB^j(t;\, q^\beta;\, y^\beta) = 0 \tag{25.38}$$

since the matrix with entries A^i_j is nonsingular. Now,

$$d(\mu_i(t)B^i(t;\, q^\beta;\, y^\beta)) = \mu_i dB^i + B^i D\mu_i dt \tag{25.39}$$

since μ_i are functions of t only. This shows that

$$\mu_i(dB^i \wedge dt) = d(\mu_i B^i) \wedge dt , \tag{25.40}$$

and hence we see that

$$\mu_i \mathcal{U} \lrcorner (dB^i \wedge dt) = \mathcal{U} \lrcorner (d(\mu_i B^i) \wedge dt) = \pounds_{\mathcal{U}}(\mu_i B^i dt) . \tag{25.41}$$

Accordingly, the fundamental integral, (23.7) becomes

$$0 = \int_a^b \Phi^* \pounds_{\mathcal{U}}\{(L + \mu_i B^i)dt\} = \delta \int_a^b (L + \mu_i B^i)dt . \tag{25.42}$$

A conservative mechanical system with Lagrangian function L *and completely integrable constraints* $\omega^i = A^i_j dB^j$ *is equivalent to a variational problem with Lagrangian*

(25.43) $$\bar{L} = L + \mu_i(t)B^i(t;\ q^\beta;\ y^\beta)$$

and integrated constraints

(25.44) $$\Phi^* B^i(t;\ q^\beta;\ y^\beta) = C^i = \text{constant}\ ,\quad i=1,\ldots,R\ .$$

This result provides access to problems with completely integrable constraints through the direct method of the calculus of variations.

26. NONINTEGRABLE CONSTRAINTS - APPLICATION OF CARTAN SYSTEMS

A general problem with extended constraints will not usually be so cooperative as to yield satisfaction of the integrability conditions

$$d\omega^i \wedge \omega^1 \wedge \omega^2 \wedge \ldots \wedge \omega^R = 0\ ,\quad i=1,\ldots,R\ .$$

We accordingly construct the R forms

(26.1) $$S^i = d\omega^i \wedge \omega^1 \wedge \ldots \wedge \omega^R$$

of degree $R+2$ and the system of Cartan equations

(26.2) $$d\omega^i = \Gamma^i_j \wedge \omega^j + \Omega^i\ ,$$

(26.3) $$d\Gamma^i_j = \Gamma^i_k \wedge \Gamma^k_j + \Theta^i_j\ ,$$

where Ω^i and Θ^i_j are collections of elements of $\Lambda^2(K)$. The 2-forms $\{\Omega^i\}$ are referred to as the *Torsion forms* of the constraints and the 2-forms $\{\Theta^i_j\}$ are referred to as the *Curvature forms* of the constraints in analogy with the Cartan structure equations. These forms satisfy the consistency conditions

(26.4) $$d\Omega^i = \Gamma^i_j \wedge \Omega^j - \Theta^i_j \wedge \omega^j\ ,\quad d\Theta^i_j = \Gamma^i_k \wedge \Theta^k_j - \Theta^i_k \wedge \Gamma^k_j$$

that obtain from (26.2) and (26.3) by exterior differentiation.

Clearly, the system (26.2), (26.3) can always be constructed for any given system of R differential forms $\{\omega^i\}$. An arbitrary construction would, however, have little purpose, and, in fact, we can eliminate most of the arbitrariness by substituting (26.2) into (26.1). This gives

$$S^i = \Omega^i \wedge \omega^1 \wedge \omega^2 \wedge \ldots \wedge \omega^R , \quad i=1,\ldots,R , \tag{26.5}$$

and hence the known S's determine the Ω's on the orthogonal complement of $\omega^1 \wedge \ldots \wedge \omega^R$. Since the projections of the Ω's on $\omega^1 \wedge \ldots \wedge \omega^R$ can be absorbed in the terms $\Gamma^i_j \wedge \omega^j$ that occur in (26.2), we may take the Ω's as uniquely determined by the S's. Accordingly, the Ω's reflect the nonintegrability of the forms ω^i as described by (26.1).

We now come to the reason for all of this. It is shown in Section II of the Appendix that a Cartan system allows one to calculate a unique, nonsingular, R-by-R matrix $((A^i_j))$ of scalar valued functions on K, a unique set $\{B^i\}$ of R scalar valued functions on K, and a unique set of R differential forms $\{H^i\}$ on K, where the H's are determined by the 2-forms $\{\Omega^i\}$, $\{\Theta^i_j\}$, and vanish with them, such that

$$\omega^i = A^i_j(dB^j + H^j) . \tag{26.6}$$

This agrees exactly with what we obtained in the last Section, since $d\omega^i \wedge \omega^1 \wedge \ldots \wedge \omega^R = 0$ implies $H^j=0$, $j=1,\ldots,R$. Since $((A^i_j))$ is nonsingular, we can accordingly replace the system of constraints $\omega^i=0$ by the equivalent system of constraints

$$dB^i + H^i = 0 , \quad i=1,\ldots,R . \tag{26.7}$$

In exactly the same fashion as in the last Section, we see that the motion Φ satisfies the constraints (26.7) if and only if

$$\begin{aligned} DB^i(t;\ \phi^\beta;\ D\phi^\beta) + H^{0i}(t;\ \phi^\beta;\ D\phi^\beta) + H^{1i}_\alpha(t;\ \phi^\beta;\ D\phi^\beta)D\phi^\alpha \\ + H^{2i}_\alpha(t;\ \phi^\beta;\ D\phi^\beta)D^2\phi^\alpha = 0 , \end{aligned} \tag{26.8}$$

and is a Newtonian motion that satisfies the constraints (26.8) if and only if

$$(26.9)\qquad \{E|L\}_{\phi^\alpha} = \lambda_i\left(H_\alpha^{1i} + \frac{\partial B^i}{\partial\phi^\alpha}\right) - D\left[\lambda_i\left(H_\alpha^{2i} + \frac{\partial B^i}{\partial D\phi^\alpha}\right)\right] .$$

The energy equation for the system is thus given by

$$(26.10)\qquad D\left[T + \hat{V} + \lambda_i\left(\frac{\partial B^i}{\partial D\phi^\alpha} + H_\alpha^{2i}\right)D\phi^\alpha\right] = -\lambda_i\left(\frac{\partial B^i}{\partial t} + H^{0i}\right) .$$

If we happen to have

$$(26.11)\qquad D\mathcal{B}^i \wedge dq^1 \wedge \ldots \wedge dq^{3N} \wedge dt = 0 ,$$

then $\mathcal{B}^i = \mathcal{B}^i(t;\, q^\beta)$. In this case we can eliminate the $\mathcal{B}$'s by an admissible mapping such that $`q^i = \mathcal{B}^i(t;\, q^\beta)$, $i=1,\ldots,R$. However, the $\mathcal{H}$'s still remain very much in evidence, as might be expected in view of the nonintegrability of the given constraints.

27. CATASTATIC CONSTRAINTS IN THE GENERAL CASE

This Section considers catastatic constraints in terms of the viewpoint arrived at in the last two Sections. In particular, we consider constraints whose defining differential forms ω^i satisfy the conditions

$$(27.1)\qquad \omega^i \wedge dq^1 \wedge \ldots \wedge dq^{3N} = 0 , \quad d\omega^i \wedge dq^1 \wedge \ldots \wedge dq^{3N} = 0 ;$$

that is $\omega^i = \omega^i(q^\beta)$ so that the support of the constraint differential forms and their closure is configuration space. Under these conditions, the Cartan equations (26.2), (26.3) become

$$(27.2)\qquad d\omega^i = \Gamma_j^i(q^\beta) \wedge \omega^j + \Omega^i(q^\beta) ,$$

$$(27.3)\qquad d\Gamma_j^i(q^\beta) = \Gamma_k^i(q^\beta) \wedge \Gamma_j^k(q^\beta) + \Theta_j^i(q^\beta) ,$$

and C is also seen to be the support space for the torsion and curvature forms of the constraints.

Now, the ω's are independent so that we have

$$\omega^1 \wedge \omega^2 \wedge \ldots \wedge \omega^R \neq 0 \tag{27.4}$$

on C. We can thus complete $\{\omega^1,\ldots,\omega^R\}$ to a basis for $\Lambda^1(C)$:

$$i:\{\omega^i\} \to \{\bar{\omega}^\alpha(q^\beta)\}|\bar{\omega}^i = \omega^i, \quad i=1,\ldots,R ; \tag{27.5}$$

$$\bar{\omega}^1 \wedge \bar{\omega}^2 \wedge \ldots \wedge \bar{\omega}^{3N} \neq 0 .$$

Since $\bar{\omega}^1 \wedge \bar{\omega}^2 \wedge \ldots \wedge \bar{\omega}^{3N} \in \Lambda^{3N}(C)$ and $\dim(C) = 3N$, we have

$$d\bar{\omega}^\alpha \wedge \bar{\omega}^1 \wedge \ldots \wedge \bar{\omega}^{3N} = 0 . \tag{27.6}$$

The Frobenius theorem thus guarantees the existence of a 3N-by-3N matrix $((A^\alpha_\beta(q^\mu)))$ of functions on C and $3N$ functions $B^\alpha(q^\mu)$, $\alpha=1,\ldots,3N$ on C such that

$$\bar{\omega}^\alpha = A^\alpha_\beta \, dB^\beta , \quad dB^\beta = \overset{-1}{A}{}^\beta_\alpha \, \bar{\omega}^\alpha . \tag{27.7}$$

Accordingly, since the $\bar{\omega}$'s are linearly independent, the B's are $3N$ functionally independent functions on C ; that is

$$dB^1 \wedge dB^2 \wedge \ldots \wedge dB^{3N} \neq 0 . \tag{27.8}$$

We now allow Φ^* to act on these results to obtain

$$\Phi^*\bar{\omega}^\alpha = A^\alpha_\beta \, DB^\beta \, dt , \quad \Phi^* dB^\beta = \overset{-1}{A}{}^\beta_\alpha \, \Phi^*\omega^\alpha . \tag{27.9}$$

Our original constraint equations thus become translated as follows:

$$0 = \Phi^*\omega^i = A^i_\beta(\phi^\mu) \, DB^\beta(\phi^\mu) dt , \quad DB^\beta \, dt = \overset{-1}{A}{}^\beta_{\dagger} \, \Phi^*\bar{\omega}^{\dagger} , \tag{27.10}$$

with $\dagger = R+1,\ldots,3N$. The governing equations are therefore

$$\{E|L\}_{\phi^\alpha} = \lambda_i A^i_\beta(\phi^\mu) \frac{\partial B^\beta(\phi^\mu)}{\partial \phi^\alpha} , \quad \alpha=1,\ldots,3N , \tag{27.11}$$

$$0 = A^i_\beta(\phi^\mu) \frac{\partial B^\beta(\phi^\mu)}{\partial \phi^\alpha} D\phi^\alpha , \quad i=1,\ldots,R , \tag{27.12}$$

and the energy equation reads

(27.13) $$D(T+\hat{V}) = \lambda_i A^i_\beta(\phi^\mu) \frac{\partial B^\beta(\phi^\mu)}{\partial \phi^\alpha} D\phi^\alpha = 0 .$$

A further simplification can be achieved if we generate a factor map from the functions $\mathcal{B}^\alpha(q^\mu)$:

(27.14) $$\Gamma: K \to {}^\backprime K | \Gamma(t) = t = {}^\backprime t , \quad {}^\backprime q^\alpha = \mathcal{B}^\alpha(q^\beta) .$$

Use of the results established in Chapter I then yields

(27.15) $$\{E|{}^\backprime L\}_{{}^\backprime\phi^\alpha} = \lambda_i {}^\backprime A^i_\alpha({}^\backprime\phi^\mu) ,$$

(27.16) $$0 = {}^\backprime A^i_\alpha({}^\backprime\phi^\mu) D{}^\backprime\phi^\alpha ,$$

(27.17) $$D({}^\backprime T + {}^\backprime\hat{V}) = 0 ,$$

where

(27.18) $${}^\backprime L = {}^\backprime\Phi^{*}\, {}^\backprime L , \quad {}^\backprime L = L\left(\bar{\mathcal{B}}^\beta({}^\backprime q^\mu) ; \frac{\partial \bar{\mathcal{B}}^\beta({}^\backprime q^\mu)}{\partial {}^\backprime q^\alpha} {}^\backprime y^\alpha\right) ,$$

$${}^\backprime y^\alpha = \frac{\partial \mathcal{B}^\alpha}{\partial q^\beta} y^\beta$$

and

(27.19) $$q^\alpha = \bar{\mathcal{B}}^\alpha({}^\backprime q^\mu)$$

are the inverse transformations of ${}^\backprime q^\alpha = \mathcal{B}^\alpha(q^\beta)$. Accordingly, even though the nonintegrability of the differential forms ω^i prevents us from a direct reduction to a problem with only 3N-R governing equations, we do obtain substantial simplifications in the specification of the forces that maintain the constraints,

(27.20) $$f^c_\alpha = \lambda_i {}^\backprime A^i_\alpha(\phi^\mu) ,$$

and in the constraints themselves.

CHAPTER V

FUNDAMENTAL DIFFERENTIAL FORMS I. VARIATIONAL SYSTEMS

28. GENERAL VARIATIONAL SYSTEMS

General variational systems can arise in two distinct ways. The first is where we have a dynamical system without constraints and forces that derive from a potential function; the second is where we have the same system together with a system of holonomic constraints on event space. In the latter case, an appropriate admissible mapping of the system can be used to eliminate the holonomic constraints, as shown in Section 25. In both cases, we wind up with a Lagrangian function $L = L(t;\ q^{\beta};\ y^{\beta})$ on kinematic space which, because of the positive definite nature of kinetic energy, satisfies the condition

$$(28.1)\qquad \det\left(\frac{\partial^2 L}{\partial y^{\alpha}\partial y^{\beta}}\right) > 0\ .$$

This Lagrangian function, in turn, leads to the following conditions in order that Φ define a Newtonian motion for the *general variational system* associated with L.

A map Φ: $[a\leq t\leq b] \to K|\Phi(t)=t$, $\Phi(q^{\alpha})=\phi^{\alpha}(t)$, $\phi^{\alpha}(a)=a^{\alpha}$, $\phi^{\alpha}(b)=b^{\alpha}$ *defines a Newtonian motion for the general variational system associated with the Lagrangian function* L *if and only if*

$$(28.2)\qquad \int_a^b \Phi^{*}\pounds_{U}(L dt) = 0$$

holds for all horizontal vector fields U *such that* $u_1^{\alpha}(a;a^{\beta}) = u_1^{\alpha}(b;b^{\beta}) = 0$. *If these conditions are met, then the functions* $\{\phi^{\alpha}(t)\}$ *satisfy the Lagrange equations*

$$(28.3) \qquad \{E|L\}_{\phi^\alpha} = 0 , \quad \alpha=1,\ldots,\text{Deg}.$$

with $L = \Phi^* L$, *and admit the energy equation*

$$(28.4) \qquad D(E) = -\frac{\partial L(t;\phi^\beta;D\phi^\beta)}{\partial t}$$

with $E = \Phi^* E$ *and*

$$(28.5) \qquad E(t;\ q^\beta;\ y^\beta) = \frac{\partial L}{\partial y^\beta}\, y^\beta - L$$

as the total energy function.

These results are fine as they stand, with the exception of one glaring drawback; the problem is phrased as a boundary value problem, $\phi^\alpha(a) = a^\alpha$, $\phi^\alpha(b) = b^\alpha$, rather than as an initial value problem. This is indeed unfortunate since most problems of interest in mechanics are initial value problems rather than boundary value problems. It would therefore be most desirable to replace the fundamental integral condition (28.2) by a different system of conditions whereby we can recover the Lagrange equations (28.3) without having to assume given boundary data. In this context we note that the Lagrange equations themselves are not representable in any simple fashion in terms of quantities defined directly on Kinematic Space, K . This follows by observing that $\{E|L\}_{\phi^\alpha}$ = $D\left(\frac{\partial L}{\partial D\phi^\alpha}\right) - \frac{\partial L}{\partial \phi^\alpha}$, and $\frac{\partial L}{\partial \phi^\alpha} = \Phi^*\left(\frac{\partial L}{\partial q^\alpha}\right)$, but $D\left(\frac{\partial L}{\partial D\phi^\alpha}\right)dt = \Phi^* d\left(\frac{\partial L}{\partial y^\alpha}\right)$. Lastly, we actually need a formulation whereby we can consider families of Newtonian motions, not just a Newtonian motion that satisfies a given system of initial data.

29. THE FUNDAMENTAL DIFFERENTIAL FORM

Consider the differential form that is defined on all of K by

$$(29.1) \qquad J = P_\alpha dq^\alpha - E dt = P_\alpha dq^\alpha - (P_\beta y^\beta - L) dt ,$$

where

$$(29.2)\qquad P_\alpha = \frac{\partial L}{\partial q^\alpha}\ ,\qquad E = P_\beta y^\beta - L\ .$$

This differential form is referred to as the *Fundamental Differential Form* associated with the variational system with Lagrangian function L for the reasons given below.

A simple rearrangement of the terms that comprise J gives

$$J = Ldt + P_\alpha(dq^\alpha - y^\alpha dt)\ .$$

Accordingly, any map $\Phi\colon \mathbb{R} \to K|\Phi(t)=t$ gives

$$(29.3)\qquad \Phi^*J = Ldt + \Phi^*(P_\alpha)\Phi^*(dq^\alpha - y^\alpha dt) = Ldt\ ,$$

and the action integral, whose variation gives the fundamental integral (28.2), is expressible directly in terms of J by

$$(29.4)\qquad A = \int_a^b Ldt = \int_a^b \Phi^*J\ .$$

Exterior differentiation of J yields

$$dJ = dP_\alpha \wedge dq^\alpha - y^\beta dP_\beta \wedge dt - P_\alpha dy^\alpha \wedge dt$$
$$+ \frac{\partial L}{\partial q^\alpha}\, dq^\alpha \wedge dt + \frac{\partial L}{\partial y^\alpha}\, dy^\alpha \wedge dt\ .$$

However, since $P_\alpha = \partial L/\partial y^\alpha$ by (29.2), the terms involving $dy^\alpha \wedge dt$ cancel out. We thus have

$$(29.5)\qquad dJ = \left(dP_\alpha - \frac{\partial L}{\partial q^\alpha}\, dt\right) \wedge dq^\alpha - y^\alpha dP_\alpha \wedge dt$$
$$= \left(dP_\alpha - \frac{\partial L}{\partial q^\alpha}\, dt\right) \wedge (dq^\alpha - y^\alpha dt)\ .$$

Next, we note that (29.1) gives

$$(29.6)\qquad U \lrcorner J = P_\alpha u_1^\alpha$$

for any horizontal vector field U on K. Thus, (29.5) and

(29.6) combine to yield

(29.7) $$\pounds_{U} J = U \rfloor dJ + d(U \rfloor J) = U \rfloor dJ + d(P_\alpha u_1^\alpha)$$

$$= -(dP_\alpha - \frac{\partial L}{\partial q^\alpha} dt) u_1^\alpha + (U \rfloor dP_\alpha)(dq^\alpha - y^\alpha dt)$$

$$+ d(P_\alpha u_1^\alpha) .$$

Since $A = \int_a^b L \, dt = \int_a^b \Phi^* J$, we obtain

(29.8) $$\delta A = \int_a^b \delta(Ldt) = \int_a^b \delta(\Phi^* J) = \int_a^b \Phi^* \pounds_{U}(Ldt)$$

while (29.7) yields

(29.9) $$\Phi^* \pounds_{U} J = -\Phi^*(dP_\alpha - \frac{\partial L}{\partial q^\alpha} dt)\Phi^*(u_1^\alpha)$$

$$+ \Phi^*(U \rfloor dP_\alpha)\Phi^*(dq^\alpha - y^\alpha dt) + \Phi^* d(P_\alpha u_1^\alpha)$$

$$= -(DP_\alpha - \frac{\partial L}{\partial q^\alpha})\delta\phi^\alpha dt + D(P_\alpha \delta\phi^\alpha) dt ,$$

where we have used $\Phi^*(u_1^\alpha) = \delta\phi^\alpha$ and

(29.10) $$P_\alpha = \Phi^*(P_\alpha) = \partial L / \partial D\phi^\alpha .$$

Thus, since $\delta\phi^\alpha(a) = \delta\phi^\alpha(b) = 0$, (29.8) and (29.9) combine to yield

(29.11) $$\int_a^b \Phi^* \pounds_{U}(Ldt) = \int_a^b \Phi^* \pounds_{U} J ,$$

and hence *the fundamental integral condition* (28.2) *can be expressed directly in terms of the differential form* J .

If this were all that there were, there would have been no point to considering the fundamental differential form J , for we would just be reproducing our previous results. This is not the case, however. Let

$$V = v_o \frac{\partial}{\partial t} + v_1^\alpha \frac{\partial}{\partial q^\alpha} + v_2^\alpha \frac{\partial}{\partial y^\alpha}$$

be an arbitrary element of $T(K)$; i.e., $v_2^\alpha = Z(v_1^\alpha - y^\alpha v_o)$. Equations (29.1) and (29.5) give

$$(29.12)\qquad V\rfloor J = -(P_\alpha y^\alpha - L)(V\rfloor dt) - P_\alpha(V\rfloor dq^\alpha) ,$$

$$(29.13)\qquad \begin{aligned} V\rfloor dJ = {} & (y^\alpha dP_\alpha - \frac{\partial L}{\partial q^\alpha} dq^\alpha)(V\rfloor dt) \\ & - (dP_\alpha - \frac{\partial L}{\partial q^\alpha} dt)(V\rfloor dq^\alpha) \\ & + (dq^\alpha - y^\alpha dt)(V\rfloor dP_\alpha) , \end{aligned}$$

so that

$$(29.14)\qquad \begin{aligned} \pounds_V J = {} & (y^\alpha dP_\alpha - \frac{\partial L}{\partial q^\alpha} dq^\alpha)(V\rfloor dt) \\ & - (dP_\alpha - \frac{\partial L}{\partial q^\alpha} dt)(V\rfloor dq^\alpha) \\ & + (dq^\alpha - y^\alpha dt)(V\rfloor dP_\alpha) \\ & + d\{P_\alpha(V\rfloor dq^\alpha) - (P_\alpha y^\alpha - L)(V\rfloor dt)\} . \end{aligned}$$

We consequently obtain

$$(29.15)\qquad \begin{aligned} \Phi^*\{\pounds_V J - d(V\rfloor J)\} & = (D\phi^\alpha DP_\alpha - \frac{\partial L}{\partial \phi^\alpha} D\phi^\alpha)V_o dt \\ & \quad - (DP_\alpha - \frac{\partial L}{\partial \phi^\alpha})V_1^\alpha dt \\ & = -\{E|L\}_{\phi^\alpha}(V_1^\alpha - V_o D\phi^\alpha)dt . \end{aligned}$$

It follows from this that any Φ such that $\{E|L\}_{\phi^\alpha} = 0$ yields $\Phi^*\{\pounds_V J - d(V\rfloor J)\} = 0$. Conversely, if

$$(29.16)\qquad 0 = \int \Phi^*\{\pounds_V J - d(V\rfloor J)\}$$

holds for all vector fields $V \varepsilon T(K)$, we obtain the Lagrange equations $\{E|L\}_{\phi^\alpha} = 0$ from the fundamental lemma of the calculus of variations without fixed endpoints.

Any map $\Phi:\mathbb{R}\to K|\Phi(t)=t$ *such that*

$$(29.17)\qquad 0 = \int \Phi^*\{\pounds_V J - d(V\rfloor J)\} = \int \Phi^* V\rfloor dJ$$

holds for all vector fields $V\in T(K)$ *gives a Newtonian motion for the variational system that is characterized by the fundamental differential form*

$$(29.18)\qquad J = P_\alpha dq^\alpha - (P_\beta y^\beta - L)dt\ ,\qquad P_\alpha = \partial L/\partial y^\alpha\ .$$

Conversely, if Φ *gives a Newtonian motion for the variational system that is characterized by the fundamental differential form* J *, then*

$$(29.19)\qquad \Phi^*\pounds_V J = \Phi^* d(V\rfloor J) = D\Phi^*(V\rfloor J)dt\ .$$

We have thus achieved our objective of replacing the boundary value formulation of mechanics by a formulation that does not depend on either given initial data or given terminal data. Consideration of families of Newtonian motions now becomes a straightforward matter. It is both interesting and ironic to note that all of this comes about through replacing the differential form Ldt on K by the differential form $J = P_\alpha(dq^\alpha - y^\alpha dt) + Ldt$, where the first terms in J are annihilated by the action of any Φ^* such that $\Phi(t)=t$. As may be expected, it is the fundamental differential form J , rather than Ldt , that is central to the remaining considerations of this tract.

30. INVARIANCE AND QUADRATURES

If we go back to (29.14) and allow Φ^* to act on this equation, we obtain

$$(30.1)\qquad \Phi^*\pounds_V J = (D\phi^\alpha DP_\alpha - \frac{\partial L}{\partial\phi^\alpha} D\phi^\alpha)V_o dt$$
$$- (DP_\alpha - \frac{\partial L}{\partial\phi^\alpha})V_1^\alpha dt$$
$$+ D(P_\alpha V_1^\alpha - (P_\alpha D\phi^\alpha - L)V_o)dt$$
$$= \Big[-\{E|L\}_{\phi^\alpha}(V_1^\alpha - V_o D\phi^\alpha)$$
$$+ D(P_\alpha V_1^\alpha - (P_\alpha D\phi^\alpha - L)V_o)\Big]dt\ .$$

The results established in Section 11 and (29.3) accordingly yield

$$(30.2)\qquad \Phi^*\pounds_V J = \Delta(V)\Phi^*(Ldt) = \Delta(V)\Phi^* J\ ,$$

where $\Delta(V)$ is the symbol for the extended variation that is generated by the vector field V ; that is

$$(30.3)\qquad \Delta(V)A = \int\Delta(V)\Phi^*(Ldt) = \int\Delta(V)\Phi^* J = \int\Phi^*\pounds_V J\ .$$

Now,

$$A(\sigma) = A(0) + \sigma\Delta(V)A(0) + o(\sigma)\ ,$$

for composition of Φ with the orbits of V ; that is, $\Phi\to\Phi_\sigma(V)$. We thus see that the condition $\pounds_V J = 0$ guarantees invariance of the action integral $A = \int\Phi^* J$ under the mapping $\Phi\to\Phi_\sigma(V)$ for any $\Phi:\mathbb{R}\to K|\Phi(t)=t$.

A vector field $V\varepsilon T(K)$ such that

$$(30.4)\qquad \pounds_V J = 0$$

is said to generate an *invariance flow* of $A = \int\Phi^* J$ for every map $\Phi:\mathbb{R}\to K|\Phi(t)=t$. This definition and (30.1) give rise to the following Noetherian theorem.

If $V\varepsilon T(K)$ *generates an invariance flow of* $A = \int\Phi^* J$ *then the following conditions hold:* (i) *if* Φ *is a Newtonian motion for the variational system characterized by* J , *then the equations* *of motion,* $\{E|L\}_{\phi^\alpha} = 0$, *admit the quadrature*

$$(30.5) \qquad \eta(V;\Phi) = P_\alpha V_1^\alpha - (P_\beta D\phi^\beta - L)V_o = \text{constant};$$

(ii) *if* Φ *is an arbitrary map of* $\mathbb{R}$ *into* K *such that* $\Phi(t)=t$, *then the functions* $\{\phi^\alpha(t)\}$ *and the functions* V_o, V_1^α *satisfy the identity*

$$(30.6) \qquad \{E|L\}_{\phi^\alpha}(V_1^\alpha - V_o D\phi^\alpha) = D[P_\alpha V_1^\alpha - (P_\alpha D\phi^\alpha - L)V_o] ;$$

(iii) *for any map* $\Phi:\mathbb{R} \to K|\Phi(t)=t$, *the quantities* $\Phi_* V$ *and* $\Phi^* L$ *stand in the relation*

$$(30.7) \qquad 0 = \Phi^*(\pounds_V J) = \frac{\partial L}{\partial t} V_o + \frac{\partial L}{\partial \phi^\alpha} V_1^\alpha + \frac{\partial L}{\partial D\phi^\alpha}(DV_1^\alpha - DV_o D\phi^\alpha) + LDV_o .$$

The last relation follows from use of (11.3) and (30.2).

The next result follows immediately from the identity

$$(30.8) \qquad \pounds_{[V_1,V_2]} = \pounds_{V_1}\pounds_{V_2} - \pounds_{V_2}\pounds_{V_1} .$$

If $V_1 \epsilon T(K)$ *and* $V_2 \epsilon T(K)$ *generate invariance flows of* $A = \int \Phi^* J$, *then the following results hold:* (i) $[V_1,V_2]$ *generates an invariance flow of* A ; (ii) *if* $\Phi:\mathbb{R} \to K|\Phi(t)=t$ *is a Newtonian motion for the variational system characterized by* J , *then the equations of motion admit the quadrature*

(30.9)

$$\eta([V_1,V_2],\Phi) = P_\alpha \Phi_*[V_1,V_2]_1^\alpha - (P_\alpha D\phi^\alpha - L)\Phi_*[V_1,V_2]_o = \text{constant},$$

(iii) *any* $\Phi:\mathbb{R} \to K|\Phi(t)=t$ *satisfies the identity*

$$(30.10) \qquad \{E|L\}_{\phi^\alpha}(\Phi_*[V_1,V_2]_1^\alpha - \Phi_*[V_1,V_2]_o D\phi^\alpha) = D\eta([V_1,V_2];\Phi)$$

where

$$(30.11)\qquad [V_1,V_2] = [V_1,V_2]_o \frac{\partial}{\partial t} + [V_1,V_2]_1^\alpha \frac{\partial}{\partial q^\alpha} + [V_1,V_2]_2^\alpha \frac{\partial}{\partial y^\alpha} .$$

The process of forming the commutators

$$(30.12)\qquad [V_1,V_2] = \pounds_{V_1} V_2 = - \pounds_{V_2} V_1$$

of any two vector fields that generate invariance flows leads to another vector field that also generates an invariance flow. This process is by no means trivial, as the simplest examples easily show. In fact, since these results obtain in a context of Lagrangian mechanics on kinematic space, K , rather than in Hamiltonian mechanics on phase space[†], the construction of vector fields that generate invariance flows by (30.12) is significantly more useful than is found to be the case in Hamiltonian mechanics. This follows very simply from a comparison between phase space and kinematic space and the fact that the vector field (30.11) in K will satisfy the nontrivial restrictions (5.17); that is,

$$(30.13)\qquad [V_1,V_2]_2^\alpha = Z([V_1,V_2]_1^\alpha - y^\alpha[V_1,V_2]_o) .$$

The results obtained above constitute the basis for quadrature theory in classical analytical dynamics[††]; namely integrals of ignorable or cyclic coordinates and integrals of energy. Suppose, for the sake of argument that ϕ^1 is an *ignorable coordinate*; that is

$$(30.14)\qquad \partial L(t;\ \phi^\alpha;\ D\phi^\alpha)/\partial\phi^1 = 0 .$$

Equation (30.7) is then satisfied by

$$(30.15)\qquad V_o = 0 ,\quad V_1^\alpha = \delta^{\alpha 1} ,$$

and (30.5) gives the ignorable coordinate quadrature

[†] Abraham (Ref. 14).

[††] Whittaker (Ref. 13); Goldstein (Ref. 12); Wintner (Ref. 18).

(30.16) $\quad P_1 = \text{constant}.$

Secondly, suppose that $L(t;\ \phi^\alpha;\ D\phi^\alpha)$ does not depend explicitly on the time variable,

(30.17) $\quad \partial L(t;\ \phi^\alpha;\ D\phi^\alpha)/\partial t = 0\ .$

In this case, (30.7) is satisfied by

(30.18) $\quad V_o = 1\ ,\quad V_1^\alpha = 0\ ,$

and (30.5) gives the energy quadrature

(30.19) $\quad P_\beta\, D\phi^\beta - L = \text{constant}.$

31. THE SEARCH FOR QUADRATURES CONTINUED

A careful examination of the results contained in the last section shows that it is the vanishing of $\Phi^*\pounds_V J$, not $\pounds_V J$, that leads to the quadratures; the vanishing of $\pounds_V J$ was just a convenient way of guaranteeing that $\Phi^*\pounds_V J = 0$. This suggests that we can generalize the procedure given in the previous section by allowing $\pounds_V J = \rho$, where ρ is a differential form on K with the property that $\Phi^*\rho \equiv 0$. This, in turn, can be generalized by allowing ρ to be a differential form on K that mimics the various terms that occur on the right-hand side of (29.14). However, $\pounds_V J$ is linear and homogeneous in V , and hence we must likewise take ρ to be linear and homogeneous in V if we wish to satisfy $\pounds_V J = \rho$. Accordingly, let $A,\ B^\alpha,\ R_\alpha,\ S$ be differential forms on K , and set

(31.1)
$$\rho(V;A,B^\alpha,R_\alpha,S) = (y^\alpha dP_\alpha - \frac{\partial L}{\partial q^\alpha}\, dq^\alpha)(V\rfloor A)$$
$$- (dP_\alpha - \frac{\partial L}{\partial q^\alpha}\, dt)(V\rfloor B^\alpha) + (dq^\alpha - y^\alpha dt)(V\rfloor R_\alpha)$$
$$+ d(V\rfloor S)\ .$$

Action by Φ^* gives

$$(31.2)\qquad \Phi^*\rho = \Big[\{E|L\}_{\phi^\alpha} D\phi^\alpha \Phi^*(V\rfloor A) - \{E|L\}_{\phi^\alpha}\Phi^*(V\rfloor B^\alpha) + D\Phi^*(V\rfloor S)\Big]\,dt\ ,$$

and we achieve the desired form for $\Phi^*\rho$. Thus, if $V\varepsilon T(K)$ is such that

$$(31.3)\qquad \pounds_V J = \rho\ ,$$

we see that (30.1) and (31.2) yield the relation

$$(31.4)\qquad -\{E|L\}_{\phi^\alpha}(V_1^\alpha - V_o D\phi^\alpha) + D(P_\alpha V_1^\alpha - (P_\alpha D\phi^\alpha - L)V_o) = -\{E|L\}_{\phi^\alpha}(\bar{B}^\alpha - \bar{A}D\phi^\alpha) + D\bar{S}$$

where we have used the notation

$$(31.5)\qquad \Phi^*(V\rfloor A) = \bar{A}\ ,\quad \Phi^*(V\rfloor B^\alpha) = \bar{B}^\alpha\ ,\quad \Phi^*(V\rfloor S) = \bar{S}\ .$$

It also follows from (29.14) and (31.1) that $\pounds_V J = \rho$ is satisfied if and only if $V\varepsilon T(K)$ is such that

$$(31.6)\qquad 0 = (y^\alpha dP_\alpha - \frac{\partial L}{\partial q^\alpha}dq^\alpha)[V\rfloor(dt-A)] - (dP_\alpha - \frac{\partial L}{\partial q^\alpha}dt)[V\rfloor(dq^\alpha - B^\alpha)] + (dq^\alpha - y^\alpha dt)[V\rfloor(dP_\alpha - R_\alpha)] + d[V\rfloor(J-S)]\ .$$

The following result is now immediate.

Any vector field $V\varepsilon T(K)$ *such that* $\pounds_V J = \rho(V;A,B^\alpha,R_\alpha,S)$ *yields the identity*

$$(31.7)\qquad \{E|L\}_{\phi^\alpha}(V_1^\alpha - \bar{B}^\alpha - (V_o - \bar{A})D\phi^\alpha) = D(P_\alpha V_1^\alpha - (P_\alpha D\phi^\alpha - L)V_o - \bar{S})$$

and any $\Phi:\mathbb{R}\to K|\Phi(t)=t$ *that generates a Newtonian motion for the variational system characterized by* J *yields the quadrature*

$$(31.8) \qquad P_\alpha V_1^\alpha - (P_\alpha D\phi^\alpha - L)V_o - \bar{S} = \text{constant.}$$

Although these results provide a wide latitude in the search for quadratures of the variational system characterized by J, they have one important drawback; namely, they do not generate new quadratures from $[V_1,V_2]$. In order to see what happens, suppose that

$$(31.9) \qquad \pounds_{V_i} J = \rho(V_i;A_i,B_i^\alpha,R_{\alpha i},S_i) .$$

Then

$$(31.10) \qquad \begin{aligned} \pounds_{[V_1,V_2]} J &= \pounds_{V_1}\pounds_{V_2} J - \pounds_{V_2}\pounds_{V_1} J \\ &= \pounds_{V_1}\rho(V_2;A_2,B_2^\alpha,R_{\alpha 2},S_2) \\ &\quad - \pounds_{V_2}\rho(V_1;A_1,B_1^\alpha,R_{\alpha 1},S_1) . \end{aligned}$$

Thus, in order to obtain new quadratures, we would need

$$(31.11) \qquad \begin{aligned} &\pounds_{V_1}\rho(V_2;A_2\ldots) - \pounds_{V_2}\rho(V_1;A_1,\ldots) \\ &\quad = \rho([V_1,V_2];A_3,\ldots) , \end{aligned}$$

which is a very complex system of conditions that must be satisfied by the form $\rho(V;A,\ldots)$. General solutions of this problem are unknown. There is, however, a special class of solutions of this problem that is of interest in its own right.

We consider the case where the vector field V is required to satisfy

$$(31.12) \qquad 0 = \pounds_V dJ .$$

A direct expansion of $\pounds_V dJ$ gives

$$(31.13)\qquad 0 = V\rfloor d^2J + d(V\rfloor dJ) = d(V\rfloor dJ) \ ,$$

and hence we infer the existence of a scalar-valued function $\chi(t; q^\beta; y^\beta)$ such that

$$(31.14)\qquad V\rfloor dJ = d\chi \ .$$

We accordingly have

$$(31.15)\qquad \pounds_V J = V\rfloor dJ + d(V\rfloor J) = d(\chi + V\rfloor J) \ .$$

Although the function $\chi(t; q^\beta; y^\beta)$ plays a heavy role in Hamiltonian mechanics†, a more useful formulation in the Lagrangian theory obtains from the identity

$$(31.16)\qquad 0 = \pounds_V dJ = d\pounds_V J \ .$$

Thus, $\pounds_V J$ is a closed differential form on K, from which we infer the existence of a scalar-valued function $Q(t; q^\beta; y^\beta)$ on K such that

$$(31.17)\qquad \pounds_V J = dQ \ .$$

This result is particularly nice, for it gives

$$(31.18)\qquad \Delta(V)A = \int_a^b \Phi^*(\pounds_V J) = \int_a^b \Phi^* dQ = \int_a^b DQ\,dt$$

$$= [\Phi^*(b) - \Phi^*(a)](Q)$$

and

$$(31.19)\qquad \Phi^*(\pounds_V J) = \Big[-\{E|L\}_{\phi^\alpha}(V_1^\alpha - V_o D\phi^\alpha)$$

$$+ D(P_\alpha V_1^\alpha - (P_\beta D\phi^\beta - L)V_o)\Big]dt$$

$$= DQ\,dt \ ,$$

† Abraham (Ref. 14).

where $Q = \Phi^* \mathcal{Q}$. We therefore give the following definition.

A vector field $V \varepsilon T(K)$ is said to generate a *mobility transformation* of K relative to the fundamental differential form J if and only if

$$(31.20) \qquad \pounds_V dJ = 0 .$$

We note for purposes of future reference that the *orbits* of a mobility transformation generated by

$$V = v_o \frac{\partial}{\partial t} + v_1^\alpha \frac{\partial}{\partial q^\alpha} + Z(v_1^\alpha - y^\alpha v_o)\frac{\partial}{\partial y^\alpha}$$

are the solutions of the system of ordinary differential equations

$$\frac{d\grave{t}}{d\sigma} = v_o(\grave{t};\grave{q}^\beta) , \quad \frac{d\grave{q}^\alpha}{d\sigma} = v_1^\alpha(\grave{t};\grave{q}^\beta) , \quad \frac{d\grave{y}^\alpha}{d\sigma} = v_2^\alpha(\grave{t};\grave{q}^\beta;\grave{y}^\beta) ,$$

where the group is considered as a group of *point transformations* acting on kinematic space; it maps the point in K with coordinates (t, q^β, y^β) onto the point in K with coordinates $(\grave{t}, \grave{q}^\beta, \grave{y}^\beta)$. Our previous calculations lead directly to the following results.

If each of the vector fields V_i *generates a mobility transformation of* K *relative to* J *, then the following results hold: (i) there exists a scalar valued function* $\mathcal{Q}_i(t; q^\beta; y^\beta)$ *for each* V_i *such that*

$$(31.21) \qquad \pounds_{V_i} J = d\mathcal{Q}_i ;$$

(ii) *if* $A(a,b) = \int_a^b \Phi^* J$, *then*

$$(31.22) \qquad \Delta(V_i)A(a,b) = [\Phi^*(b) - \Phi^*(a)](\mathcal{Q}_i) ;$$

(iii) *every invariance flow generates a mobility transformation with* $\mathcal{Q}_i = 0$; (iv) *any map* $\Phi: \mathbb{R} \to K|\Phi(t)=t$ *satisfies the identity*

$$(31.23) \qquad \{E|L\}_{\phi^\alpha}(V_{1i}^\alpha - V_{0i}D\phi^\alpha) \equiv - D(Q_i - P_\alpha V_{1i}^\alpha + (P_\alpha D\phi^\alpha - L)V_{0i}) ;$$

(v) *if* Φ *generates a Newtonian motion for the variational system characterized by* J , *then the system admits the quadrature*

$$(31.24)\qquad \xi(V_i,\Phi) = P_\alpha V^\alpha_{1i} - (P_\alpha D\phi^\alpha - L)V_{0i} - Q_i = \text{constant};$$

(vi) *the vector fields* $[V_i,V_j]$ *generate mobility transformations of* K *relative to* J *and*

$$(31.25)\qquad \pounds_{[V_i,V_j]}J = dQ_{ij} = d\left(\pounds_{V_i}Q_j - \pounds_{V_j}Q_i\right) .$$

Part (vi) of the above theorem yields an immediate corollary that is of sufficient importance to be explicitly noted.

The collection of mobility transformations of K *relative to the fundamental differential form* J *forms a group under composition, and the generating vector fields of this group form a Lie algebra with the multiplication* $[V_1,V_2]$.

32. GROUP PROPERTIES OF KINEMATIC SPACE

It should be evident from the results just established, that the mobility group of Kinematic space relative to a fundamental differential form J depends in a very intimate way on the structure of the variational system characterized by the differential form J . Different variational systems will, in general, admit entirely different mobility groups. One would expect, however, that all mobility groups should be realized as subgroups of some master universal mobility group. In the case of arbitrary variational systems, the Lagrangian $L(t;\, q^\beta;\, y^\beta)$ can depend on the variables $(t;\, q^\beta;\, y^\beta)$ in quite an arbitrary manner, and, indeed, there is no simple classification of all such Lagrangian functions. Our interests, on the other hand, are in mechanics, and Lagrangian function for a dynamical system of N particles in the posited inertial frame $(t;\, q^\beta;\, y^\beta)$ has the explicit form

$$(32.1)\qquad L(t;\, q^\beta;\, y^\beta) = T - \hat{V} = \frac{1}{2}\, y^\alpha m_{\alpha\beta} y^\beta - \hat{V}(t;\, q^\beta) .$$

Now, it is clear physically, that the mobility group of K , for a mechanical system, is maximal when there are no interactions among the N particles of the system; that is, when the N particles undergo free motion. This situation is characterized by

$$L_F = T = \frac{1}{2} y^{\alpha} m_{\alpha\beta} y^{\beta} , \tag{32.2}$$

in which case, the fundamental differential form becomes

$$J_F = m_{\alpha\beta}(y^{\alpha}dq^{\beta} - \frac{1}{2}y^{\alpha}y^{\beta}dt) .$$

The *maximal mobility group*, M_N , of a mechanical system of N particles is the mobility group of K relative to the fundamental differential form

$$J_F = m_{\alpha\beta}(y^{\alpha}dq^{\beta} - \frac{1}{2}y^{\alpha}y^{\beta}dt) . \tag{32.3}$$

There are several subgroups of M_N that prove to be important, so we may as well define them now as later.

The *mass-universal group*, U_N , of a mechanical system of N particles is the maximal subgroup of M_N that is invariant under changes in the masses, m_i , $i=1,\ldots,N$, that define the components of the mass tensor

$$((m_{\alpha\beta})) = \text{diag}(m_1,m_1,m_1,m_2,\ldots,m_{N-1},m_N,m_N,m_N) \tag{32.4}$$

in the global inertial coordinate cover $(t;\ q^{\beta};\ y^{\beta})$ of K .

The *inertial group*, I_N , of a mechanical system of N particles is the maximal subgroup of U_N that leaves time invariant; that is $\pounds_v dt = 0$.

The *space group*, S_N , of a mechanical system of N particles is the maximal subgroup of U_N that can be generated by the action of a group on $\mathbb{R}\times E_3$. The space group is thus a group that can be realized by a group acting on E_3 which is populated by all N particles with arbitrarily given masses.

The *kinematic group*, K_N , of a mechanical system of N

particles is the maximal subgroup of S_N that leaves time invariant; that is, $\pounds_{\upsilon} dt = 0$. It is also the maximal subgroup of I_N that can be generated by the action of a group on $\mathbb{R}\times E_3$.

The proof of the basic theorem on maximal mobility groups is of sufficient length that it seems preferable to state the theorem and its corollaries at this point, and then follow along with the proof.

The maximal mobility group, M_N, *of a mechanical system with* N *particles is a* $\frac{3}{2}(N(3N-1)+4N+2)$*-parameter group. The generating vectors of this group are given by*

$$(32.5)\qquad \upsilon_o = \ell + rt + st^2 ,$$

$$(32.6)\qquad \upsilon_1^\beta = \frac{1}{2} rq^\beta + stq^\beta + k^\beta t + a_\alpha^\beta q^\alpha + c^\beta ,$$

$$(32.7)\qquad \upsilon_2^\beta = -\frac{1}{2} ry^\beta + s(q^\beta - ty^\beta) + k^\beta + a_\alpha^\beta y^\alpha ,$$

where ℓ, r, s, c^β, k^β *and the independent* a_α^β *such that*

$$(32.8)\qquad m_{\mu\beta} a_\eta^\beta = - m_{\eta\beta} a_\mu^\beta$$

are the $\frac{3}{2}(N(3N-1)+4N+2)$*-group parameters.* Since $\upsilon_2^\beta = Z(\upsilon_1^\beta - y^\beta \upsilon_o)$, we shall not list the entries of these functions in the following corollaries.

The mass universal group, U_N, *of a mechanical system with* N *particles is a* $3(3N+1)$*-parameter group. The generating vectors of this group are given by*

$$(32.9)\qquad \upsilon_o = \ell + rt + st^2$$

$$(32.10)\qquad \upsilon_1^\beta = \frac{1}{2} rq^\beta + stq^\beta + k^\beta t + j_\alpha^\beta q^\alpha + c^\beta ,$$

with

$$(32.11)\qquad ((j_\beta^\alpha)) = \operatorname{diag}(S_1, S_2, \ldots, S_N)$$

and S_i *are skew symmetric* 3-by-3 *matrices.*

The inertial group, I_N *, of a mechanical system with* N *particles is a* (9N+1)*-parameter group. The generating vectors of this group are given by* (32.9) *and* (32.10) *with* r=s=0 .

The space group, S_N *, of a mechanical system with* N *particles is a* 12*-parameter gorup. The generating vectors of this group are given by*

$$(32.12)\qquad v_o = \ell + rt + st^2$$

$$(32.13)\qquad v_1^\beta = \frac{1}{2}rq^\beta + stq^\beta + \bar{k}^\beta t + \bar{j}_\alpha^\beta q^\alpha + \bar{c}^\beta$$

with

$$(32.14)\qquad ((\bar{j}_\beta^\alpha)) = \mathrm{diag}(S,S,\ldots,S) ,$$

S *a skew symmetric* 3-by-3 *matrix, and*

$$(32.15)\qquad \begin{matrix} \bar{k}^{3i-2} = k_1 , & \bar{k}^{3i-1} = k_2 , & \bar{k}^{3i} = k_3 , \\ \bar{c}^{3i-2} = c_1 , & \bar{c}^{3i-1} = c_2 , & \bar{c}^{3i} = c_3 , \end{matrix} \qquad i=1,\ldots,N .$$

The kinematic group, K_N *, of a mechanical system with* N *particles is a* 10*-parameter group. The generating vectors of this group are given by* (32.12) *through* (32.15) *with* r=s=0 .

The proofs of these corollaries constitute straightforward applications of the definitions of the subgroups involved, and follow immediately from the general theorem. We therefore leave the proofs of the corollaries to the reader if he is interested, and embark on the proof of the fundamental theorem concerning M_N .

33. PROOF OF THE FUNDAMENTAL THEOREM ON MAXIMAL MOBILITY GROUPS

The maximal mobility group is obtained by constructing the general solution of

$$(33.1)\qquad d\pounds_{v} J_F = 0$$

with the fundamental differential form

$$(33.2)\qquad J_F = P_\alpha dq^\alpha - Tdt = m_{\alpha\beta}(y^\alpha dq^\beta - \frac{1}{2} y^\alpha y^\beta dt)$$

that obtains from

$$(33.3)\qquad L_F = T = \frac{1}{2} y^\alpha m_{\alpha\beta} y^\beta$$

and (29.1), (29.2). A straightforward calculation based on (32.2) and use of $v_2^\beta = Z(v_1^\beta - y^\beta v_o)$ yields

$$(33.4)\qquad \pounds_v J_F = Adt + B_\mu dq^\mu ,$$

where

$$(33.5)\qquad A = T\frac{\partial v_o}{\partial t} + 2Ty^\mu \frac{\partial v_o}{\partial q^\mu} - m_{\alpha\beta} y^\alpha y^\beta \frac{\partial v_1^\beta}{\partial q^\mu} ,$$

$$(33.6)\qquad B_\mu = m_{\mu\beta}\frac{\partial v_1^\beta}{\partial t} + m_{\mu\beta} y^\nu \frac{\partial v_1^\beta}{\partial q^\nu} - m_{\mu\nu} y^\nu \frac{\partial v_o}{\partial t}$$

$$+ m_{\beta\nu} y^\nu \frac{\partial v_1^\beta}{\partial q^\mu} - m_{\mu\nu} y^\nu y^\sigma \frac{\partial v_o}{\partial q^\sigma} - T\frac{\partial v_o}{\partial q^\mu} ,$$

and the functions v_o , v_1^β are functions of t and $\{q^\beta\}$ only. Now, (33.4) and (33.1) give the conditions

$$(33.7)\qquad 0 = d\pounds_v J_F = \frac{\partial A}{\partial q^\eta} dq^\eta \wedge dt + \frac{\partial A}{\partial y^\eta} dy^\eta \wedge dt + \frac{\partial B_\mu}{\partial t} dt \wedge dq^\mu$$

$$+ \frac{\partial B_\mu}{\partial q^\eta} dq^\eta \wedge dq^\mu + \frac{\partial B_\mu}{\partial y^\eta} dy^\eta \wedge dq^\mu ,$$

so that we obtain the conditions

$$(33.8)\qquad \frac{\partial A}{\partial y^\eta} = 0 ,$$

$$(33.9)\qquad \frac{\partial B_\mu}{\partial y^\eta} = 0 ,$$

$$\text{(33.10)} \qquad \frac{\partial B_\mu}{\partial q^\eta} - \frac{\partial B_\eta}{\partial q^\mu} = 0 ,$$

$$\text{(33.11)} \qquad \frac{\partial A}{\partial q^\eta} - \frac{\partial B_\eta}{\partial t} = 0 .$$

These four sets of conditions will be studied sequentially.

We start by substituting (33.5) into (33.8). This gives, after arrangement of certain terms,

$$\text{(33.12)} \qquad 0 = y^\alpha\left(m_{\alpha\eta}\frac{\partial \upsilon_o}{\partial t} - m_{\eta\beta}\frac{\partial \upsilon_1^\beta}{\partial q^\alpha} - m_{\alpha\beta}\frac{\partial \upsilon_1^\beta}{\partial q^\eta}\right) + y^\alpha y^\mu\left(2m_{\alpha\eta}\frac{\partial \upsilon_o}{\partial q^\mu} + m_{\alpha\mu}\frac{\partial \upsilon_o}{\partial q^\eta}\right) .$$

Since υ_o and υ_1^α do not depend on the y's , (33.12) can be satisfied if and only if the symmetrized coefficients of each power of the y's vanishes separately; that is, υ_o , υ_1^α must satisfy

$$\text{(33.13)} \qquad 0 = m_{\alpha\eta}\frac{\partial \upsilon_o}{\partial t} - m_{\eta\beta}\frac{\partial \upsilon_1^\beta}{\partial q^\alpha} - m_{\alpha\beta}\frac{\partial \upsilon_1^\beta}{\partial q^\eta} ,$$

$$\text{(33.14)} \qquad 0 = m_{\alpha\eta}\frac{\partial \upsilon_o}{\partial q^\mu} + m_{\mu\eta}\frac{\partial \upsilon_o}{\partial q^\alpha} + m_{\alpha\mu}\frac{\partial \upsilon_o}{\partial q^\eta} .$$

If we substitute (33.6) into (33.9) and use the same argument, a straightforward calculation yields the same system of conditions (33.13) and (33.14). Thus, the conditions (33.8) and (33.9) are satisfied if and only if υ_o and υ_1^α satisfy the system (33.13), (33.14).

Since $\{m_{\alpha\beta}\}$ vanishes for $\alpha \neq \beta$, it is easily seen that the system (33.14) is satisfied if and only if

$$\text{(33.15)} \qquad \upsilon_o = s(t) .$$

When (33.15) is substituted into (33.13), we obtain

$$\text{(33.16)} \qquad 0 = m_{\mu\beta}\frac{\partial \upsilon_1^\beta}{\partial q^\eta} + m_{\eta\beta}\frac{\partial \upsilon_1^\beta}{\partial q^\mu} - m_{\mu\eta}\frac{ds}{dt}$$

and (33.5), (33.6) yield

$$(33.17)\qquad A = T\frac{ds}{dt} - m_{\alpha\beta}\dot{y}^{\alpha}y^{\mu}\frac{\partial v_1^{\beta}}{\partial q^{\mu}},$$

$$(33.18)\qquad B_{\mu} = m_{\mu\beta}\frac{\partial v_1^{\beta}}{\partial t}.$$

Thus, putting (33.18) into (33.10) gives

$$(33.19)\qquad 0 = \frac{\partial}{\partial t}\left(m_{\mu\beta}\frac{\partial v_1^{\beta}}{\partial q^{\eta}} - m_{\eta\beta}\frac{\partial v_1^{\beta}}{\partial q^{\mu}}\right),$$

while (33.17) and (33.11) yield

$$(33.20)\qquad 0 = -m_{\alpha\beta}y^{\alpha}y^{\mu}\frac{\partial^2 v_1^{\beta}}{\partial q^{\mu}\partial q^{\eta}} - m_{\eta\beta}\frac{\partial^2 v_1^{\beta}}{\partial t^2}.$$

Using the fact that v_1^{β} is independent of the variables y^{β}, (33.20) can be satisfied if and only if

$$(33.21)\qquad 0 = \partial^2 v_1^{\beta}/\partial t^2,$$

$$(33.22)\qquad 0 = \frac{\partial}{\partial q^{\eta}}\left(m_{\alpha\beta}\frac{\partial v_1^{\beta}}{\partial q^{\mu}} + m_{\mu\beta}\frac{\partial v_1^{\beta}}{\partial q^{\alpha}}\right)$$

hold. Now, (33.21) is satisfied if and only if

$$(33.23)\qquad v_1^{\beta} = a^{\beta}(q^{\alpha})t + b^{\beta}(q^{\alpha}).$$

However, (33.23) and (33.16) combine to give

$$(33.24)\qquad 0 = m_{\mu\beta}t\frac{\partial a^{\beta}}{\partial q^{\eta}} + m_{\eta\beta}t\frac{\partial a^{\beta}}{\partial q^{\mu}} + m_{\mu\beta}\frac{\partial b^{\beta}}{\partial q^{\eta}} + m_{\eta\beta}\frac{\partial b^{\beta}}{\partial q^{\mu}} - m_{\mu\eta}\frac{ds}{dt}.$$

If we now use the fact that s is a function of t only and the a's and b's are functions of the q's only, all of the above equations reduce to the following system:

$$(33.25)\qquad v_o = \ell + rt + st^2,$$

$$(33.26)\qquad v_1^\alpha = a^\alpha(q^\beta)t + b^\alpha(q^\beta)\ ,$$

$$(33.27)\qquad 0 = m_{\mu\beta}\frac{\partial a^\beta}{\partial q^\eta} + m_{\eta\beta}\frac{\partial a^\beta}{\partial q^\mu} - 2m_{\mu\eta}s\ ,$$

$$(33.28)\qquad 0 = m_{\mu\beta}\frac{\partial a^\beta}{\partial q^\eta} - m_{\eta\beta}\frac{\partial a^\beta}{\partial q^\mu}\ ,$$

$$(33.29)\qquad 0 = m_{\mu\beta}\frac{\partial b^\beta}{\partial q^\eta} + m_{\eta\beta}\frac{\partial b^\beta}{\partial q^\mu} - m_{\mu\eta}r\ .$$

Progress from this point is most easily made by changing to matrix notation. If we set $\underset{\sim}{M} = ((m_{\alpha\beta}))$, $\underset{\sim}{A} = ((\partial a^\beta/\partial q^\alpha))$, $\underset{\sim}{B} = ((\partial b^\beta/\partial q^\alpha))$, $\underset{\sim}{E} = ((\delta_{\alpha\beta}))$, (33.27)-(33.29) become

$$(33.30)\qquad \underset{\sim}{0} = \underset{\sim}{M}\underset{\sim}{A} + \underset{\sim}{A}^T\underset{\sim}{M} - 2s\underset{\sim}{M}\ ,$$

$$(33.31)\qquad \underset{\sim}{0} = \underset{\sim}{M}\underset{\sim}{A} - \underset{\sim}{A}^T\underset{\sim}{M}\ ,$$

$$(33.32)\qquad \underset{\sim}{0} = \underset{\sim}{M}\underset{\sim}{B} + \underset{\sim}{B}^T\underset{\sim}{M} - r\underset{\sim}{M}\ .$$

It is then an easy matter to see that the general solution of this system of matrix equations is given by

$$(33.33)\qquad \underset{\sim}{A} = s\underset{\sim}{E}$$

$$(33.34)\qquad \underset{\sim}{B} = \frac{1}{2}r\underset{\sim}{E} + \underset{\sim}{J}\ ,$$

where

$$(33.35)\qquad \underset{\sim}{J} = \underset{\sim}{M}^{-\frac{1}{2}}\underset{\sim}{S}\underset{\sim}{M}^{\frac{1}{2}}\ ,\qquad \underset{\sim}{S}^T = -\underset{\sim}{S}$$

is the general solution of

$$(33.36)\qquad \underset{\sim}{M}\underset{\sim}{J} + \underset{\sim}{J}^T\underset{\sim}{M} = \underset{\sim}{0}$$

and $\underset{\sim}{M}^{\frac{1}{2}}$ is the positive square root of the positive definite matrix $\underset{\sim}{M}$.

We now convert the solutions (33.33)-(33.36) back in terms of the a's and the b's by means of the definitions $\underset{\sim}{A} = ((\partial a^\beta/\partial q^\alpha))$, $\underset{\sim}{B} = ((\partial b^\beta/\partial q^\alpha))$. This gives

$$\frac{\partial a^\beta}{\partial q^\alpha} = s\delta^\beta_\alpha , \tag{33.37}$$

$$\frac{\partial b^\beta}{\partial q^\alpha} = \frac{1}{2}r\delta^\beta_\alpha + J^\beta_\alpha , \tag{33.38}$$

$$m_{\mu\beta}J^\beta_\eta = - m_{\eta\beta}J^\beta_\mu . \tag{33.39}$$

The system (33.37) integrates directly to give

$$a^\beta = sq^\beta + k^\beta , \tag{33.40}$$

where the k's are constants. If we define the differential forms $\mathcal{B}^\beta$ by

$$\mathcal{B}^\beta = db^\beta , \tag{33.41}$$

then (33.38) gives

$$\mathcal{B}^\beta = \frac{1}{2}rdq^\beta + J^\beta_\alpha dq^\alpha . \tag{33.42}$$

Since the $\mathcal{B}$'s are exact, it follows that

$$0 = d\mathcal{B}^\beta = \frac{\partial J^\beta_\alpha}{\partial q^\gamma} dq^\gamma \wedge dq^\alpha ,$$

and hence

$$\frac{\partial J^\beta_\alpha}{\partial q^\gamma} = \frac{\partial J^\beta_\gamma}{\partial q^\alpha} ;$$

that is, there exist 3N functions $j^\mu(q^\beta)$ such that

$$J^\mu_\eta = \frac{\partial j^\mu}{\partial q^\eta} . \tag{33.43}$$

Accordingly, the system of conditions (33.39) becomes

(33.44) $$m_{\mu\beta}\frac{\partial j^\beta}{\partial q^\eta} = -m_{\eta\beta}\frac{\partial j^\beta}{\partial q^\mu}.$$

Thus

$$m_{\mu\beta}\frac{\partial^2 j^\beta}{\partial q^\eta \partial q^\gamma} = -m_{\eta\beta}\frac{\partial^2 j^\beta}{\partial q^\mu \partial q^\gamma} = -\frac{\partial}{\partial q^\mu}\left(m_{\eta\beta}\frac{\partial j^\beta}{\partial q^\gamma}\right)$$

$$m_{\mu\beta}\frac{\partial^2 j^\beta}{\partial q^\eta \partial q^\gamma} = m_{\mu\beta}\frac{\partial^2 j^\beta}{\partial q^\gamma \partial q^\eta} = -m_{\gamma\beta}\frac{\partial^2 j^\beta}{\partial q^\mu \partial q^\eta} = -\frac{\partial}{\partial q^\mu}\left(m_{\gamma\beta}\frac{\partial j^\beta}{\partial q^\eta}\right),$$

and (33.44) may be combined with this to yield

(33.45) $$\frac{\partial}{\partial q^\mu}\left(m_{\eta\beta}\frac{\partial j^\beta}{\partial q^\gamma} + m_{\gamma\beta}\frac{\partial j^\beta}{\partial q^\eta}\right) = 0,$$
$$\frac{\partial}{\partial q^\mu}\left(m_{\eta\beta}\frac{\partial j^\beta}{\partial q^\gamma} - m_{\gamma\beta}\frac{\partial j^\beta}{\partial q^\eta}\right) = 0.$$

These equations imply that

$$\frac{\partial}{\partial q^\mu} m_{\eta\beta}\frac{\partial j^\beta}{\partial q^\gamma} = 0,$$

and hence, since $\det(m_{\alpha\beta}) \neq 0$, we finally obtain

(33.46) $$J^\beta_\alpha = \partial j^\beta/\partial q^\alpha = a^\beta_\alpha = \text{constants}.$$

Since the J's must satisfy the equations (33.39), the constants a^β_α are required to satisfy

(33.47) $$m_{\mu\beta}a^\beta_\eta = -m_{\eta\beta}a^\beta_\mu.$$

We note that the general solution of this system is given in matrix notation by

(33.48) $$\underset{\sim}{a} = ((a^\beta_\alpha)) = \underset{\sim}{M}^{-\frac{1}{2}}\underset{\sim}{S}\underset{\sim}{M}^{\frac{1}{2}}, \quad \underset{\sim}{S}^T = -\underset{\sim}{S}.$$

We now combine all of the results to obtain

(33.49) $$\upsilon_o = \ell + rt + \mathscr{s}t^2,$$

$$(33.50) \qquad v_1^\beta = \tfrac{1}{2}rq^\beta + \delta tq^\beta + tk^\beta + a_\alpha^\beta q^\alpha + c^\beta ,$$

while v_2^β is given by $v_2^\beta = Z(v_1^\beta - v_o y^\beta)$. This establishes equations (32.5) through (32.8) of the basic theorem on maximal mobility groups. It is now an easy matter to check that the vectors generated by the possible values of the constants ℓ, r, s, c^β, k^β and the independent a_β^α satisfying (33.47) are linearly independent with constant coefficients and form a Lie algebra under the product $[v_1, v_2] = \pounds_{v_1} v_2$. The set of all transformations generated by these vectors thus forms a group under composition. Finally, we note that there are 3 independent parameters ℓ, r, s, 6N independent parameters c^β, k^β, $\beta=1,\ldots,N$, and $3N(3N-1)/2$ independent a_β^α that satisfy the conditions (33.47), as follows immediately from the solution (33.48). This gives a total of $\frac{3}{2}[N(3N-1)+4N+2]$ independent parameters, and the proof of the fundamental theorem is complete.

We note for future reference that (33.4), (33.17), (33.18), (33.49) and (33.50) give

$$(33.51) \qquad \pounds_v J_F = m_{\mu\beta}(sq^\beta + k^\beta)dq^\mu = d\left[m_{\mu\beta}(\tfrac{s}{2}q^\beta + k^\beta)q^\mu\right] ,$$

for vector fields that generate the maximal mobility group.

34. QUADRATURES, GROUP ORBITS AND RELATED TOPICS

The first thing that comes to mind in obtaining a full understanding of the maximal mobility group is to use the theorem in Section 31 to obtain the quadratures that result from the maximal mobility group. The dynamical system under study is that characterized by the fundamental differential form (see (33.2))

$$(34.1) \qquad J_F = P_\alpha dq^\alpha - Tdt = m_{\alpha\beta}(y^\alpha dq^\beta - \tfrac{1}{2}y^\alpha y^\beta dt) .$$

The last section has shown that any vector field that generates the maximal mobility group gives (see (33.51))

(34.2) $$\pounds_v J_F = d\left[m_{\mu\beta}(\frac{s}{2}q^\beta + k^\beta)q^\mu\right] .$$

and hence (31.21) yields

(34.3) $$\mathcal{Q} = m_{\mu\beta}(\frac{s}{2}q^\beta + k^\beta)q^\mu .$$

A substitution of (34.1) and (34.3) into (31.24) shows that any Φ that gives a Newtonian motion of the system admits the quadratures

(34.4) $$\xi(a^\beta_\alpha, c^\beta, k^\beta, \ell, r, s; \Phi) = P_\alpha V^\alpha_1 - TV_o - Q = \text{constant}$$

for each of the independent parameters in the set $(a^\beta_\alpha,\ c^\beta,\ k^\beta,\ \ell,\ r,\ s)$ that occur in the explicit representation of V_o, V^α_1 obtained by allowing Φ_* to act on (33.49) and (33.50). When this is written out and we use the fact that (33.47) shows $a^\beta_\gamma m_{\beta\alpha}$ to be skewsymmetric in (γ,α), we obtain

(34.5) $$\begin{aligned}\xi = {} & \ell[\tfrac{1}{2}m_{\alpha\beta}D\phi^\alpha D\phi^\beta] + r[\tfrac{1}{2}m_{\alpha\beta}\phi^\beta D\phi^\alpha - \tfrac{1}{2}tm_{\alpha\beta}D\phi^\alpha D\phi^\beta] \\ & + s[tm_{\alpha\beta}\phi^\alpha D\phi^\beta - \tfrac{1}{2}t^2 m_{\alpha\beta}D\phi^\alpha D\phi^\beta - \tfrac{1}{2}m_{\alpha\beta}\phi^\alpha\phi^\beta] \\ & + k^\beta[tm_{\alpha\beta}D\phi^\alpha - m_{\alpha\beta}\phi^\alpha] + c^\beta[m_{\alpha\beta}D\phi^\alpha] \\ & + \tfrac{1}{2}a^\beta_\gamma m_{\beta\alpha}(\phi^\gamma D\phi^\alpha - \phi^\alpha D\phi^\gamma) .\end{aligned}$$

This allows us to read off the following list of quadratures: (1) *conservation of generalized angular momentum*

(34.6) $$\xi^\gamma_{1\beta} = m_{\beta\alpha}(\phi^\gamma D\phi^\alpha - \phi^\alpha D\phi^\gamma) , \quad \gamma,\beta = 1,\dots,3N$$

from (a^β_γ) ; (2) *conservation of linear momentum*

(34.7) $$\xi_{2\beta} = m_{\beta\alpha}D\phi^\alpha , \quad \beta = 1,\dots,3N$$

from (c^β) ; (3) *conservation of generalized mass centers*

(34.8) $$\xi_{3\beta} = m_{\beta\alpha}(tD\phi^\alpha - \phi^\alpha) , \quad \beta = 1,\dots,3N$$

from (k^β) ; (4) *conservation of energy*

$$\xi_4 = \frac{1}{2} m_{\alpha\beta} D\phi^\alpha D\phi^\beta \tag{34.9}$$

from ℓ ; (5) *conservation of moment of generalized mass centers*

$$\xi_5 = m_{\alpha\beta}\{(tD\phi^\alpha - \phi^\alpha)D\phi^\beta + (tD\phi^\beta - \phi^\beta)D\phi^\alpha\} \tag{34.10}$$

from r ; (6) *conservation of moment of inertial about mass center*

$$\xi_6 = m_{\alpha\beta}(tD\phi^\alpha - \phi^\alpha)(tD\phi^\beta - \phi^\beta) \tag{34.11}$$

from s . All of these conserved quantities are physically well defined and constitute an extension of the classic quadratures of total angular momentum, total linear momentum, total mass center, and total energy that obtain from invariance under the Kinematic group K_N . We note in this context, that *the Kinematic group is actually just the Galilean group* of classical mechanics, as follows immediately from the corollary given in Section 32.

The next thing to do is to obtain the finite transformations that are generated by the generating vectors of the maximal mobility group. Since the maximal mobility group is a group of point transformations on Kinematic space, its orbits are obtained by solving the orbit equations

$$\frac{d\grave{t}}{d\sigma} = v_o(\grave{t};\grave{q}^\beta) , \quad \frac{d\grave{q}^\alpha}{d\sigma} = v_1^\alpha(\grave{t};\grave{q}^\beta) , \tag{34.12}$$

$$\frac{d\grave{y}^\alpha}{d\sigma} = v_2^\alpha(\grave{t};\grave{q}^\beta) = \grave{Z}(v_1^\alpha - v_o\grave{y}^\alpha) .$$

However, the meaning of these transformations can be deduced from solving only the equations

$$\frac{d\grave{t}}{d\sigma} = v_o(\grave{t};\grave{q}^\beta) , \quad \frac{d\grave{q}^\alpha}{d\sigma} = v_1^\alpha(\grave{t};\grave{q}^\beta) \tag{34.13}$$

that define the orbits of Event space in view of the group extension structure of K (i.e., $\grave{y}^\alpha = Z(\grave{q}^\alpha)/Z(\grave{t})$ by (6.6)). By itself, this is useful, but significantly more information can be obtained if we correlate the finite transformations with the quadrature that obtain from them and the quantities that are invariant. In the context of invariance, we note from (34.2) that the transformations generated by all parameters but s and k^β leave J_F invariant, while the transformations generated from s and k^β leave dJ_F invariant.

The simplest point transformation is that generated by the parameter ℓ. With all parameters but ℓ set equal to zero, $\ell=1$, (34.13), (33.49) and (33.50) yield

$$\text{(34.14)} \qquad \grave{t} = t + \sigma\,, \qquad \grave{q}^\alpha = q^\alpha\,;$$

that is, ℓ generates the group of all constant time translations. Accordingly, invariance of J_F under the group of all constant time translations gives the quadrature (34.9) of conservation of energy. Next, if we set all parameters but c^i equal to zero, $c^i=1$, we obtain

$$\text{(34.15)} \qquad \grave{t} = t\,, \qquad \grave{q}^\alpha = q^\alpha + \delta^\alpha_{\cdot i}\sigma$$

that is, c^i generates the group of all constant translations in the q^i direction. Invariance of J_F under the group of all constant translations in the q^i direction gives the quadrature (34.7) (with $\beta=i$) of conservation of linear momentum in the q^i direction. If we set all of the parameters but k^i equal to zero, $k^i=1$, we obtain

$$\text{(34.16)} \qquad \grave{t} = t\,, \qquad \grave{q}^\alpha = q^\alpha + t\delta^\alpha_{\cdot i}\sigma$$

that is; k^i generates the group of all translations in the q^i direction with constant velocity equal to σ. Invariance of dJ_F under the group of all translations in the q^i direction with constant velocity gives the quadrature (34.8) (with $\beta=i$) of conservation of the ith generalized mass center. If we use the general solution for the independent a^β_α given by (33.48), we obtain

$$(34.17) \qquad `t = t \ , \quad `q^{\alpha} = A^{\alpha}_{\beta} q^{\beta}$$

where

$$(34.18) \qquad A^{\alpha}_{\beta} = (m^{-\frac{1}{2}})^{\alpha\mu} O^{\nu}_{\mu} (m^{\frac{1}{2}})_{\nu\beta}$$

and $\underset{\sim}{O} = \exp(\underset{\sim}{S})$ is orthogonal. Thus, the finite transformations generated by the independent a^{β}_{α} constitute all mass-weighted rotations of configuration space. Invariance of J_F under the group of all mass-weighted rotations on configuration space gives the quadratures (34.6) of conservation of generalized angular momentum.

We now come to the two remaining sets of transformations of M_N, namely, those generated by r and by s. If we set all parameters but r equal to zero, $r=1$, the orbit equations are

$$(34.19) \qquad \frac{d`t}{d\sigma} = `t \ , \quad \frac{d`q^{\alpha}}{d\sigma} = \frac{1}{2}`q^{\alpha} \ , \quad `y^{\alpha} = Z(`q^{\alpha})/Z(`t)$$

and we obtain the finite transformations

$$(34.20) \qquad `t = e^{\sigma} t \ , \quad `q^{\alpha} = e^{\sigma/2} q^{\alpha} \ , \quad `y^{\alpha} = e^{-\sigma/2} y^{\alpha} \ .$$

The group of transformations generated by r is thus the extension of the group of all scale transformations $t \rightarrowtail e^{\sigma} t$, $q^{\alpha} \rightarrowtail e^{\sigma/2} q^{\alpha}$ of Event space. Invariance of J_F under this group of scale transformations gives the quadrature (34.10) of conservation of moment of generalized mass center.

If we set all of the parameters but s equal to zero, $s=1$, the orbit equatons are

$$(34.21) \qquad \frac{d`t}{d\sigma} = `t^2 \ , \quad \frac{d`q^{\alpha}}{d\sigma} = `t`q^{\alpha} \ , \quad `y^{\alpha}(\sigma) = Z(`q^{\alpha})/Z(`t) \ ,$$

and we obtain the finite transformations

$$(34.22) \qquad `t = \frac{t}{1-\sigma t} \ , \quad `q^{\alpha} = \frac{q^{\alpha}}{1-\sigma t} \ , \quad `y^{\alpha} = y^{\alpha} + \sigma(q^{\alpha} - t y^{\alpha}) \ .$$

The first of these transformations suggests that s generates translations of the point at infinity on the time line. Invariance of dJ under this group of all translations of the point at infinity of time gives the quadrature (34.11) of conservation of moment of inertia about the mass center. Although the transformations generated by s are of an entirely different nature from all of the rest of the transformations of M_N, they are still reasonable in that the motion of a system of N free particles is unaware of the location of the point at infinity on the time line. On the other hand, as soon as the particles cease to be free particles, the interaction introduces a characteristic length or a characteristic time for the system and the dynamic relations render the system aware of the distinction between motion for finite time and motion for infinite time. This is a point that is often glossed over or not mentioned at all in the classic treatment of mechanics.

Further understanding of the group generated by s can be achieved by inquiring into what kinds of variational interactions will preserve this group as a mobility group of the interacting system. We could do this for the other transformations of M_N, but these transformations constitute a subgroup of the general linear group, and as such, are well understood. Variational systems that arise in mechanics have a Lagrangian function given by $L = T - \hat{V}(t;\, q^\beta)$, and hence their fundamental differential forms are

$$(34.23)\qquad J = P_\alpha dq^\alpha - (P_\alpha y^\alpha - T + \hat{V})dt\ ,$$

with

$$(34.24)\qquad P_\alpha = \frac{\partial L}{\partial y^\alpha} = \frac{\partial T}{\partial y^\alpha}\ .$$

However, $J_F = P_\alpha dq^\alpha - (P_\alpha y^\alpha - T)dt$ with $P_\alpha = \partial T/\partial y^\alpha$, so that we can write

$$(34.25)\qquad J = J_F + \hat{V}dt\ .$$

This implies that

$$(34.26)\qquad \pounds_V J = \pounds_V J_F + \pounds_V(\hat{V}dt)\ .$$

Thus, since $\pounds_V dJ_F = 0$ for the V generated by s, we can have $\pounds_V dJ = 0$ if and only if

$$(34.27)\qquad 0 = \pounds_V d(\hat{V}dt) = d\pounds_V(\hat{V}dt)\ .$$

Since $\hat{V}$ is independent of the variables y^α, we can dispense with the quantities v_2^α and take

$$(34.28)\qquad V = t^2 \frac{\partial}{\partial t} + tq^\alpha \frac{\partial}{\partial q^\alpha}\ .$$

A straightforward calculation then gives

$$(34.29)\qquad \pounds_V(\hat{V}dt) = \left(t^2 \frac{\partial \hat{V}}{\partial t} + tq^\alpha \frac{\partial \hat{V}}{\partial q^\alpha} + 2t\hat{V}\right) dt\ ,$$

and consequently (34.27) gives the condition

$$0 = \frac{\partial}{\partial q^\beta}\left(t^2 \frac{\partial \hat{V}}{\partial t} + tq^\alpha \frac{\partial \hat{V}}{\partial q^\alpha} + 2t\hat{V}\right) dq^\beta \wedge dt\ .$$

Thus dJ is invariant under the transformations generated by s if and only if the potential function $\hat{V}(t;\ q^\beta)$ satisfies

$$(34.30)\qquad t^2 \frac{\partial \hat{V}}{\partial t} + tq^\alpha \frac{\partial \hat{V}}{\partial q^\alpha} + 2t\hat{V} = h(t)$$

for any function $h(t)$. We note that $h(t) = 0$ gives $\pounds_V(\hat{V}dt) = 0$, as follows from (34.29), and that (34.30) has nontrivial solutions for $h(t) = 0$; namely, all functions of q^β that are homogeneous of degree -2. Accordingly,

$$(34.31)\qquad \hat{V} = \theta(q^\beta)\ ,\qquad q^\alpha \frac{\partial \theta}{\partial q^\alpha} = -\,2\theta$$

yields an interacting dynamical system such that J is invariant under the group generated by s.

CHAPTER VI

FUNDAMENTAL DIFFERENTIAL FORMS 2. NONCONSERVATIVE NONHOLONOMIC SYSTEMS

35. NONCONSERVATIVE, NONHOLONOMIC DYNAMICAL SYSTEMS

The structure of a general nonconservative, nonholonomic dynamical system is obtained by combining the results established in Chapters III and IV. The formulation of this structure occurs over Kinematic space K referred to the global inertial coordinate cover $(t;\, q^\beta;\, y^\beta)$, and is obtained through the assignment of the following information: (i) a differential form $W(t;\, q^\beta;\, y^\beta)$ of work such that

$$(35.1)\qquad W \wedge dq^1 \wedge dq^2 \wedge \ldots \wedge dq^{3N} = 0 ;$$

(ii) R independent differential forms $\omega^i(t;\, q^\beta;\, y^\beta)$ that specify the constraints

$$(35.2)\qquad \omega^i(t;\, q^\beta;\, y^\beta) = 0 , \quad i=1,\ldots,R ,$$

the independence being stated by

$$(35.3)\qquad \omega^1 \wedge \omega^2 \wedge \ldots \wedge \omega^R \neq 0 ;$$

(iii) the kinetic energy scalar

$$(35.4)\qquad T = \frac{1}{2}\, y^\alpha m_{\alpha\beta} y^\beta ,$$

where $\{m_{\alpha\beta}\}$ is the mass tensor that is given by

$$(35.5)\qquad ((m_{\alpha\beta})) = \mathrm{diag}(m_1, m_1, m_1, m_2, \ldots, m_{N-1}, m_N, m_N, m_N)$$

in the given coordinate cover $(t;\ q^\beta;\ y^\beta)$. A straightforward combination of the results established in Chapters III and IV allows us to draw the following conclusions.

A map $\Phi:[a\leq t\leq b]\to K|\Phi(t)=t$, $\Phi(q^\alpha)=\phi^\alpha(t)$, $\Phi(y^\alpha)=D\phi^\alpha(t)$, $\phi^\alpha(a)=a^\alpha$, $\phi^\alpha(b)=b^\alpha$ *defines a Newtonian motion for the dynamical system that is acted upon by the system of forces that arise from the work* 1-*form* $W(t;\ q^\beta;\ y^\beta)$ *and from the* R *independent constraints* $\omega^i(t;\ q^\beta;\ y^\beta)=0$ *(assumed to be compatible with the boundary data* $\phi^\alpha(a)=a^\alpha$, $\phi^\alpha(b)=b^\alpha$ *) if and only if the* 3N+R *functions* $\{\phi^\alpha(t),\ \lambda_i(t)\}$ *satisfy*

$$(35.6)\qquad \Phi^*\omega^i = 0\ ,$$

and

$$(35.7)\qquad 0 = \int_a^b \Phi^*\{\mathcal{L}_U(T dt) + U\rfloor(W+\lambda_i\omega^i)\wedge dt\}$$

for all horizontal vector fields

$$(35.8)\qquad U = u_1^\alpha(t;\ q^\beta)\frac{\partial}{\partial q^\alpha} + \left(\frac{\partial u_1^\alpha(t;q^\beta)}{\partial t} + \frac{\partial u_1^\alpha(t;q^\beta)}{\partial q^\mu}\, y^\mu\right)\frac{\partial}{\partial y^\alpha}$$

such that

$$(35.9)\qquad u_1^\alpha(a;a^\beta) = u_1^\alpha(b;b^\beta) = 0\ .$$

Although a specific representation is not required at this point for the constraint 1-forms ω^i, a specific representation for the work 1-form W is essential. The representation that is most useful is that which was obtained in Section 21:

$$(35.10)\qquad W = -\,dp(t;\ q^\beta;\ y^\beta) - \mathcal{Q}(t;\ q^\beta;\ y^\beta)\ ,$$

where

$$(35.11)\qquad p(t;\ q^\beta;\ y^\beta) = -\,(HW)(t;\ q^\beta;\ y^\beta)$$

is the unique potential of W,

$$(35.12)\qquad \mathcal{Q}(t;\ q^\beta;\ y^\beta) = -(H d\mathcal{W})(t;\ q^\beta;\ y^\beta)\ ,$$

and H is the homotopy operator introduced in Section 10 of the Appendix. The exact same line of reasoning as used in Section 21 shows that the fundamental integral conditon (35.7) is equivalent to the integral condition

$$(35.13)\qquad 0 = \int_a^b \Phi^*\{\pounds_{\mathcal{U}}(L dt) + \mathcal{U}\rfloor(-\mathcal{Q}+\lambda_i\omega^i)\wedge dt\}\ ,$$

where

$$(35.14)\qquad L(t;\ q^\beta;\ y^\beta) = T(y^\beta) - p(t;\ q^\beta;\ y^\beta)$$

is the *Lagrangian function* of the system.

The explicit form of the governing equations of a nonconservative nonholonomic system follows directly from (35.6) and (35.13) once we agree on the following explicit representations of ω^i and $\mathcal{Q}$

$$(35.15)\qquad \omega^i = w^{0i}(t;\ q^\beta;\ y^\beta)dt + w^{1i}_\alpha(t;\ q^\beta;\ y^\beta)dq^\alpha + w^{2i}_\alpha(t;\ q^\beta;\ y^\beta)dy^\alpha$$

$$(35.16)\qquad \mathcal{Q} = \mathcal{Q}^0(t;\ q^\beta;\ y^\beta)dt + \mathcal{Q}^1_\alpha(t;\ q^\beta;\ y^\beta)dq^\alpha + \mathcal{Q}^2_\alpha(t;\ q^\beta;\ y^\beta)dy^\alpha\ .$$

We thus obtain the governing system

$$(35.17)\qquad w^{0i}(t;\phi^\beta;D\phi^\beta) + w^{1i}_\alpha(t;\phi^\beta;D\phi^\beta)D\phi^\alpha + w^{2i}_\alpha(t;\phi^\beta;D\phi^\beta)D^2\phi^\alpha = 0\ ,\quad i=1,\ldots,R$$

$$(35.18)\qquad \{E|L\}_{\phi^\alpha} = -D(\lambda_i w^{2i}_\alpha - Q^{2i}_\alpha) + \lambda_i w^{1i}_\alpha - Q^{1i}_\alpha\ ,\quad \alpha=1,\ldots,3N$$

of 3N+R equations that are to be satisfied by the 3N+R functions

$\{\phi^\alpha(t),\ \lambda_i(t)\}$ for $a \le t \le b$ subject to the boundary data $\phi^\alpha(a)=a^\alpha$, $\phi^\alpha(b)=b^\alpha$.

Much of the mathematical literature gives the impression that classical mechanics is just a special (and uninteresting) case of the calculus of variations. This is easily seen to be false for nonconservative nonholonomic dynamical systems. Granted, we have

$$(35.19) \qquad \int_a^b \Phi^* \pounds_{\mathcal{U}}(L dt) = \delta \int_a^b \Phi^* L dt = \delta \int_a^b L dt ,$$

but (35.13) shows that

$$(35.20) \qquad \delta \int_a^b L dt = \int_a^b \Phi^* \{\mathcal{U} \rfloor (\mathcal{Q} - \lambda_i \omega^i) \wedge dt\} .$$

Thus, we have a classic variational problem $\delta \int_a^b L dt = 0$ only when the system is both conservative and holonomic with constraints that obtain only on event space. This latter generality in the constraints follows from Section 25, where it is shown that an appropriately chosen admissible map can reduce the problem to one in which we have 3N-R degrees of freedom without constraints.

The formulation of nonconservative nonholonomic systems given above suffer from the same difficiencies as noted in Section 28 for variational systems; namely, the formulation is in terms of a boundary value problem and does not provide easy access to problems in which one wishes to consider families of Newtonian motions. We therefore proceed in search of a reformulation of the problem in terms of differential forms since this proved to be successful in connection with variational systems.

36. THE FUNDAMENTAL DIFFERENTIAL FORM

As may be expected, the problem of finding one or more appropriate differential forms for the formulation of nonconservative nonholonomic systems is more difficult than the corresponding problem for variational systems that was solved in the last chapter. The results of the last chapter do, however, provide us with a firm starting point. In order to take full advantage of what we already know, it is useful to introduce a notational simplification by setting

$$N(t;\ q^\beta;\ y^\beta) = -\mathcal{Q}(t;\ q^\beta;\ y^\beta) + \lambda_i(t)\omega^i(t;\ q^\beta;\ y^\beta)\ , \tag{36.1}$$

in which case we write

$$N = N^0 dt + N^1_\alpha dq^\alpha + N^2_\alpha dy^\alpha\ . \tag{36.2}$$

The fundamental integral condition (35.13) thus becomes

$$0 = \int_a^b \Phi^*\{\pounds_U(L dt) + U \rfloor (N \wedge dt)\}\ . \tag{36.3}$$

The 1-form N is referred to as *the* 1-*form of non*-C-H of the dynamical system since it gives all of the effects that result in both the nonconservative and the nonholonomic aspects of the system.

Let us proceed exactly as in the previous chapter and introduce the 1-form $J(t;\ q^\beta;\ y^\beta)$ by

$$J = P_\alpha dq^\alpha - (P_\beta y^\beta - L)dt\ , \tag{36.4}$$

with

$$P_\alpha = \partial L/\partial y^\alpha\ . \tag{36.5}$$

The results of the last chapter show that

$$\int_a^b \Phi^* \pounds_U(L dt) = \int_a^b \Phi^* \pounds_U J$$

and hence the integral condition (36.3) becomes

$$0 = \int_a^b \Phi^*\{\pounds_U J + U \rfloor (N \wedge dt)\}\ . \tag{36.6}$$

The basic reason why this works is that

$$J = P_\alpha(dq^\alpha - y^\alpha dt) + L dt = \frac{\partial L}{\partial y^\alpha}(dq^\alpha - y^\alpha dt) + L dt\ , \tag{36.7}$$

and hence

$$(36.8) \qquad \Phi^* J = \Phi^*(L\,dt) ;$$

that is

$$(36.9) \qquad \Phi^*\left\{\frac{\partial L}{\partial y^\alpha}(dq^\alpha - y^\alpha dt)\right\} \equiv 0 .$$

This suggests that we do the same thing by comparing N with dL. We thus define the 1-form $j(t;\, q^\beta;\, y^\beta)$ by

$$(36.10) \qquad j(t;\, q^\beta;\, y^\beta) = N^2_\alpha(t;\, q^\beta;\, y^\beta)(dq^\alpha - y^\alpha dt) .$$

This gives

$$(36.11) \qquad \Phi^* j \equiv 0 .$$

Further, for U a horizontal vector field, we have

$$\begin{aligned}(36.12) \qquad \pounds_U j &= U \rfloor dj + d(U \rfloor j) \\ &= (U \rfloor dN^2_\alpha)(dq^\alpha - y^\alpha dt) \\ &\quad + N^2_\alpha\{d(U \rfloor dq^\alpha) - (U \rfloor dy^\alpha)dt\} \\ &= (U \rfloor dN^2_\alpha)(dq^\alpha - y^\alpha dt) \\ &\quad + N^2_\alpha\{d(u^\alpha_1) - u^\alpha_2 dt\} .\end{aligned}$$

Thus, since $\Phi^*(dq^\alpha - y^\alpha dt) = 0$, $\Phi^* du^\alpha_1 = D\delta\phi^\alpha dt$, $\Phi^* u^\alpha_2 = D\delta\phi^\alpha$, we see that

$$(36.13) \qquad \Phi^* \pounds_U j = 0 .$$

Accordingly, the integral condition (36.6) assumes the more symmetric form

$$(36.14) \qquad 0 = \int_a^b \Phi^*\{\pounds_U(J+j) + U \rfloor (N \wedge dt)\}$$

where

$$(36.15) \qquad J + j = (P_\alpha + N^2_\alpha)(dq^\alpha - y^\alpha dt) + L\,dt .$$

This motivates the following definition.

The *fundamental differential form* of a nonconservative nonholonomic dynamic system with Lagrangian $L = T-p$, and non-C-H form N is given by

$$(36.16) \qquad J + j = (P_\alpha + N^2_\alpha)(dq^\alpha - y^\alpha dt) + L dt .$$

However, it is clear from (36.14) that the fundamental differential form of a nonconservative nonholonomic system does not fully characterize such a system. We must thus proceed further.

37. THE FUNDAMENTAL 2-FORM

We take up the search for the additional forms needed in order to characterize a nonconservative nonholonomic system by writing out the integral condition (36.14):

$$(37.1) \qquad 0 = \int_a^b \Phi^* \{U \rfloor \{d(J+j) + N \wedge dt\} + d\{U \rfloor (J+j)\}\} .$$

This leads us to make the following definition.

The *fundamental 2-form* of a nonconservative nonholonomic dynamical system with Lagrangian L, fundamental differential form $J+j$, and non-C-H form N, is given by

$$(37.2) \qquad F = d(J+j) + N \wedge dt .$$

It then follows directly from (37.1) that the fundamental integral condition can be expressed directly in terms of the fundamental forms of the system by

$$(37.3) \qquad 0 = \int_a^b \Phi^* \{U \rfloor F + d\{U \rfloor (J+j)\}\} .$$

The entire dynamic interplay of the system is thus characterized by the 1-form

$$(37.4) \qquad \Omega(U) = U \rfloor F + d\{U \rfloor (J+j)\}$$

on K. It is exactly this ability to characterize the dynamic interplay of the system in terms of a differential form on K that allows us to rid ourselves of the formulation in terms of boundary value problems and to consider families of Newtonian motions, as we shall now show.

Let us replace the horizontal vector field U in (37.4) by an arbitrary tangent vector field

$$(37.5)\qquad V = v_o \frac{\partial}{\partial t} + v_1^\alpha \frac{\partial}{\partial q^\alpha} + Z(v_1^\alpha - y^\alpha v_o) \frac{\partial}{\partial y^\alpha} .$$

This gives us

$$(37.6)\qquad \Omega(V) = \pounds_V (J+j) + V \rfloor (N \wedge dt)$$
$$= V \rfloor F + d\{V \rfloor (J+j)\} .$$

However, (37.2) and (36.16) give

$$(37.7)\qquad F = d[(P_\alpha + N_\alpha^2)(dq^\alpha - y^\alpha dt) + L dt] + N \wedge dt$$
$$= d(P_\alpha + N_\alpha^2) \wedge (dq^\alpha - y^\alpha dt) + \left(\frac{\partial L}{\partial q^\alpha} + N_\alpha^1\right) dq^\alpha \wedge dt$$
$$= \left[d(P_\alpha + N_\alpha^2) - \left(\frac{\partial L}{\partial q^\alpha} + N_\alpha^1\right) dt\right] \wedge (dq^\alpha - y^\alpha dt) .$$

A comparison of (36.16) with (29.1) and (37.7) with (29.5) shows that all of the formula from Chapter V can be taken over directly by the correspondence $P_\alpha \rightarrowtail P_\alpha + N_\alpha^2$, $\frac{\partial L}{\partial q^\alpha} \rightarrowtail \frac{\partial L}{\partial q^\alpha} + N_\alpha^1$. This gives the evaluation

$$(37.8)\qquad \Omega(V) = -\left\{d(P_\alpha + N_\alpha^2) - \left(\frac{\partial L}{\partial q^\alpha} + N_\alpha^1\right) dt\right\} (V \rfloor dq^\alpha - y^\alpha V \rfloor dt)$$
$$+ (dq^\alpha - y^\alpha dt)\left\{V \rfloor d(P_\alpha + N_\alpha^2) - \left(\frac{\partial L}{\partial q^\alpha} + N_\alpha^1\right) V \rfloor dt\right\}$$
$$+ d\left\{(P_\alpha + N_\alpha^2) V \rfloor dq^\alpha - (P_\alpha y^\alpha + N_\alpha^2 y^\alpha - L) V \rfloor dt\right\} ,$$

which we record for use in the next Section. The important thing at this point is that (37.7) yields

$$(37.9)\qquad V\rfloor F = (dq^\alpha - y^\alpha dt)V\rfloor\left[d(P_\alpha + N^2_\alpha) - \left(\frac{\partial L}{\partial q^\alpha} + N^1_\alpha\right)dt\right]$$
$$- \left\{d(P_\alpha + N^2_\alpha) - \left(\frac{\partial L}{\partial q^\alpha} + N^1_\alpha\right)dt\right\} V\rfloor(dq^\alpha - y^\alpha dt) ,$$

and hence

$$(37.10)\qquad \Phi^* V\rfloor F = -\left[D(P_\alpha + N^2_\alpha) - \frac{\partial L}{\partial\phi^\alpha} - N^1_\alpha\right]\Phi^*(v^\alpha_1 - y^\alpha v_c)dt .$$

Now, the equations of motion (35.18) in this notation are

$$(37.11)\qquad D(P_\alpha + N^2_\alpha) - \frac{\partial L}{\partial\phi^\alpha} - N^1_\alpha = 0 ,$$

so that any Φ that defines a Newtonian motion yields

$$(37.12)\qquad \Phi^* V\rfloor F = 0 .$$

Conversely, if $\int\Phi^* V\rfloor F = 0$ for all vector fields V, the fundamental lemma of the calculus of variations without fixed endpoints shows that the functions $\{\phi^\alpha\}$ that define Φ satisfy the equations of motion (37.10).

Any map $\Phi: \mathbb{R} \to K|\Phi(t)=t$, $\Phi^*\omega^i = 0$, $i=1,\dots,R$ *and*

$$(37.13)\qquad 0 = \int\Phi^* V\rfloor F$$

holds for all vector fields V, *gives a Newtonian motion of a nonconservative nonholonomic system with Lagrangian* L *and non-C-H form* N ; *the fundamental 2-form*

$$(37.14)\qquad F = d(J+j) + N\wedge dt$$

and the constraint forms ω^i *give a complete characterization of such dynamical systems.*

This achieves our aim of ridding ourselves of a formulation in terms of boundary value problems. We also note that elimination of the boundary value problem formulation removes the requirement of checking that the given boundary values are compatible with the given constraints: a nontrivial result in light of intrinsic difficulty in effecting such checks for nonholonomic constraints.

38. QUADRATURES

The development of quadrature theory for nonconservative nonholonomic systems proceeds along lines parallel to those established in Sections 30 and 31 for variational systems. We start with the 1-form $\Omega(V)$ given by (37.6) and (37.8);

$$(38.1)\quad \Omega(V) = V\rfloor F + d\{V\rfloor(J+j)\}$$
$$= -\left\{d(P_\alpha + N^2_\alpha) - \left(\frac{\partial L}{\partial q^\alpha} + N^1_\alpha\right)dt\right\}(V\rfloor dq^\alpha - y^\alpha V\rfloor dt)$$
$$+ (dq^\alpha - y^\alpha dt)\left\{V\rfloor d(P_\alpha + N^2_\alpha) - \left(\frac{\partial L}{\partial q^\alpha} + N^1_\alpha\right)V\rfloor dt\right\}$$
$$+ d\left\{(P_\alpha + N^2_\alpha)V\rfloor dq^\alpha - (P_\alpha y^\alpha + N^2_\alpha y^\alpha - L)V\rfloor dt\right\},$$

$$(38.2)\quad d\{V\rfloor(J+j)\} = d\{(P_\alpha + N^2_\alpha)V\rfloor dq^\alpha - (P_\alpha y^\alpha + N^2_\alpha y^\alpha - L)V\rfloor dt\}.$$

When we allow Φ^* to act on (38.1), we obtain

$$(38.3)\qquad \Phi^*\Omega(V) = \Phi^*(V\rfloor F) + D\Phi^*(V\rfloor(J+j))dt.$$

This result motivates the following definition.

A vector field $V\varepsilon T(K)$ is said to constitute a *quadrature field* of the dynamical system characterized by the forms $J+j$ and F if and only if

$$(38.4)\qquad \Omega(V) = 0.$$

It is now a simple matter to combine (37.12) and (38.1) through (38.3) to arrive at the following results.

If $V\varepsilon T(K)$ *constitutes a quadrature field of the dynamical system characterized by the forms* $J+j$ *and* F, *then the following results hold:* (i) *if* Φ *defines a Newtonian motion of the system then the equations of motion* (37.11) *admit the quadrature*

$$(38.5) \qquad \eta(V;\Phi) = (P_\alpha + N_\alpha^2)V_1^\alpha - (P_\alpha + N_\alpha^2)V_o D\phi^\alpha + V_o L = \text{constant}$$

(ii) *if* Φ *is an arbitrary map of* $\mathbb{R}$ *into* K *such that* $\Phi(t)=t$, *then the functions* $\{\phi^\alpha(t)\}$ *and the functions* $\{V_o, V_1^\alpha\}$ *satisfy the identity*

$$(38.6) \qquad \left\{D(P_\alpha + N_\alpha^2) - \frac{\partial L}{\partial\phi^\alpha} - N_\alpha^1\right\}(V_1^\alpha - V_o D\phi^\alpha)$$

$$= D\{(P_\alpha + N_\alpha^2)V_1^\alpha - (P_\alpha + N_\alpha^2)D\phi^\alpha V_o + V_1 L\} .$$

In contrast with the results for variational systems obtained in the previous chapter, we can not form new quandrature fields by constructing the vector fields $[V_1, V_2]$. The simplest way to see this is to use (37.6) to note that

$$(38.7) \qquad \Omega(V_i) = \pounds_{V_i}(J+j) + V_i \rfloor (N \wedge dt) .$$

Thus

$$(38.8) \qquad \Omega([V_i, V_j]) = \pounds_{[V_i, V_j]}(J+j) + [V_i, V_j]\rfloor(N \wedge dt)$$

$$= \pounds_{V_i}\pounds_{V_j}(J+j) - \pounds_{V_j}\pounds_{V_i}(J+j)$$

$$+ [V_i, V_j]\rfloor(N \wedge dt)$$

$$= \pounds_{V_i}(\Omega(V_j) - V_j\rfloor(N \wedge dt))$$

$$- \pounds_{V_j}(\Omega(V_i) - V_i\rfloor(N \wedge dt))$$

$$+ [V_i, V_j]\rfloor(N \wedge dt) ,$$

so that $\Omega(V_i) = \Omega(V_j) = 0$ yields

$$(38.9) \qquad \Omega([V_i, V_j]) = \pounds_{V_j}(V_i\rfloor(N \wedge dt)) - \pounds_{V_i}(V_j\rfloor(N \wedge dt))$$

$$+ [V_i, V_j]\rfloor(N \wedge dt)$$

which does not vanish in general.

Again, we note that the important thing is that we have $\Phi^*\Omega(V) = 0$, and that $\Omega(V) = 0$ is a convenient way of obtaining $\Phi^*\Omega(V) = 0$. Thus, we can obtain significant generalizations by allowing V to satisfy

$$(38.10) \qquad \Omega(V) = \rho$$

with ρ a differential form that is also a homogeneous functional of degree one in V and satisfies

$$(38.11) \qquad \Phi^*\rho = 0 .$$

An important subcase of such efforts yields the extension of the notion of a mobility transformation to a nonconservative nonholonomic system.

A vector field $V\varepsilon T(K)$ is said to generate a *mobility transformation* of the dynamical system characterized by the forms $J+j$ and F if and only if V satisfies

$$(38.12) \qquad d\Omega(V) = 0 .$$

Since satisfaction of (38.12) shows that $\Omega(V)$ is closed, there accordingly exists a scalar-valued function $\mathcal{Q}(t;\ q^\beta;\ y^\beta)$ such that

$$(38.13) \qquad \Omega(V) = d\mathcal{Q} .$$

This leads to the following results similar to those established in Section 31.

If the vector field $V\varepsilon T(K)$ *generates a mobility transformation of the dynamical system characterized by the forms* $J+j$ *and* F *, then the following results hold:* (i) *there exists a scalar valued function* $\mathcal{Q}(t;\ q^\beta;\ y^\beta)$ *such that*

$$(38.14) \qquad \Omega(V) = d\mathcal{Q} ;$$

(ii) *every quadrature field of the system generates a mobility transformation with* $Q=0$; (iii) *any map* $\Phi: \mathbb{R} \to K|\Phi(t)=t$ *satisfies the identity*

$$(38.15) \qquad \left\{D(P_\alpha + N_\alpha^1) - \frac{\partial L}{\partial \phi^\alpha} - N_\alpha^1\right\}(V_1^\alpha - V_o D\phi^\alpha)$$

$$= D\{(P_\alpha + N_\alpha^2)V_1^\alpha - (P_\alpha + N_\alpha^2)D\phi^\alpha V_o + V_o L - Q\} ,$$

(iv) *if* Φ *generates a Newtonian motion of the system then the equations of motion* (37.11) *admit the quadrature*

$$(38.16) \quad \eta(V,\Phi) = (P_\alpha + N_\alpha^2)V_1^\alpha - (P_\alpha + N_\alpha^2)V_o D\phi^\alpha + V_o L - Q = \text{constant}.$$

We note that we lose the group property of mobility transformations for nonconservative nonholonomic dynamical systems for the same reason as noted above; that is $[V_i, V_j]$ does not generate a mobility transformation if both V_i and V_j do. The maximal mobility group constructed in Section 32 can still be used, however, as an underlying group whose symmetries are broken by the forces and the constraints, for any nonholonomic nonconservative system of N particles can be thought of as a dynamical system that starts out as a free system and then has the forces and the constraints "switched on" so to speak.

39. A UNIFORM REFORMULATION

The results obtained in Section 37 show that $\Phi: \mathbb{R} \to K|\Phi(t)=t$ is a Newtonian motion of a dynamical system characterized by the forms $J+j$ and K if and only if

$$(39.1) \qquad \Phi^* \omega^i = 0 , \quad i=1,\dots,R$$

and

$$(39.2) \qquad \int \Phi^* V \rfloor F = 0$$

holds for all vector fields $V \varepsilon T(K)$. There are, however, certain instances, approximation theory in particular, when we would like to have a single integral condition lead to both the equations of motion and the constraints. We obtain such a formulation in this section.

We start with kinematic space, K , referred to its standard inertial coordinate cover $(t;\ q^{\beta};\ y^{\beta})$. From this we build a larger space of $6N+R+1$ dimensions

$$(39.3)\qquad \bar{K} = K\times E_R$$

with standard coordinate cover $(t;\ q^{\beta};\ y^{\beta};\ \mu_i)$. A general vector field on $\bar{V}$ is defined on $\bar{K}$ by

$$(39.4)\qquad \bar{V} = v_o(t;\ q^{\beta})\frac{\partial}{\partial t} + v_1^{\alpha}(t;\ q^{\beta})\frac{\partial}{\partial q^{\alpha}} + Z(v_1^{\alpha} - v_o y^{\alpha})\frac{\partial}{\partial y^{\alpha}} + v_i(\mu_j)\frac{\partial}{\partial \mu_i}$$

and a horizontal vector field $\bar{U}$ is defined on $\bar{K}$ by

$$(39.5)\qquad \bar{U} = u_1^{\alpha}(t;\ q^{\beta})\frac{\partial}{\partial q^{\alpha}} + Z(u_1^{\alpha})\frac{\partial}{\partial y^{\alpha}} + u_i(\mu_j)\frac{\partial}{\partial \mu_i}\ .$$

This allows us to define a map

$$(39.6)\qquad \bar{\Phi}:\ \mathbb{R} \to \bar{K}\,|\,\Phi(t) = t\ ,\quad \Phi(q^{\alpha}) = \phi^{\alpha}(t)\ ,\quad \Phi(y^{\alpha}) = D\phi^{\alpha}(t)\ ,$$
$$\Phi(\mu_i) = \lambda_i(t)\ ,$$

that plays the same role as played by the map Φ in the previous formulations. We now go back and replace the integral condition (35.7) with the new condition

$$(39.7)\qquad 0 = \int_a^b \bar{\Phi}^{*}\{\pounds_{\bar{U}}(Tdt) + \bar{U}\rfloor(W + \mu_i\omega^i)\wedge dt + \omega^i\bar{U}\rfloor d\mu_i\}$$

for all horizontal vector fields $\bar{U}$ such that $u_1^{\alpha}(a;\phi^{\beta}(a)) = u_1^{\alpha}(b;\phi^{\beta}(b)) = 0$. It is easily seen that (39.5) and (39.6) yield

$$(39.8)\qquad \int_a^b \bar{\Phi}^{*}\{\pounds_{\bar{U}}(Tdt) + \bar{U}\rfloor(W + \mu_i\omega^i)\wedge dt\} =$$

$$= \int_a^b \Phi^*\{\mathcal{L}_{U}(Tdt) + U \rfloor (W + \lambda_i \omega^i) \wedge dt\} ,$$

while

$$(39.9) \qquad \int_a^b \bar{\Phi}^*\{\omega^i \bar{U} \rfloor d\mu_i\} = \int_a^b (\Phi^*\omega^i) u_i(\lambda_j(t)) .$$

Accordingly, if we set

$$(39.10) \qquad \delta\lambda_i(t) = u_i(\lambda_j(t))$$

and use the result established in the first part of this chapter, we see that (39.7) can hold for all horizontal vector fields $\bar{U}$ such that $\delta\phi^{\alpha}(a) = u_1^{\alpha}(a;\phi^{\beta}(a)) = \delta\phi^{\alpha}(b) = 0$ if and only if

$$(39.11) \qquad \int_a^b \left[-D(P_{\alpha} + N_{\alpha}^2) + \frac{\partial L}{\partial \phi^{\alpha}} + N_{\alpha}^2\right]\delta\phi^{\alpha}(t)dt + \int_a^b \Phi^*\omega^i \delta\lambda_i = 0 .$$

The fundamental lemma of the calculus of variations shows that this can be the case if and only if

$$(39.12) \qquad D(P_{\alpha} + N_{\alpha}^2) - \frac{\partial L}{\partial \phi^{\alpha}} - N_{\alpha}^1 = 0 , \quad \alpha=1,\ldots,3N$$

$$(39.13) \qquad \Phi^*\omega^i = 0 , \quad i=1,\ldots,R .$$

A map $\Phi\colon [a\le t\le b] \to K|\Phi(t)=t$, $\phi^{\alpha}(a) = a^{\alpha}$, $\phi^{\alpha}(b) = b^{\alpha}$ *defines a Newtonian motion of a dynamical system with work* 1-*form* W *and constraints* $\omega^i = 0$, $i=1,\ldots,R$ *if and only if*

$$(39.14) \qquad 0 = \int_a^b \bar{\Phi}^*\{\mathcal{L}_{\bar{U}}(Tdt) + \bar{U} \rfloor (W + \mu_i\omega^i) \wedge dt + \omega^i \bar{U} \rfloor d\mu_i\}$$

holds for all horizontal vector fields $\bar{U}$ *on* $\bar{K} = K\times E_R$ *of the form* (39.5) *such that* $u_1^{\alpha}(a;a^{\beta}) = u_1^{\alpha}(b;b^{\beta}) = 0$.

This accomplishes our task and we leave this subject at this point since it is not overly germain to problems in mechanics. The interested reader can easily go back and reformulate the whole thing so as to eliminate the dependence of this formulation on the boundary value problem if he wishes; just follow what was done in Section 37.

CHAPTER VII

INTEGRAL FORMS OF CONSERVATION AND BALANCE

40. FAMILIES OF NEWTONIAN MOTIONS FOR VARIATIONAL SYSTEMS

The basic idea underlying the analysis of this chapter is that families of Newtonian motions contain almost all of the information needed in order to reconstruct the full setting of the problem in Kinematic Space. This sounds obvious on first reading, but is not. In order to make this distinction clear, let us restrict attention to the variational systems treated in Chapter V. The full setting of a variational system in Kinematic Space requires specification of a Lagrangian function, $L(t;\, q^\beta;\, y^\beta)$, whose domain is Kinematic space, K , and whose range is some subset of the real line. On the other hand, a Newtonian motion is a single map

$$\Phi\colon J \to K \,|\, \Phi(t) = t \;,\quad \Phi(q^\alpha) = \phi^\alpha(t) \;,\quad \Phi(y^\alpha) = D\phi^\alpha(t) \tag{40.1}$$

that satisfies the equations of motion

$$D\left(\frac{\partial L(t;\phi^\beta(t);D\phi^\beta(t))}{\partial(D\phi^\alpha(t))}\right) = \frac{\partial L(t;\phi^\beta(t);D\phi^\beta(t))}{\partial\phi^\alpha(t)} \;, \quad \alpha=1,\ldots,3N \;. \tag{40.2}$$

These can also be written as

$$D\left(\frac{\partial(L\circ\Phi)}{\partial(D\phi^\alpha)}\right) = \frac{\partial(L\circ\Phi)}{\partial\phi^\alpha} \;,\quad \alpha=1,\ldots,3N \;. \tag{40.3}$$

Accordingly, a Newtonian motion only gives information about $L \circ \Phi$, which is a function of the single variable t, rather than information about the function L as a function defined on the $(6N+1)$-dimension space K. Put another way, a Newtonian motion samples L only along the 1-dimensional orbit of that Newtonian motion. The question thus naturally arises as to what extent can a full description of the dynamical system be obtained by the study of a family of Newtonian motions rather than just one such motion.

Let X_r denote r-dimensional number space whose points are labeled with respect to a fixed coordinate cover by $(\eta_1, \ldots, \eta_r)$. This space will be used as a space of parameters that will distinguish between members of an r-parameter family of curves in K. To this end, let n be a given, simply connected, r-dimensional region of X_r, and let $a(\eta_i)$ and $b(\eta_i)$ be two functions that are defined on n with values on the real line and such that

$$(40.4) \qquad a(\eta_i) < b(\eta_i) .$$

We then construct the simply connected $(r+1)$-dimensional region M of $R \times X_r$ with coordinates $(t; \eta_i)$ with $(\eta_i) \epsilon n$ and

$$(40.5) \qquad a(\eta_i) \leq t \leq b(\eta_i) .$$

The region M is shown diagramatically in Figure 5-1. An r-parameter family of curves in K is defined by the map

$$(40.6) \quad \Psi : M \to K | \Psi(t) = t , \quad \Psi(q^\alpha) = \psi^\alpha(t;\eta_i) , \quad \Phi(y^\alpha) = \partial\psi^\alpha(t;\eta_i)/\partial t .$$

Let us agree to denote the range of Ψ by S_{r+1}. S_{r+1} is thus an $(r+1)$-dimensional surface (manifold) in K that is the image of the $(r+1)$-dimensional region M of $\mathbb{R} \times X_r$ under the map Ψ. All that now remains is implementation of the condition that each of the curves in K that is obtained under Ψ for fixed values of the parameters η_i shall be a Newtonion motion of a variational system with Lagrangian function L.

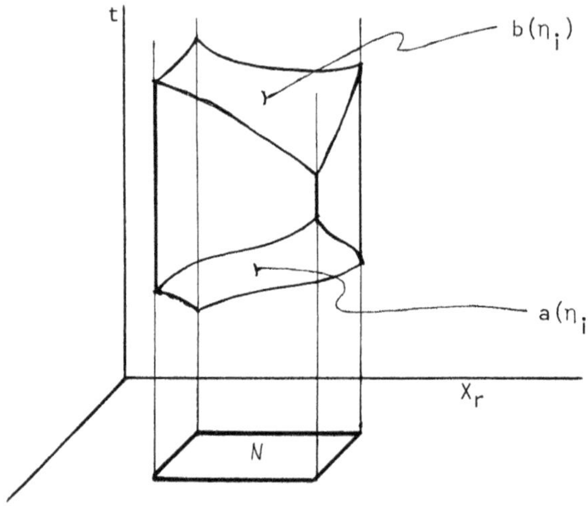

Fig. 5-1. The region M

The results established in Chapter V show that all of the dynamical information of a variational system with Lagrangian L is carried by the fundamental 1-form

$$(40.7) \qquad J = P_\alpha dq^\alpha - (P_\alpha y^\alpha - L)dt ,$$

where P_α is defined by

$$(40.8) \qquad P_\alpha = \partial L(t;\, q^\beta;\, y^\beta)/\partial y^\alpha .$$

In fact, we know if $\Phi\colon J \to K|\Phi(t)=t$ defines a Newtonian motion, then (see (29.17) and (29.19)) $\Phi^* V \rfloor dJ = 0$ for any vector field V on K, where

$$(49.9) \qquad dJ = \left(dP_\alpha - \frac{\partial L}{\partial q^\alpha}dt\right) \wedge (dq^\alpha - y^\alpha dt) .$$

Thus, *if each of the maps* $\Psi|_\eta$, *that obtains from the map* Ψ *by fixing the values of the parameters* (η_i) , *is a Newtonian motion, then*

$$(40.10)\qquad (\Psi|_\eta)^* V \rfloor dJ = 0$$

for any vector field V *on* K. This result provides the information contained in the assumption that S_{r+1}, the image of M under the map Ψ, is comprised of an r-parameter family of Newtonian motions.

The rest of this Section is devoted to a lengthy calculation that yields the basis for the subsequent study of variational systems. If we apply Ψ^* to (40.9), we obtain

$$(40.11)\qquad \Psi^* dJ = \Psi^*(dP_\alpha - \frac{\partial L}{\partial q^\alpha} dt) \wedge \Psi^*(dq^\alpha - y^\alpha dt) .$$

However, (40.6) shows that

$$(40.12)\qquad \Psi^*(dq^\alpha - y^\alpha dt) = \frac{\partial \psi^\alpha}{\partial \eta_i} d\eta_i ,$$

and hence (40.11) becomes

$$(40.13)\qquad \Psi^* dJ = \Psi^* \left[\left(dP_\alpha - \frac{\partial L}{\partial q^\alpha} dt \right) \frac{\partial \psi^\alpha}{\partial \eta_i} \right] \wedge d\eta_i .$$

Further progress is significantly simplified by introducing the r vector fields V_i on S_{r+1} by

$$(40.14)\qquad V_i = \frac{\partial \psi^\alpha}{\partial \eta_i} \frac{\partial}{\partial q^\alpha} + \frac{\partial^2 \psi^\alpha}{\partial t \partial \eta_i} \frac{\partial}{\partial y^\alpha} .$$

If we consider all quantities to be restricted to the (r+1)-dimensional manifold S_{r+1} in K, we see that (40.12) implies

$$\begin{aligned} V_i \rfloor dJ &= \left[V_i \rfloor \left(dP_\alpha - \frac{\partial L}{\partial q^\alpha} dt \right) \right] (dq^\alpha - y^\alpha dt) \\ &\quad - \left[V_i \rfloor (dq^\alpha - y^\alpha dt) \right] \left(dP_\alpha - \frac{\partial L}{\partial q^\alpha} dt \right) \\ &= \left[V_i \rfloor (dP_\alpha) \right] (dq^\alpha - y^\alpha dt) \\ &\quad - \left(dP_\alpha - \frac{\partial L}{\partial q^\alpha} dt \right) \frac{\partial \psi^\alpha}{\partial \eta_i} . \end{aligned}$$

We thus have

$$\left(dP_\alpha - \frac{\partial L}{\partial q^\alpha}dt\right)\frac{\partial\psi^\alpha}{\partial\eta_i} = \left[V_i\lrcorner(dP_\alpha)\right](dq^\alpha - y^\alpha dt) - (V_i\lrcorner dJ) \ ,$$

and hence a substitution of this result into (40.13) yields

$$(40.15)\quad \Psi^* dJ = \Psi^*\left[-(V_i\lrcorner dJ) + (V_i\lrcorner dP_\alpha)(dq^\alpha - y^\alpha dt)\right]\wedge d\eta_i$$

$$= -\Psi^*(V_i\lrcorner dJ)\wedge d\eta_i + \Psi^*(V_i\lrcorner dP_\alpha)\frac{\partial\psi^\alpha}{\partial\eta_j}d\eta_j\wedge d\eta_i \ .$$

Now, $(V_i\lrcorner dJ)$ is a differential form, so we may write

$$\Psi^*(V_i\lrcorner dJ) = (\Psi|_\eta)^*(V_i\lrcorner dJ) + (\Psi|_t)^*(V_i\lrcorner dJ) \ ,$$

where $|_\eta$ and $|_t$ mean at constant (η_i) and at constant t, respectively. The evaluation (40.15) accordingly becomes

$$(40.16)\quad \Psi^* dJ = -(\Psi|_\eta)^*(V_i\lrcorner dJ)\wedge d\eta_i - (\Psi|_t)^*(V_i\lrcorner dJ)\wedge d\eta_i$$

$$+ \Psi^*(V_i\lrcorner dP_\alpha)\frac{\partial\psi^\alpha}{\partial\eta_j}d\eta_j\wedge d\eta_i \ .$$

The next step is to note that

$$(40.17)\quad (\Psi|_t)^*(V_i\lrcorner dJ) = (\Psi|_t)^*(V_i\lrcorner(dJ)|_t)$$

$$= (\Psi|_t)^*(V_i\lrcorner(dP_\alpha\wedge dq^\alpha)|_t)$$

$$= (\Psi|_t)^*(V_i\lrcorner(dP_\alpha\wedge dq^\alpha))$$

$$= (\Psi|_t)^*\{(V_i\lrcorner dP_\alpha)dq^\alpha - dP_\alpha\frac{\partial\psi^\alpha}{\partial\eta_i}\}$$

$$= (\Psi|_t)^*(V_i\lrcorner dP_\alpha)\frac{\partial\psi^\alpha}{\partial\eta_j}d\eta_j - (\Psi|_t)^*(dP_\alpha)\frac{\partial\psi^\alpha}{\partial\eta_i} \ .$$

When (40.17) is substituted into (40.17), we accordingly obtain

$$(40.18)\quad \Psi^* dJ = -(\Psi|_\eta)^*(V_i\lrcorner dJ)\wedge d\eta_i$$

$$+ (\Psi|_t)^*(dP_\alpha)\frac{\partial\psi^\alpha}{\partial\eta_i}\wedge d\eta_i \ .$$

Now, $(\Psi|_t)^* dP_\alpha = d(P_\alpha)|_t$, where $P_\alpha = P_\alpha \circ \Psi$ in conformity with previous usage of the relation between P_α and P_α . Thus,

$$(40.19)\qquad (\Psi|_t)^*(dP_\alpha)\frac{\partial\psi^\alpha}{\partial\eta_i}\wedge d\eta_i = d(P_\alpha)|_t \frac{\partial\psi^\alpha}{\partial\eta_i}\wedge d\eta_i$$

$$= d\left(P_\alpha \frac{\partial\psi^\alpha}{\partial\eta_i}\right)\Big|_t \wedge d\eta_i - P_\alpha d\left(\frac{\partial\psi^\alpha}{\partial\eta_i}\right)\Big|_t \wedge d\eta_i$$

$$= d\left(P_\alpha \frac{\partial\psi^\alpha}{\partial\eta_i}\right)\Big|_t \wedge d\eta_i \ .$$

On the other hand, (40.7) and (40.14) give $V_i \lrcorner J = P_\alpha \frac{\partial\psi^\alpha}{\partial\eta_i}$, and hence

$$(40.20)\qquad \Psi^*(V_i \lrcorner J) = P_\alpha \frac{\partial\psi^\alpha}{\partial\eta_i} \ .$$

Thus, a combination of (40.19) and (40.20) yields

$$(40.21)\qquad (\Psi|_t)^*(dP_\alpha)\frac{\partial\psi^\alpha}{\partial\eta_i}\wedge d\eta_i = d\left[\Psi^*(V_i \lrcorner J)\right]|_t \wedge d\eta_i$$

$$= (\Psi|_t)^* d(V_i \lrcorner J)\wedge d\eta_i$$

so that (40.18) gives

$$(40.22)\qquad \Psi^* dJ = -(\Psi|_\eta)^*(V_i \lrcorner dJ)\wedge d\eta_i$$

$$+ (\Psi|_t)^* d(V_i \lrcorner J)\wedge d\eta_i \ .$$

We have seen, however, that if $\Psi|_\eta$ is a Newtonian motion for the variational system with Lagrangian function L for all $(\eta_i)\varepsilon n$ (S_{r+1} is comprised of an r-parameter family of Newtonian motions of the variational system with Lagrangian function L), then $(\Psi|_\eta)^*(V \lrcorner dJ) = 0$ for any vector field V . Under these circumstances, (40.22) reduces to

$$(40.23)\qquad \Psi^*(dJ) = (\Psi|_t)^* d(V_i \lrcorner J)\wedge d\eta_i \ ,$$

and hence

$$(40.24)\qquad \Psi^*(dJ) = \frac{\partial \xi_i(t;\eta_k)}{\partial \eta_j} d\eta_j \wedge d\eta_i \ .$$

Finally, since $\Psi^*(dJ) = d\Psi^*(J)$, (40.24) yields

$$0 = d\Psi^*(dJ) = \frac{\partial^2 \xi_i(t;\eta_k)}{\partial t\, \partial \eta_j} dt \wedge d\eta_j \wedge d\eta_i \ .$$

These conditions can be satisfied if and only if

$$\frac{\partial}{\partial t}\left(\frac{\partial \xi_i(t;\eta_k)}{\partial \eta_j}\right) = 0 \ ,$$

and (40.24) shows that

$$(40.25)\qquad \Psi^*(dJ) = \bar{\xi}_{ji}(\eta_k) d\eta_j \wedge d\eta_k \ .$$

Thus, the differential form $\Psi^*(dJ)$ is a differential form on the region $\mathcal{R}$ of X_r ; that is, it contains no dt term and its coefficients are independent of t as well.

If a map Ψ *maps a region* $M(a(\eta_i)\le t\le b(\eta_i),(\eta_i)\in\mathcal{R})$ *of* $\mathbb{R}\times X_r$ *onto an* $(r+1)$*-dimensional manifold* S_{r+1} *in* K *such that* S_{r+1} *consists of an* r*-parameter family of Newtonian motions of a variational system with Lagrangian function* L *, then* Ψ^* *maps* dJ *onto an exterior form of degree two on* $\mathcal{R}$ *; that is,* $\Psi^*(dJ)$ *is independent of* t *and* dt *.*

A simpler way of stating this is to say that $\Psi^*(dJ)$ is invariant under all mappings

$$\rho : (t;\eta_i) \rightarrowtail (\rho(t);\eta_i)$$

that can be generated by all strictly increasing functions $\rho(t)$. This means that

$$(40.26)\qquad \overset{-1}{\rho}{}^*(\Psi^*(dJ)) = \Psi^*(dJ) \ .$$

Still another way is to say that

$$(40.27)\qquad \Psi^*(dJ)\big|_t = \Psi^*(dJ) \ .$$

41. ABSOLUTE AND RELATIVE INTEGRAL INVARIANTS OF VARIATIONAL SYSTEMS

The purpose of all of the calculation in the previous section is that it allows us to obtain almost immediate implementations of the classic definition of absolute integral invariants.

An exterior form ω of degree k is an *absolute integral invariant* of a system of differential equations with independent variable t if and only if (i)

$$d\omega = 0 ,$$

and (ii) $\Psi^*(\omega)$ is independent of t and dt for any map Ψ of any region of $\mathbb{R}\times X_r$ onto $S_{r+1} \subset K$ such that S_{r+1} is comprised of an r-parameter family of solutions of the system of differential equations.

Every entry in the following list of exterior forms is an absolute integral invariant of the variational system with Lagrangian function L :

$$(41.1) \qquad dJ,\ dJ \wedge dJ,\ (dJ)^{(3)},\ (dJ)^{(4)},\ \ldots,\ (dJ)^{(3N)},$$

where $(dJ)^{(k)}$ *means the* k*-fold exterior product of* dJ .

The proof of this assertion follows directly from the results established in the last section on noting that $d(dJ)^{(k)} = 0$ and that $\Psi^*[(dJ)^{(k)}] = [\Psi^*(dJ)]^{(k)}$.

We now proceed to relate integral invariants as differential forms on K to invariant integrals over submanifolds in K . We first consider a 2-dimensional manifold S_2 that is comprised of a 1-parameter family of Newtonian motions of a variational system with Lagrangian function L . This means that we have a mapping Ψ that maps the 2-dimensional region $M_2(a(\eta) \le t \le b(\eta)\ ;\ \eta_1 \le \eta \le \eta_2)$ onto $S_2 \subset K$ that is defined by

$$(41.2) \quad t=t,\quad q^\alpha = \psi^\alpha(t;\eta),\quad y^\alpha = \partial\psi^\alpha(t;\eta)/\partial t,\quad (\Psi|_\eta)^*(V \lrcorner dJ) = 0 ,$$

where the latter equality holds for all vector fields V. We know, however, that $\Psi^*(dJ)$ is an exterior form on the 1-dimensional subset $\eta_1 \le \eta \le \eta_2$ of the 1-dimensional space X_1 with coordinate η because dJ is an absolute integral invariant of the variational system. Accordingly,

$$(41.3) \qquad \Psi^*(dJ) = 0$$

since all forms of degree higher than the dimension of the space vanish. This in turn gives

$$(41.4) \qquad 0 = \int_M \Psi^*(dJ) = \int_{S_2} dJ \ .$$

Conversely, *if* (41.4) *holds for every* S_2 *that is comprised of a 1-parameter family of Newtonian motions, then* dJ *is an absolute integral invariant.*

Now, let S_3 be a 3-dimensional manifold in K that is comprised of a 2-parameter family of Newtonian motions of a variational system with Lagrangian function L; that is, M_3 is given by

$$(41.5) \qquad \begin{cases} a(\eta_1,\eta_2) \le t \le b(\eta_1,\eta_2) \ , & (\eta_1,\eta_2)\varepsilon n \ , \\ q^\alpha = \psi^\alpha(t;\ \eta_1,\ \eta_2) \ , & y^\alpha = \partial\psi^\alpha(t;\ \eta_1,\eta_2)/\partial t \ , \end{cases}$$

and

$$(41.6) \qquad (\Psi|_\eta)^*(V \rfloor dJ) = 0 \ , \qquad V\varepsilon T(K) \ ,$$

where (41.6) means that the functions $\psi^\alpha(t;\ \eta_i)$ satisfy Lagranges equations for each fixed $(\eta_1,\ \eta_2)\varepsilon n$:

$$\frac{d}{dt}\left\{ \frac{\partial L(t;\psi^\beta;\partial\psi^\beta/\partial t)}{\partial(\partial\psi^\alpha/\partial t)}\bigg|_\eta \right\} = \frac{\partial L(t;\psi^\beta,\partial\psi^\beta/\partial t)}{\partial\psi^\alpha}\bigg|_\eta \ , \quad \alpha=1,\dots,3N \ .$$

Now, the boundary, ∂S_3, of the manifold S_3 consists of three 2-dimensional manifolds:

$$\Sigma_o : t = a(\eta_i),\ q^\alpha = \psi^\alpha(a(\eta_i);\eta_i),\ y^\alpha = \frac{\partial\psi^\alpha(a(\eta_i);\eta_i)}{\partial a(\eta_i)},\ (\eta_i)\varepsilon n\ ;$$

(41.7) $$\Sigma_1 : t = b(\eta_i),\ q^\alpha = \psi^\alpha(b(\eta_i);\eta_i),\ y^\alpha = \frac{\partial\psi^\alpha(b(\eta_i);\eta_i)}{\partial b(\eta_i)},\ (\eta_i)\varepsilon n\ ;$$

$$S_2 : a(\eta_i) \le t \le b(\eta_i),\ q^\alpha = \psi^\alpha(t;\eta_i),\ y^\alpha = \frac{\partial\psi^\alpha(t;\eta_i)}{\partial t},\ (\eta_i)\varepsilon\ \partial n\ ,$$

where ∂n denotes the 1-dimensional boundary of the 2-dimensional region n in X_2. Thus, in particular, S_2 is a 2-dimensional manifold in K that is comprised of the 1-parameter family of Newtonian motions of the variational system with Lagrangian function L. Since $d(dJ) = 0$, Stokes' theorem and $\int_M \Psi^* d(dJ) = \int_{S_3} d(dJ)$, $\int_{\partial M} \Psi^* dJ = \int_{\partial S_3} dJ$ (see Appendix, Section 12) provide us with the following result:

(41.8) $$0 = \int_{S_3} d(dJ) = \int_{\partial S_3} dJ = \int_{\Sigma_1} dJ - \int_{\Sigma_0} dJ + \int_{S_2} dJ\ .$$

We have already seen, however that the integral of dJ over S_2 vanishes, and hence (41.8) yields

(41.9) $$\int_{\Sigma_1} dJ = \int_{\Sigma_0} dJ\ .$$

The exact same reasoning can be used to conclude that

(41.10)
$$\int_{\Sigma_{1,2k}} (dJ)^{(k)} = \int_{\Sigma_{0,2k}} (dJ)^{(k)}\ ,$$
$$\int_{S_{2k}} (dJ)^{(k)} = 0\ ,$$

where the boundary of $S_{2k+1} = \Psi^*(\partial M_{2k+1})$ is given by

$$\Sigma_{0,2k}: a(\eta_i) = t, \quad q^\alpha = \psi^\alpha(a(\eta_i);\eta_i),$$
$$y^\alpha = \partial\psi^\alpha(a(\eta_i);\eta_i)/\partial a, \quad (\eta_i)\varepsilon n\subset X_{2k} ;$$

(41.11)a
$$\Sigma_{1,2k}: b(\eta_i) = t, \quad q^\alpha = \psi^\alpha(b(\eta_i);\eta_i),$$
$$y^\alpha = \partial\psi^\alpha(b(\eta_i);\eta_i)/\partial b, \quad (\eta_i)\varepsilon n\subset X_{2k} ;$$

$$S_{2k}: a(\eta_i)\le t\le b(\eta_i), \quad q^\alpha = \psi^\alpha(t,\eta_i),$$
$$y^\alpha = \partial\psi^\alpha(t,\eta_i)/\partial t, \quad (\eta_i)\varepsilon\ \partial n\subset X_{2k} .$$

Careful note should be taken of the fact that the manifolds $\Sigma_{0,2k}$ and $\Sigma_{1,2k}$ are not required to be intersections of a timewise extension of S_{2k+1} with surfaces of constant time, as is easily seen from the parametric equations (41.11)a that define $\Sigma_{0,2k}$ and $\Sigma_{1,2k}$. However, $\Sigma_{0,2k}$ and $\Sigma_{1,2k}$ can be taken as constant time surfaces if we take $a(\eta_i) = t_0$, $b(\eta_i) = t_1$ in (41.11)a. In this instance, we have

$$\Sigma_{0,2k}: t = t_0, \quad q^\alpha = \psi^\alpha(t_0,\eta_i),$$
$$y^\alpha = \partial\psi^\alpha(t_0,\eta_i)/\partial t_0, \quad (\eta_i)\varepsilon n\subset X_{2k}$$

(41.11)b
$$\Sigma_{1,2k}: t = t_1, \quad q^\alpha = \psi^\alpha(t_1,\eta_i),$$
$$y^\alpha = \partial\psi^\alpha(t_1,\eta_i)/\partial t_1, \quad (\eta_i)\varepsilon n\subset X_{2k} .$$

These equations show that $\Sigma_{1,2k}$ is the image in K of the 2k-dimensional region $\Sigma_{0,2k}$ under the map $\Psi|_{t=t_1}$. Thus, since S_{2k+1} is comprised of Newtonian motions of the given variational system, we have

$$\Sigma_{1,2k} = \Psi|_{t=t_1}(\Sigma_{0,2k}) \tag{41.12}$$

and (41.11) becomes

$$(41.13)\qquad \int_{\Psi|_{t=t_1}(\Sigma_{0,2k})} (dJ)^{(k)} = \int_{\Sigma_{0,2k}} (dJ)^{(k)} .$$

Now, since the time t_1 is quite arbitrary, (41.13) gives

$$(41.14)\qquad \int_{\Psi(\Sigma_{0,2k})} (dJ)^{(k)} = \int_{\Sigma_{0,2k}} (dJ)^{(k)}$$

for every value of t for which the map Ψ is defined. Thus, in particular, we conclude that

$$(41.15)\qquad \frac{d}{dt}\int_{\Psi(\Sigma_{0,2k})} (dJ)^{(k)} = 0 ;$$

the integral

$$(41.16)\qquad I(\Sigma_{0,2k}) = \int_{\Psi(\Sigma_{0,2k})} (dJ)^{(k)}$$

is conserved under the motion of a variational system with Lagrangian function L *for any simply connected* 2k*-dimensional region* $\Sigma_{0,2k}$: $t=t_0$, $q^\alpha = \psi^\alpha(t_0,\eta_i)$, $y^\alpha = \partial\psi^\alpha(t_0,\eta_i)/\partial t_0$, $(\eta_i)\,\varepsilon\, n \subset X_{2k}$.

A case of particular importance occurs when $k=3N$. In this instance, we have

$$\begin{aligned} (dJ)^{(3N)}\Big|_{\Psi(\Sigma_{0,6N})} &= (\Psi|_{t=t_0})^*(dJ)^{(3N)} = (\Psi|_{t=t_0})^*(dP_\alpha \wedge dq^\alpha)^{(3N)} \\ &= (\Psi|_{t=t_0})^*\left[(3N)!\,dP_1\wedge dq^1\wedge dP_2\wedge dq^2\wedge\ldots\wedge dP_{3N}\wedge dq^{3N}\right] \\ &= (\Psi|_{t=t_0})^*\left[(3N)!\det\left(\frac{\partial P_\alpha}{\partial y^\beta}\right)dy^1\wedge dq^1\wedge\ldots\wedge dy^{3N}\wedge dq^{3N}\right] . \end{aligned}$$

Accordingly, we conclude that *the* 6N*-dimensional volume measure*

$$(41.17)\qquad V(\Sigma_{6N}) = \int_{\Psi(\Sigma_{0,6N})} \det\left(\frac{\partial P_\alpha}{\partial y^\beta}\right)dy^1\wedge dq^1\wedge\ldots\wedge dy^{3N}\wedge dq^{3N}$$

is conserved under Newtonian motions of variational systems. This theorem is the Lagrangian mechanics analog of Liouville's theorem in Hamiltonian mechanics, and, as such, constitutes a basis for a Lagrangian formulation of statistical mechanics of variational systems.

An exterior form ω of degree k is said to be a *relative integral invariant* of a system of differential equations if and only if $d\omega$ is an absolute integral invariant of the same system of differential equations.

The following theorem results immediately upon noting that $(dJ)^{(k)}$ is an absolute integral invariant and that $d(J \wedge (dJ)^{(k)}) = (dJ)^{(k+1)}$.

Each of the following exterior forms is a relative integral invariant of a variational system with Lagrangian function L :

$$J,\ J \wedge dJ,\ J \wedge (dJ)^{(2)},\ \ldots,\ J \wedge (dJ)^{(3N-1)} . \tag{41.18}$$

The relation between relative integral invariants and time invariant integrals is obtained directly from Stokes' theorem $\int_S d\omega = \int_{\partial S} \omega$. Thus, if we consider the situation described by (41.10) and (41.11)a, we have

$$\begin{aligned} \int_{\partial\Sigma_{0,2k}} J \wedge (dJ)^{(k-1)} &= \int_{\Sigma_{0,2k}} d(J \wedge (dJ)^{(k-1)}) \\ &= \int_{\Sigma_{0,2k}} (dJ)^{(k)} = \int_{\Sigma_{1,2k}} (dJ)^{(k)} \\ &= \int_{\Sigma_{1,2k}} d(J \wedge (dJ)^{(k-1)} = \int_{\partial\Sigma_{1,2k}} J \wedge (dJ)^{(k-1)} . \end{aligned} \tag{41.19}$$

If we take the special case of constant time slices, as described by (41.11)b, then (41.14) holds and we conclude from (40.19) that

$$\int_{\partial\Psi(\Sigma_{0,2k})} J \wedge (dJ)^{(k-1)} = \int_{\partial\Sigma_{0,2k}} J \wedge (dJ)^{(k-1)} , \tag{41.20}$$

or

$$\frac{d}{dt} \int_{\partial\Psi(\Sigma_{0,2k})} J \wedge (dJ)^{(k-1)} = 0 ; \tag{41.21}$$

the boundary integrals

$$(41.22)\qquad I(\partial\Sigma_{0,2k}) = \int_{\partial\Psi(\Sigma_{0,2k})} J \wedge (dJ)^{(k-1)}, \quad k=1,\dots,3N-1$$

are conserved under the Newtonian motion of a variational system with Lagrangian function L .

The analysis given in Chapter V showed that a variational system is completely characterized by its fundamental differential form J . The above results allow us to view this characterization from a different perspective; namely, that associated with the sequence of exterior forms that is generated by the given differential form for purposes of rank and class classification via the Darboux theorem. This sequence is generated recursively from J by means of the recursion relations

$$(41.23)\qquad \begin{aligned} I_1 &= J, & I_{2k+1} &= I_1 \wedge I_{2k}, \\ I_2 &= dJ, & I_{2k+2} &= I_2 \wedge I_{2k} = d(I_{2k+1}). \end{aligned}$$

Now, $K(J)$, the class of J , is equal to the degree of last nonzero exterior form in this sequence, and $2\rho(J)$, the rank of J , is equal to the largest even integer less than or equal to $K(J)$. We have seen, however, that

$$(41.24)\qquad I_{6N}\Big|_t = (dJ)^{(3N)}\Big|_t = (3N)!\det\left(\frac{\partial P_\alpha}{\partial y^\beta}\right) dy^1 \wedge dq^1 \wedge \dots \wedge dy^{3N} \wedge dq^{3N} \neq 0$$

and hence

$$(41.25)\qquad K(J) \geq 6N, \quad 2\rho(J) = 6N.$$

Thus, since dimension $(K) = 6N+1$, *the exterior form* J *has maximal rank.* We have also established that *each of the elements* $I_2, I_4, \dots, I_{6N}$ *is an absolute integral invariant and each of the elements* $I_1, I_3, \ldots, I_{6N-1}$ *is a relative integral invariant.* Thus, *each element in the sequence* (41.23) *gives a volume or a surface integral that is conserved under Newtonian motions of the variational system characterized by* J .

42. DERIVATION OF LAGRANGE'S EQUATIONS FROM A RELATIVE INTEGRAL INVARIANT

We saw in Chapter V that a map $\Phi\colon \mathbb{R} \to K|\Phi(t)=t$, $\Phi(q^\alpha) = \phi^\alpha(t)$ defined a Newtonian motion of a variational system with Lagrangian function L if and only if $\Phi^*(V\rfloor dJ) = 0$ holds for all vector fields $V\varepsilon T(K)$; that is, if and only if the functions $\phi^\alpha(t)$ satisfy Lagrange's equations

$$(42.1) \qquad DP_\alpha - \partial L/\partial\phi^\alpha = 0\ , \quad P_\alpha = \partial L/\partial(D\phi^\alpha)\ .$$

The results on absolute and relative integral invariants established in the last Section provide yet another way of obtaining Lagrange's equations that is the Lagrangian mechanics analog of the Poincaré theorem† in Hamiltonian mechanics.

If the system of second order differential equations

$$(42.2) \qquad P_\alpha = p_\alpha(t;\ q^\beta;\ y^\beta)\ , \quad \Psi^* dP_\alpha = \Psi^* r_\alpha(t;\ q^\beta;\ y^\beta)\ , \qquad \alpha=1,\ldots,3N$$

that defines the map

$$\Psi\colon \mathbb{R} \to K|\Psi(t) = t\ , \quad \Psi(q^\alpha) = \psi^\alpha(t)\ , \quad \Psi(y^\alpha) = D\psi^\alpha(t)\ ,$$

admits the relative integral invariant

$$(42.3) \qquad J = P_\alpha dq^\alpha - (P_\alpha y^\alpha - L(t;\ q^\beta;\ y^\beta))dt\ ,$$

then

† Abraham (Ref. 14); Whittaker (Ref. 13).

$$(42.4)\qquad p_\alpha(t;\ q^\beta;\ y^\beta) \;=\; \partial L(t;\ q^\beta;\ y^\beta)/\partial y^\alpha\ ,\quad \alpha=1,\ldots,3N\ ,$$

$$(42.5)\qquad r_\alpha(t;\ q^\beta;\ y^\beta) \;=\; \partial L(t;\ q^\beta;\ y^\beta)/\partial q^\alpha\ ,\quad \alpha=1,\ldots,3N\ ;$$

that is, the system (42.2) is the system of Lagrange's equations for a variational system with Lagrangian function L .

The proof of this result is as follows. Since J is a relative integral invariant of the system (42.2),

$$(42.6)\quad dJ = \left(dP_\alpha - \frac{\partial L}{\partial q^\alpha}dt\right)\wedge dq^\alpha + \left(\frac{\partial L}{\partial y^\alpha} - P_\alpha\right)dy^\alpha\wedge dt - y^\alpha dP_\alpha\wedge dt$$

is an absolute integral invariant of the system (42.2). Now, a complete solution of the system of second order differential equations (42.2) is a map

$$(42.7)\qquad \Psi\colon\ \mathbb{R}\times X_{6N}\to K\,|\,\Psi(t) = t\ ,\quad \Psi(q^\alpha) = f^\alpha(t;\eta_1,\ldots,\eta_{6N})\ ,$$

$$\Psi(y^\alpha) = \partial f^\alpha/\partial t\ ,$$

where the $6N$ parameters (η_i) constitute the $6N$ integration constants of the second order system (42.2) in the $3N$ unknown ψ^α , $\alpha=1,\ldots,3N$. (42.7) gives $\Psi^* dt = dt$,

$$(42.8)\qquad \Psi^* dq^\alpha \;=\; \frac{\partial f^\alpha}{\partial t}dt + \frac{\partial f^\alpha}{\partial \eta_i}d\eta_i\ ,$$

$$(42.9)\qquad \Psi^* dy^\alpha \;=\; \frac{\partial^2 f^\alpha}{\partial t^2}dt + \frac{\partial^2 f^\alpha}{\partial t\partial\eta_i}d\eta_i\ ,$$

while (42.2) gives

$$(42.10)\qquad \Psi^* dP_\alpha \;=\; \frac{\partial p_\alpha}{\partial t}dt + \frac{\partial p_\alpha}{\partial f^\beta}\left(\frac{\partial f^\beta}{\partial t}dt + \frac{\partial f^\beta}{\partial\eta_i}d\eta_i\right)$$

$$+ \frac{\partial p_\alpha}{\partial\left(\frac{\partial f^\beta}{\partial t}\right)}\left(\frac{\partial^2 f^\beta}{\partial t^2}dt + \frac{\partial^2 f^\beta}{\partial t\partial\eta_i}d\eta_i\right) \;=$$

$$= \left(\frac{\partial p_\alpha}{\partial t} + \frac{\partial p_\alpha}{\partial \mathfrak{f}^\beta}\frac{\partial \mathfrak{f}^\beta}{\partial t} + \frac{\partial p_\alpha}{\partial (D\mathfrak{f}^\beta)}\frac{\partial^2 \mathfrak{f}^\beta}{\partial t^2}\right)dt$$

$$+ \left(\frac{\partial p_\alpha}{\partial \mathfrak{f}^\beta}\frac{\partial \mathfrak{f}^\beta}{\partial \eta_i} + \frac{\partial p_\alpha}{\partial\left(\frac{\partial \mathfrak{f}^\beta}{\partial t}\right)}\frac{\partial^2 \mathfrak{f}^\beta}{\partial t\partial \eta_i}\right)d\eta_i$$

$$= Dp_\alpha dt + \frac{\partial p_\alpha}{\partial \eta_i}d\eta_i = r_\alpha dt + \frac{\partial p_\alpha}{\partial \eta_i}d\eta_i \ .$$

We thus have, from (42.6), (42.8)-(42.10),

$$(42.11)\quad \Psi^* dJ = \left(r_\alpha dt + \frac{\partial p_\alpha}{\partial \eta_i}d\eta_i - \frac{\partial L}{\partial \mathfrak{f}^\alpha}dt\right)\wedge\left(\frac{\partial \mathfrak{f}^\alpha}{\partial t}dt + \frac{\partial \mathfrak{f}^\alpha}{\partial \eta_j}d\eta_j\right)$$

$$+ \left(\frac{\partial L}{\partial\left(\frac{\partial \mathfrak{f}^\alpha}{\partial t}\right)} - p_\alpha\right)\left(\frac{\partial^2 \mathfrak{f}^\alpha}{\partial t^2}dt + \frac{\partial^2 \mathfrak{f}^\alpha}{\partial t\partial \eta_i}d\eta_i\right)\wedge dt$$

$$- \frac{\partial \mathfrak{f}^\alpha}{\partial t}\left(r_\alpha dt + \frac{\partial p_\alpha}{\partial \eta_i}d\eta_i\right)\wedge dt$$

$$= \left(r_\alpha - \frac{\partial L}{\partial \mathfrak{f}^\alpha}\right)dt\wedge\frac{\partial \mathfrak{f}^\alpha}{\partial \eta_i}d\eta_i + \left(\frac{\partial L}{\partial\left(\frac{\partial \mathfrak{f}^\alpha}{\partial t}\right)} - p_\alpha\right)\frac{\partial^2 \mathfrak{f}^\alpha}{\partial t\partial \eta_i}d\eta_i\wedge dt$$

$$+ \frac{\partial p_\alpha}{\partial \eta_i}\frac{\partial \mathfrak{f}^\alpha}{\partial \eta_j}d\eta_i\wedge d\eta_j \ .$$

Since an absolute integral invariant can not contain terms involving dt, we conclude from (42.11) that

(42.12)

$$0 = \left(r_\alpha - \frac{\partial L}{\partial \mathfrak{f}^\alpha}\right)dt\wedge\frac{\partial \mathfrak{f}^\alpha}{\partial \eta_i}d\eta_i + \left(p_\alpha - \frac{\partial L}{\partial\left(\frac{\partial \mathfrak{f}^\alpha}{\partial t}\right)}\right)dt\wedge\frac{\partial^2 \mathfrak{f}^\alpha}{\partial t\partial \eta_i}d\eta_i$$

$$= \left(r_\alpha - \frac{\partial L}{\partial \mathfrak{f}^\alpha}\right)dt\wedge(\Psi|_t)^*(dq^\alpha) + \left(p_\alpha - \frac{\partial L}{\partial\left(\frac{\partial \mathfrak{f}^\alpha}{\partial t}\right)}\right)dt\wedge(\Psi|_t)^*(dy^\alpha) \ .$$

Thus, since $(\Psi|_t)^*(dq^\alpha)$ and $(\Psi|_t)^*(dy^\alpha)$ can be chosen independently by choice of the 6N integration constants $\eta_1,\ldots,\eta_{6N}$ that enter into the map Ψ, we conclude that

$$(42.13) \qquad \begin{aligned} r_\alpha(t;\ \mathfrak{f}^\alpha;\ \partial\mathfrak{f}^\alpha/\partial t) &= \partial L(t;\ \mathfrak{f}^\alpha;\ \partial\mathfrak{f}^\alpha/\partial t)/\partial\mathfrak{f}^\alpha , \\ p_\alpha(t;\ \mathfrak{f}^\alpha;\ \partial\mathfrak{f}^\alpha/\partial t) &= \partial L(t;\ \mathfrak{f}^\alpha;\ \partial\mathfrak{f}^\alpha/\partial t)/\partial(\partial\mathfrak{f}^\alpha/\partial t) . \end{aligned}$$

However, (42.7) gives $\Psi(q^\alpha) = \mathfrak{f}^\alpha(t;\ \eta_i)$, $\Psi(y^\alpha) = \partial\mathfrak{f}^\alpha(t;\ \eta_i)/\partial t$, and the independence of the 6N parameters (η_i) allows us to choose the parameters η_i so that Ψ maps the point $t = t_0$ in $\mathbb{R}$ into any generic point in K. This has the effect of allowing us to replace the $\mathfrak{f}$'s by the q's and the $(\partial\mathfrak{f}/\partial t)$'s by the y's ; that is, (42.13) yields

$$(42.14) \qquad \begin{aligned} r_\alpha(t;\ q^\beta;\ y^\beta) &= \partial L(t;\ q^\beta;\ y^\beta)/\partial q^\alpha , \\ p_\alpha(t;\ q^\beta;\ y^\beta) &= \partial L(t;\ q^\beta;\ y^\beta)/\partial y^\beta . \end{aligned}$$

This establishes (42.4) and (42.5). A substitution of these results back into the system (42.2) now gives

$$(42.15) \qquad P_\alpha = \partial L/\partial(D\psi^\alpha) , \quad DP_\alpha - \partial L/\partial\psi^\alpha = 0 ,$$

which are Lagrange's equations for a variational system with Lagrangian function L.

There is one very important point that should be noted about the above proof. The full properties of an absolute integral invariant were not used, for we only needed the fact that $\Psi^*(dJ)$ contained no term with a dt. In particular, we did not use the fact that $\Psi^*(dJ)$ is also independent of the variable t. This is an important point, for it will allow us to draw a similar conclusion concerning nonconservative nonholonomic systems wherein the quantity corresponding to $\Psi^*(dJ)$ can be shown to contain no term involving dt, but its coefficients can be functions of the variable t.

43. FAMILIES OF NEWTONIAN MOTIONS IN THE GENERAL CASE

We now turn to the general case of a nonconservative, nonholonomic dynamical system. The results established in Chapter VI show that all of the dynamical information about such systems is contained in the second degree fundamental exterior form

$$(43.1)\qquad F = d(J+j) + N \wedge dt$$

and the constraint forms ω^i, where

$$(43.2)\qquad J + j = (P_\alpha + N^2_\alpha)(dq^\alpha - y^\alpha dt) + L dt ,$$

and

$$(43.3)\qquad N = - Q + \lambda_i \omega^i = N^0 dt + N^1_\alpha dq^\alpha + N^2_\alpha dy^\alpha$$

is the non-C-H form that characterizes the nonconservative, nonholonomic aspects of the system. For the purposes of this Section, we will be interested primarily in the equations of motion and their solutions, so we will suppose that any solutions that are considered will be such as to satisfy the constraints. This simply says that the Lagrange multipliers, λ_i, that are associated with the constraints, will be determined by substituting the solutions of the equations of motion into the constraint equations. Thus, if $\Phi: \mathbb{R} \to K | \Phi(t) = t$ gives a Newtonian motion, we have

$$(43.4)\qquad \Phi^*(V \lrcorner F) = 0$$

for any vector field $V \epsilon T(K)$.

We now proceed exactly as in Section 40, and construct an r-parameter family of curves in K by means of the map

$$(43.5)\qquad \Psi: M \to S_{r+1} \subset K | \Psi(t) = t , \quad \Psi(q^\alpha) = \psi^\alpha(t; \eta_i) ,$$

$$\Psi(y^\alpha) = \partial\psi^\alpha(t; \eta_i)/\partial t .$$

Since F contains the term dJ additively, as shown by (43.1), the results obtained for variational systems suggests that we evaluate Ψ^*F. To this end, we first use (37.7) to write F in the equivalent form

$$(43.6)\qquad F = \left[d(P_\alpha + N^2_\alpha) - \left(\frac{\partial L}{\partial q^\alpha} + N^1_\alpha\right)dt\right]\wedge(dq^\alpha - y^\alpha dt)\ .$$

A comparison of (43.6) with (40.9) shows that we can take over all of the results established in Section 40 if we make the replacements $dJ \rightarrowtail F$, $P_\alpha \rightarrowtail (P_\alpha + N^2_\alpha)$, $\partial L/\partial q^\alpha \rightarrowtail (\partial L/\partial q^\alpha + N^1_\alpha)$. Thus, (40.18) yields

$$(43.7)\qquad \Psi^*(F) = -(\Psi|_\eta)^*(V_i \lrcorner F)\wedge d\eta_i + (\Psi|_t)^* d(P_\alpha + N^2_\alpha)\frac{\partial\psi^\alpha}{\partial\eta_i}\wedge d\eta_i\ .$$

Now,

$$\begin{aligned}(\Psi|_t)^* d(P_\alpha + N^2_\alpha)\frac{\partial\psi^\alpha}{\partial\eta_i}\wedge d\eta_i &= d(P_\alpha + N^2_\alpha)|_t\frac{\partial\psi^\alpha}{\partial\eta_i}\wedge d\eta_i\\ &= d\left\{(P_\alpha + N^2_\alpha)\frac{\partial\psi^\alpha}{\partial\eta_i}\right\}\Big|_t\wedge d\eta_i - (P_\alpha + N^2_\alpha)d\left(\frac{\partial\psi^\alpha}{\partial\eta_i}\right)\Big|_t\wedge d\eta_i\\ &= d\left\{(P_\alpha + N^2_\alpha)\frac{\partial\psi^\alpha}{\partial\eta_i}\right\}\Big|_t\wedge d\eta_i = (\Psi|_t)^* d\left\{(P_\alpha + N^2_\alpha)\frac{\partial\psi^\alpha}{\partial\eta_i}\right\}\wedge d\eta_i\ ,\end{aligned}$$

and hence (43.7) yields

$$(43.8)\qquad \Psi^*(F) = -(\Psi|_\eta)^*(V_i \lrcorner F)\wedge d\eta_i + (\Psi|_t)^* d\left\{(P_\alpha + N^2_\alpha)\frac{\partial\psi^\alpha}{\partial\eta_i}\right\}\wedge d\eta_i\ .$$

We now restrict the map Ψ so that each curve generated by Ψ for fixed (η_i) is a Newtonian motion. This means that

$$(43.9)\qquad (\Psi|_\eta)^*(V \lrcorner F) = 0$$

for any $V \in T(K)$, and hence (43.8) gives

$$(43.10)\qquad \Psi^*(F) = (\Psi|_t)^* d\left\{(P_\alpha + N_\alpha^2)\frac{\partial\psi^\alpha}{\partial\eta_i}\right\}\wedge d\eta_i$$

$$= \frac{\partial\bar{\mu}_i(t;\eta_k)}{\partial\eta_j}\, d\eta_j \wedge d\eta_i \ .$$

Thus, as with the case of variational systems, *the second degree exterior form* $\Psi^*(F)$ *on* M *contains no* dt *term.*

The similarity between variational systems and nonholonomic, nonconservative systems ends at this point, for (43.1) shows that

$$(43.11)\qquad dF = dN \wedge dt \neq 0$$

since $dN \wedge dt$ vanishes only when the system is a variational system, as seen from (43.3) and the results established in previous chapters. We can make further headway starting with (43.11) since $\Psi^*(dF) = d\Psi^*(F)$; that is,

$$(43.12)\qquad d\Psi^*(F) = \Psi^*(dN)\wedge dt = d\Psi^*(N)\wedge dt$$

since $\Psi^*(dt) = dt$. Accordingly, (43.10) gives

$$(43.13)\qquad \frac{\partial}{\partial t}\left[\frac{\partial\bar{\mu}_i(t;\eta_k)}{\partial\eta_j}\right] dt\wedge d\eta_j\wedge d\eta_i = d\Psi^*(N)\wedge dt \ ,$$

so that $\Psi^*(F)$ is not independent of the variable t .

If a map Ψ *maps a region* M $(a(\eta_i) \le t \le b(\eta_i)\ ;\ (\eta_i)\epsilon n)$ *of* $\mathbb{R}\times X_r$ *onto an* $(r+1)$*-dimensional manifold* S_{r+1} *in* K *such that* S_{r+1} *consists of an* r*-parameter family of Newtonian motions of a general dynamical system with fundamental 2-form* F *, then* Ψ^* *maps* F *onto a 2-form on* M *such that*

$$(43.14)\qquad \Psi^*(F) = \frac{\partial\bar{\mu}_i(t;\eta_k)}{\partial\eta_j}\, d\eta_j\wedge d\eta_i = d(\bar{\mu}_i(t;\eta_k))|_t \wedge d\eta_i \ .$$

44. BALANCE FORMS AND INTEGRAL BALANCE LAWS

It is clear at once that F is not an integral invariant of the associated nonconservative, nonholonomic system, for $dF \neq 0$ and $\Psi^*(F)$ depends on t, though it does not contain any term dt. Now, integral invariants lead to integrals over subsets of K that are conserved under Newtonian motions of the associated variational system. We might thus expect that the independence of $\Psi^*(F)$ on dt, but $dF \neq 0$ and $\partial\Psi^*(F)/\partial t \neq 0$, would lead to integral laws of balance rather than laws of conservation. This is indeed the case, as we shall now show.

We first consider a 2-dimensional manifold S_2 in K that is comprised of a 1-parameter family of Newtonian motions. This means that we have a mapping Ψ that maps the 2-dimensional region $M(a(\eta) \leq t \leq b(\eta)\ ;\ \eta_1 \leq \eta \leq \eta_2)$ onto S_2 that is defined by

$$(44.1)\qquad t = t\ ,\quad q^\alpha = \psi^\alpha(t;\eta)\ ,\quad y^\alpha = \partial\psi(t;\eta)/\partial t\ ,$$

$$(\Psi|_\eta)^*(V \lrcorner F) = 0\ ,$$

where the latter equality holds for all $V \varepsilon T(K)$. Now, we have

$$(44.2)\qquad \Psi^*(F) = 0$$

since $\Psi^*(F)$ is a 2-form on the two dimensional space M and contains no dt term. This in turn gives

$$(44.3)\qquad 0 = \int_M \Psi^*(F) = \int_{S_2} F\ .$$

Now, let S_3 be a 3-dimensional manifold in K that is comprised of a 2-parameter family of Newtonian motions of the system characterized by the fundamental form F; that is, (41.5) holds while (41.6) is replaced by $(\Psi|_\eta)^*(V \lrcorner F) = 0$ for all $V \varepsilon T(K)$. The boundary of S_3 thus consists of the same three pieces Σ_0, Σ_1, S_2 given by (41.7) and S_2 is a 2-dimensional manifold that is comprised of a 1-parameter family of Newtonian motions of the system. At this point, we must deviate from the considerations given in Section 41 since (43.11) gives

$$(44.4)\qquad dF = dN \wedge dt \ , \quad d(\Psi^* F) = d(\Psi^* N) \wedge dt \ .$$

Stokes' theorem accordingly yields

$$(44.5)\qquad \int_{S_3} dN \wedge dt = \int_{S_3} dF = \int_{\partial S_3} F = \int_{\Sigma_1} F - \int_{\Sigma_0} F + \int_{S_2} F \ ,$$

and use of (44.3) shows that

$$(44.6)\qquad \int_{\Sigma_1} F - \int_{\Sigma_0} F = \int_{S_3} dN \wedge dt \ .$$

We now specialize the above result to the case where the surfaces Σ_0 and Σ_1 are constant time surfaces. This means that $a(\eta_i) = t_0$, $b(\eta_i) = t_1$; that is

$$(44.7)\qquad \begin{aligned} &\Sigma_0 : \ t = t_0 , \ q^\alpha = \psi^\alpha(t_0;\eta_i) , \ y^\alpha = \partial\psi^\alpha(t_0;\eta_i)/\partial t , \ (\eta_i)\varepsilon n \subset X_2 \ , \\ &\Sigma_1 : \ t = t_1 , \ q^\alpha = \psi^\alpha(t_1;\eta_i) , \ y^\alpha = \partial\psi^\alpha(t_1;\eta_i)/\partial t , \ (\eta_i)\varepsilon n \subset X_2 \ . \end{aligned}$$

In this case, we have

$$(44.8)\qquad \Sigma_1 = \Psi|_{t=t_1}(\Sigma_0) \ ,$$

and (44.6) becomes

$$(44.9)\qquad \int_{\Psi|_{t=t_1}(\Sigma_0)} F - \int_{\Sigma_0} F = \int_{t_0}^{t_1} \left(\int_n \Psi^* dN \right) dt \ .$$

Accordingly, since t_0 and t_1 are arbitrary, (44.9) gives us the integral law of balance

$$(44.10)\qquad \frac{d}{dt} \int_{\Psi|_t(\Sigma_0)} F = \int_n (\Psi^* dn) \ .$$

It thus seems appropriate to refer to F as a balance form. The general definition is as follows.

An exterior form ω is said to be a *balance form* of a general dynamical system characterized by the fundamental form F if and only if $\Psi^*(\omega)$ is independent of dt for any map Ψ of any region of $\mathbb{R}\times X_r$ onto $S_{r+1}\subset K$ such that S_{r+1} is comprised of an r-parameter family of Newtonian motions of the system.

The results established in the previous section allow us to draw the following immediate conclusion.

Each of the following list of exterior forms is a balance form of the general dynamical system that is characterized by the fundamental form F :

$$F,\ F\wedge F,\ F\wedge F\wedge F,\ \dots,\ (F)^{(3N)}\ . \tag{44.11}$$

Integral laws of balance can be obtained for each of the integrals of these balance forms as follows. We start by considering the 2k-dimensional manifold S_{2k} that is comprised of a (2k-1)-parameter family of Newtonian motions of the system. We take the equations that define S_{2k} to be given by the map

$$\Psi : t_0 \le t \le t_1\ ,\quad q^\alpha = \psi^\alpha(t;\eta_i)\ , \tag{44.12}$$

$$y^\alpha = \partial\psi^\alpha(t;\eta_i)/\partial t\ ,\quad (\eta_i)\varepsilon n\subset X_{2k-1}\ ,$$

and M is given by $[t_0,t_1]\times n$. We then have

$$\int_{S_{2k}} (F)^{(k)} = \int_M \Psi^*(F^{(k)}) = \int_M (\Psi^* F)^{(k)}\ . \tag{44.13}$$

However, $(F)^{(k)}$ is a balance form, so that $\Psi^*(F^{(k)})$ is a 2k-form on the 2k-dimensional region M of $\mathbb{R}\times X_{2k-1}$ that contains no factor dt . Thus, since the basis elements of all exterior forms on M can be constructed from the elements $dt, d\eta_1,\dots, d\eta_{2k-1}$ all forms of degree higher than $2k-1$ that contain no factor dt vanish identically. We thus obtain

$$0 = \int_{S_{2k}} (F)^{(k)}\ . \tag{44.14}$$

We next note that F is a 2-form, from which we conclude that

$$d[(F)^{(k)}] = k(F)^{(k-1)}\wedge dF = k(F)^{(k-1)}\wedge dN\wedge dt\ . \tag{44.15}$$

Integrating both sides of (44.15) over an S_{2k+1} that is comprised of a 2k-parameter family of Newtonian motions gives us

(44.16)

$$k\int_{S_{2k+1}} (F)^{(k-1)} \wedge dN \wedge dt = \int_{S_{2k+1}} d[(F)^{(k)}] = \int_{\partial S_{2k+1}} (F)^{(k)}$$

$$= \int_{\Sigma_{1,2k}} (F)^{(k)} - \int_{\Sigma_{0,2k}} (F)^{(k)} + \int_{S_{2k}} (F)^{(k)}$$

where

$$\Sigma_{0,2k}: \ t=t_0,\ q^\alpha = \psi^\alpha(t_0;\eta_i),\ y^\alpha = \partial\psi^\alpha(t;\eta_i)/\partial t,\ (\eta_i)\varepsilon n \subset X_{2k}$$

(44.17) $$\Sigma_{1,2k}: \ t=t_1,\ q^\alpha = \psi^\alpha(t_1;\eta_i),\ y^\alpha = \partial\psi^\alpha(t;\eta_i)/\partial t,\ (\eta_i)\varepsilon n \subset X_{2k}$$

$$S_{2k}: \ t_0 \le t \le t_1,\ q^\alpha = \psi^\alpha(t;\eta_i),\ y^\alpha = \partial\psi^\alpha(t;\eta_i)/\partial t,\ (\eta_i)\varepsilon \partial n \subset X_{2k}$$

Since S_{2k} is a 2k-dimensional manifold in K that is comprised of a (2k-1)-parameter family of Newtonian motions, (44.14) gives $\int_{S_{2k}} (F)^{(k)} = 0$. We also have the obvious result

(44.18) $$\int_{S_{2k+1}} (F)^{(k-1)} \wedge dN \wedge dt = \int_{t_0}^{t_1}\int_{n} \Psi^*[(F)^{(k-1)} \wedge dN]dt \ .$$

A substitution of these results into (44.16) and use of $\Sigma_{1,2k} = \Psi|_{t_1}(\Sigma_{0,2k})$ gives us

(44.19)

$$\int_{\Psi|_{t_1}(\Sigma_{0,2k})} (F)^{(k)} - \int_{\Sigma_0} (F)^{(k)} = k\int_{t_0}^{t_1}\left\{\int_{n} \Psi^*[(F)^{(k-1)} \wedge dN]\right\} dt \ .$$

Thus, since t_0 and t_1 are arbitrary, we obtain the integral law of balance

(44.20) $$\frac{d}{dt}\int_{\Psi(\Sigma_{0,2k})} (F)^{(k)} = k\int_{n} \Psi^*[(F)^{(k-1)} \wedge dN] \ .$$

There is still a further simplification that can be made, for the domain of integration of the integral on the left-hand side of (44.20) is a 2k-dimensional subset of a constant time hypersurface in K. We can thus replace $(F)^{(k)}$ by $(F|_t)^{(k)}$. However, (43.6) gives

$$F|_t = d(P_\alpha + N^2_\alpha)|_t \wedge dq^\alpha|_t ,$$

and (44.20) becomes

$$(44.21)\quad \frac{d}{dt}\int_{\Psi(\Sigma_{0,2k})} \{d(P_\alpha + N^2_\alpha)|_t \wedge dq^\alpha|_t\}^{(k)} = k\int_{\mathcal{N}} \Psi^*[(F)^{(k-1)} \wedge dN] .$$

As before, the case of particular importance is that for which $k=3N$. In this instance, since $dP_\alpha|_t \wedge dq^\alpha|_t$ and $dN^2|_t \wedge dq^\alpha|_t$ are 2-forms, we have

$$\begin{aligned}\{d(P_\alpha + N^2_\alpha) \wedge dq^\alpha\}|_t^{(3N)} &= \{dP_\alpha \wedge dq^\alpha\}|_t^{(3N)} \\ &+ \sum_{\ell=1}^{3N}\binom{3N}{\ell}\{dP_\alpha \wedge dq^\alpha\}|_t^{(3N-\ell)} \wedge \{dN^2_\alpha \wedge dq^\alpha\}|_t^{(\ell)} \\ &= (3N)!\det\left(\frac{\partial P_\alpha}{\partial y^\beta}\right)dy^1 \wedge dq^1 \wedge \ldots \wedge dy^{3N} \wedge dq^{3N} \\ &+ \sum_{\ell=1}^{3N}\binom{3N}{\ell}\{dP_\alpha \wedge dq^\alpha\}|_t^{(3N-\ell)} \wedge \{dN^2_\alpha \wedge dq^\alpha\}|_t^{(\ell)} .\end{aligned}$$

Accordingly, *the following form of Liouville's theorem obtains for nonconservative, nonholonomic systems:*

(44.22)

$$(3N)!\frac{d}{dt}\int_{\Psi(\Sigma_{0,3N})} \det\left(\frac{\partial P_\alpha}{\partial y^\beta}\right)dy^1 \wedge dq^1 \wedge \ldots \wedge dy^{3N} \wedge dq^{3N}$$

$$= -\frac{d}{dt}\int_{\Psi(\Sigma_{0,3N})} \sum_{\ell=1}^{3N}\binom{3N}{\ell}\{dP_\alpha \wedge dq^\alpha\}|_t^{(3N-\ell)} \quad \{dN^2_\alpha \wedge dq^\alpha\}|_t^{(\ell)}$$

$$+ 3N\int_{\mathcal{N}} \Psi^*[(F)^{(3N-1)} \wedge dN] .$$

This result can be used as a starting point for construction of a statistical mechanics of nonconservative, nonholonomic systems, although such a discipline would involve very messy formulas as (44.22) indicates. Of course, if $N^2_\alpha = 0$, a significant economy results.

45. DERIVATION OF THE EQUATIONS OF MOTION FROM A BALANCE FORM

This section establishes the Poincaré theorem for nonconservative, nonholonomic dynamical systems.

If the system of second order differential equations

(45.1) $$P_\alpha = p_\alpha(t;\, q^\beta;\, y^\beta) \ , \quad \Psi^* dP_\alpha = \Psi^* r_\alpha(t;\, q^\beta;\, y^\beta)$$

that defines the map

(45.2) $$\Psi:\ \mathbb{R} \to K \,|\, \Psi(t) = t \ , \quad \Psi(q^\alpha) = \psi^\alpha(t) \ , \quad \Psi(y^\alpha) = D\psi^\alpha(t)$$

admits the balance form

(45.3) $$F = d(J + j) + N \wedge dt$$

with

(45.4) $$J = P_\alpha dq^\alpha - (P_\alpha y^\alpha - L(t;\, q^\beta;\, y^\beta))dt$$

(45.5) $$j = N^2_\alpha dq^\alpha - N^2_\alpha y^\alpha dt \ , \quad N = N^0 dt + N^1_\alpha dq^\alpha + N^2_\alpha dy^\alpha$$

then

(45.6) $$p_\alpha(t;\, q^\beta;\, y^\beta) = \partial L(t;\, q^\beta;\, y^\beta)/\partial y^\alpha \ ,$$

(45.7) $$\Psi^* r_\alpha(t;\, q^\beta;\, y^\beta) = N^1_\alpha - DN^2_\alpha + \partial L/\partial\psi^\alpha \ ;$$

that is, the system (45.1) *constitutes the equations of motion of a nonconservative, nonholonomic system that is characterized by the fundamental form* F .

The proof proceeds along exactly the same lines as the proof given in Section 42 in the case of a variational system. Starting with the general solution of the system (45.1) in the form given by (42.7), we obtain

$$\text{(45.8)} \qquad \Psi^*(F) = \left(r_\alpha - \frac{\partial L}{\partial \mathfrak{f}^\alpha} - N^1_\alpha + DN^2_\alpha\right) dt \wedge \frac{\partial \mathfrak{f}^\alpha}{\partial \eta_i} d\eta_i$$

$$+ \left(\frac{\partial L}{\partial\left(\frac{\partial \mathfrak{f}^\alpha}{\partial t}\right)} - p_\alpha\right) \frac{\partial^2 \mathfrak{f}^\alpha}{\partial t \partial \eta_i} d\eta_i \wedge dt$$

$$+ \frac{\partial (P_\alpha + N^2_\alpha)}{\partial \eta_i} \frac{\partial \mathfrak{f}^\alpha}{\partial \eta_j} d\eta_i \wedge d\eta_j \; .$$

Now, $\Psi^*(F)$ can not contain terms with a factor dt since F is a balance form. We thus have the conditions

$$\text{(45.9)} \qquad 0 = \left(r_\alpha - \frac{\partial L}{\partial \mathfrak{f}^\alpha} - N^1_\alpha + DN^2_\alpha\right) dt \wedge (\Psi|_t)^*(dq^\alpha)$$

$$+ \left(p_\alpha - \frac{\partial L}{\partial\left(\frac{\partial \mathfrak{f}^\alpha}{\partial t}\right)}\right) dt \wedge (\Psi|_t)^*(dy^\alpha) \; .$$

Noting that $(\Psi|_t)^*(dq^\alpha)$ and $(\Psi|_t)^*(dy^\alpha)$ are independent by the independence of the $6N$ parameters (η_i) , we obtain (45.6) and (45.7), and the result is established.

This theorem and the corresponding theorem for variational systems established in Section 42 provide the basis for the construction of a transformation-theoretic approach to mechanics that is the Lagrangian mechanics analog of Hamilton-Jacobi theory.

CHAPTER VIII

A TRANSFORMATION-THEORETIC APPROACH

46. LAGRANGE TRANSFORMATIONS OF VARIATIONAL SYSTEMS

The results obtained in Section 15 show that an admissible map

$$(46.1)\qquad \Gamma\colon K \to {}^{\backprime}K \,|\, {}^{\backprime}t = t,\ {}^{\backprime}q^{\alpha} = \gamma^{\alpha}(t;\, q^{\beta}),\ {}^{\backprime}y^{\alpha} = (Z\gamma^{\alpha})(t;\, q^{\beta};\, y^{\beta})\ ,$$

where Z is the linear operator $\partial/\partial t + y^{\beta}\partial/\partial q^{\beta}$, maps a variational system with Lagrangian function L onto a variational system with Lagrangian function ${}^{\backprime}L = L\circ\Gamma$ and maps a general system of Lagrange's equations $\{E|T\}_{\phi^{\alpha}} - f^{\alpha} = 0$ onto a system of Lagrange's equations $\{E|{}^{\backprime}T\}_{{}^{\backprime}\phi^{\alpha}} - {}^{\backprime}f^{\alpha} = 0$. Now, admissible maps, as maps from a $(6N+1)$-dimensional space with coordinate functions $(t;\, q^{\beta};\, y^{\beta})$ to a $(6N+1)$-dimensional space with coordinate functions $({}^{\backprime}t;{}^{\backprime}q^{\beta};\, {}^{\backprime}y^{\beta})$, are quite special, as witnessed by the lack of occurrence of any dependence of the ${}^{\backprime}q$'s on the y's, even for the class of maps that preserves time $({}^{\backprime}t=t)$. It is also well known that classes of transformations more general than admissible ones are of central importance in mechanics: the theory of contact transformations and Hamilton-Jacobi theory in the Hamiltonian formulation of variational systems. It thus appears both useful and necessary to investigate what happens when we replace admissible maps by a more general class of maps that also preserves time.

The case of general variational systems will be considered first since this is the simpler case and its results will provide a springboard for the corresponding analysis of nonconservative, nonholonomic systems. Because the ultimate aim is to obtain a transformation-theoretic approach to nonconservative, nonholonomic systems, we shall stay in the Lagrangian formulation of mechanics. The ensuing analysis of variational systems is thus not simply a redevelopment of the classic theory, for the classic theory presupposes a Hamiltonian formulation.

We start with a general class of invertible, time preserving maps of the form

(46.2) $\rho\colon K \to {}^{\backprime}K \mid {}^{\backprime}t = t,\ {}^{\backprime}q^{\alpha} = \rho_1^{\alpha}(t;\, q^{\beta};\, y^{\beta}),\ {}^{\backprime}y^{\alpha} = \rho_2^{\alpha}(t;\, q^{\beta};\, y^{\beta})$.

Suppose that we are given a variational system on K with Lagrangian function $L(t;\, q^{\beta};\, y^{\beta})$. The Newtonian motions of such a system consist of all maps $\Phi\colon \mathbb{R} \to K \mid \Phi(t) = t,\ \Phi(q^{\alpha}) = \phi^{\alpha}(t),\ \Phi(y^{\alpha}) = D\phi^{\alpha}(t)$ that satisfy the equations of motion

(46.3) $$\Phi^{*}(V \lrcorner\, dJ) = 0 ,$$

for all $V \in T(K)$, where we now write

(46.4) $$J(L) = P_{\alpha}(dq^{\alpha} - y^{\alpha}dt) + L\,dt ,$$

(46.5) $$P_{\alpha} = \partial L/\partial y^{\alpha} .$$

For any map Φ that satisfies these conditions, a composition with the map ρ induces a map

(46.6) $${}^{\backprime}\Phi = \rho\circ\Phi\colon \mathbb{R} \to {}^{\backprime}K \mid {}^{\backprime}\Phi(t) = t$$

that defines a Newtonian motion in ${}^{\backprime}K$. The equations of motion in the new space ${}^{\backprime}K$ will be quite a complex system of equations, however, with little or no resemblance to the equations of motion of a variational system. It thus appears reasonable to place conditions on the map ρ so that the new system of equations of motion is again identifiable with the equations of motion of a variational system; that is, the map ρ is to be restricted in such a manner that it preserves the form of the variational equations of motion.

The simplest and most direct way of placing condition on the map ρ so that it will preserve the form of the equations of motion of a variational system is to use the results established in Section 42. These results tell us that any system of second order differential equations that admits the absolute integral invariant $dJ(L)$ (in the notation introduced by (46.4)) is a system of equations of motion of a variational system with Lagrangian function L. Further, the theorem occurring just after equation (41.4) says that if

$$(46.7) \qquad 0 = \int_{S_2} dJ(L)$$

for every S_2 in K that is comprised of a 1-parameter family of Newtonian motions of the variational system, then $dJ(L)$ is an absolute integral invariant of the system. This is the key to the problem, for suppose we demand that ρ be such that $\rho^* dJ({}^{\backprime}L) = dJ(L)$ for some ${}^{\backprime}L = {}^{\backprime}L(L;\rho)$. If this is true, then

$$(46.8) \qquad 0 = \int_{S_2} dJ(L) = \int_M \Psi^* dJ(L) = \int_M \Psi^*(\rho^* dJ({}^{\backprime}L))$$

$$= \int_M (\rho\circ\Psi)^* dJ({}^{\backprime}L) = \int_M {}^{\backprime}\Psi^* dJ({}^{\backprime}L) = \int_{\rho\circ S_2} dJ({}^{\backprime}L) .$$

Accordingly, the equations of motion in ${}^{\backprime}K$ admit $dJ({}^{\backprime}L)$ as an absolute integral invariant, and the equations of motion in ${}^{\backprime}K$ are the equations of motion of a variational system with Lagrangian function ${}^{\backprime}L$. There is now just one further restriction to be placed on the map ρ. We do not want ρ to depend on the choice of the Lagrangian function for such a dependence would preclude compositions of such maps - the resulting Lagrangian function ${}^{\backprime}L$ would not necessarily be usable with another such map ρ if ρ were to depend on the Lagrangian function of the system. Stated another way, the kinds of maps ρ that we require should map all variational systems onto variational systems. These considerations give us a specific set of requirements that are combined in the following definition.

An invertible map

$$(46.9)\quad \rho : K \to {}^{`}K \mid {}^{`}t = t\ ,\ {}^{`}q^{\alpha} = \rho_1^{\alpha}(t;\, q^{\beta};\, y^{\beta}),\ {}^{`}y^{\alpha} = \rho_2^{\alpha}(t;\, q^{\beta};\, y^{\beta})$$

is a *Lagrange transformation* if and only if there exists an ${}^{`}L(L;\rho)$ for each Lagrangian function L such that

$$(46.10)\qquad \rho^{-1*}(dJ(L)) = dJ({}^{`}L)\ .$$

The reason for writing the conditions in the form (46.10), rather than as $dJ(L) = \rho^{*}dJ({}^{`}L)$, is because we start with the form $dJ(L)$ in any specific problem; that is, (46.10) leads to a simplification in the resulting computations. The two ways of writing the conditions are equivalent, however, for we have assumed that ρ is an invertible map.

47. PROPERTIES OF LAGRANGE TRANSFORMATIONS

Now that we have singled out the class of Lagrange transformations from K to ${}^{`}K$ as a "preferred" class of transformations, the next step is to obtain the properties of these transformations.

The class of all Lagrange transformations forms a group under composition.

First off, it is clear that the identity map is a Lagrange transformation. Suppose now, that we have two Lagrange transformations, ρ_1 and ρ_2 : that is, $\rho_1 : K \to {}^{`}K$ and $\rho_2 : {}^{`}K \to {}^{``}K$ . We want to show that $\rho_2 \circ \rho_1 : K \to {}^{``}K$ is also a Lagrange transformation. It is immediate that $((\rho_2 \circ \rho_1)^{-1})^{*} = (\rho_1^{-1} \circ \rho_2^{-1})^{*} = \rho_2^{-1*} \circ \rho_1^{-1*}$. Accordingly, since ρ_1 and ρ_2 satisfy (46.10) for all Lagrangian functions, we see that

$$(47.1)\quad ((\rho_2 \circ \rho_1)^{-1})^{*} dJ(L) = \rho_2^{-1*} \circ \rho_1^{-1*} dJ(L) = \rho_2^{-1*} dJ({}^{`}L) = dJ({}^{``}L)\ ,$$

and the result is established.

An invertible map $\rho : K \to \grave{}K \mid \grave{}t = t$ *defines a Lagrange transformation if and only if there exists an* $\grave{}L(L;\rho)$ *for each* L *and a scalar-valued function* η *, called the generating function of* ρ *, such that*

$$(47.2) \qquad \rho^{-1*}J(L) = J(\grave{}L) + d\eta .$$

Since $\Phi^* d = d\Phi^*$ for any map Φ , (46.10) yields

$$(47.3) \qquad \rho^{-1*}dJ(L) = d\rho^{-1*}J(L) = dJ(\grave{}L) ,$$

and we conclude that $\rho^{-1*}J(L) - J(\grave{}L)$ is a closed form. Thus, since K and $\grave{}K$ are (6N+1)-dimensional Euclidean spaces by the assumption that there is an inertial coordinate cover of these spaces, the Poincaré lemma shows that $\rho^{-1*}J(L) - J(\grave{}L)$ is an exact form. The converse is obvious and the result follows.

This theorem sits at the heart of the problem of generating Lagrange transformations, as we shall now show. The definition of $J(L)$ given by (46.4) yields

$$(47.4) \qquad J(L) = P_\alpha (dq^\alpha - y^\alpha dt) + L dt ,$$

$$(47.5) \qquad J(\grave{}L) = \grave{}P_\alpha (d\grave{}q^\alpha - \grave{}y^\alpha d\grave{}t) + \grave{}L d\grave{}t .$$

Now, because ρ is assumed to be invertible, we can take t and any 6N of the 12N variables q^α, y^α, $\grave{}q^\alpha$, $\grave{}y^\alpha$ as the arguments of the generating function η . An inspection of (47.4) and (47.5) shows that an obvious first choice, at least from the standpoint of simplicity, is to take

$$(47.6) \qquad \eta = u(t; q^\alpha; \grave{}q^\alpha) .$$

It is, of course, assumed that y^α and $\grave{}y^\alpha$ are then expressed in terms of $(t; q^\beta; \grave{}q^\beta)$ by means of ρ and ρ^{-1} . With this accomplished, a substitution of (47.4) through (47.6) into (47.2) gives us (recall that $\rho(t)=t$)

$$(47.7)\qquad P_\alpha(dq^\alpha - y^\alpha dt) + L dt - {}^{\backprime}P_\alpha(d{}^{\backprime}q^\alpha - {}^{\backprime}y^\alpha dt) - {}^{\backprime}L dt$$

$$= P_\alpha dq^\alpha - {}^{\backprime}P_\alpha d{}^{\backprime}q^\alpha + ({}^{\backprime}P_\alpha {}^{\backprime}y^\alpha - P_\alpha y^\alpha + L - {}^{\backprime}L)dt$$

$$= d\eta = \frac{\partial u}{\partial t} dt + \frac{\partial u}{\partial q^\alpha} dq^\alpha + \frac{\partial u}{\partial {}^{\backprime}q^\alpha} d{}^{\backprime}q^\alpha .$$

Since t, q^α and ${}^{\backprime}q^\alpha$ are independent, (47.7) is satisfied if and only if

$$(47.8)\qquad P_\alpha(t;\, q^\beta;\, y^\beta) = \partial u(t;\, q^\beta;\, {}^{\backprime}q^\beta)/\partial q^\alpha ,$$

$$(47.9)\qquad {}^{\backprime}P_\alpha(t;\, {}^{\backprime}q^\beta;\, {}^{\backprime}y^\beta) = -\, \partial u(t;\, q^\beta;\, {}^{\backprime}q^\beta)/\partial {}^{\backprime}q^\alpha ,$$

$$(47.10)\qquad {}^{\backprime}L - {}^{\backprime}P_\alpha {}^{\backprime}y^\alpha = L - P_\alpha y^\alpha - \partial u(t;\, q^\beta;\, {}^{\backprime}q^\beta)/\partial t .$$

It now remains to show that we can actually obtain the equations of transformation that define ρ from the above system of relations. Actually, we will solve for the equations that determine ρ^{-1}, since it is these equations that are needed in the applications to be given in Section 49. We first note that (47.9) can be solved for the q's as functions of t, ${}^{\backprime}q^\beta$, ${}^{\backprime}y^\beta$ provided u satisfies the determinental condition

$$(47.11)\qquad \det(\partial^2 u/\partial q^\alpha \partial {}^{\backprime}q^\beta) \neq 0 .$$

Under satisfaction of this condition, we accordingly obtain

$$(47.12)\qquad q^\alpha = \overset{-1}{\rho}{}^\alpha_1(t;\, {}^{\backprime}q^\beta;\, {}^{\backprime}y^\beta) .$$

Further, since $\det(\partial^2 L/\partial y^\alpha \partial y^\beta) \neq 0$ for any given variational system that represents a mechanical problem (i.e., the kinetic energy is a positive definite scalar-valued form in the y's), we can solve $P_\alpha = \partial L(t;\, q^\beta;\, y^\beta)/\partial y^\alpha$ for the y's so as to obtain

$$(47.13)\qquad y^\alpha = Y^\alpha(t;\, q^\beta;\, P_\beta) .$$

An elimination of P_α between (47.13) and (47.8) yields the relations

$$(47.14) \qquad y^\alpha \;=\; Y^\alpha(t;\; q^\beta; \partial u(t;\; q^\gamma;\; {}^{\backprime}q^\gamma)/\partial q^\beta) \;.$$

Thus, when (47.12) is used to eliminate the q's from these relations, we obtain $y^\alpha = \overset{-1}{\rho}{}_2^{\alpha}(t;\; {}^{\backprime}q^\beta;\; {}^{\backprime}y^\beta)$. The actual equations of transformation are indeed obtainable provided the determinental condition (47.11) is satisfied.

A second class of Lagrange transformations is obtained by setting

$$(47.15) \qquad \eta \;=\; \gamma(t;\; q^\beta;\; {}^{\backprime}P_\beta) - {}^{\backprime}P_\alpha {}^{\backprime}q^\alpha \;.$$

We assume that the equations of transformation and their inverses are used so that all quantities are expressed in terms of the 6N+1 variables $(t,\; q^\beta,\; {}^{\backprime}P_\beta)$. It then follows, upon substituting (47.15) into (47.2), that

$$P_\alpha(dq^\alpha - y^\alpha dt) + L dt - {}^{\backprime}P_\alpha(d{}^{\backprime}q^\alpha - {}^{\backprime}y^\alpha dt) - {}^{\backprime}L dt \;=\; d\eta$$

$$= \frac{\partial\gamma}{\partial t}\,dt + \frac{\partial\gamma}{\partial q^\alpha}dq^\alpha + \frac{\partial\gamma}{\partial {}^{\backprime}P_\alpha}\,d{}^{\backprime}P_\alpha - {}^{\backprime}P_\alpha d{}^{\backprime}q^\alpha - {}^{\backprime}q^\alpha d{}^{\backprime}P_\alpha \;;$$

that is

$$(47.16) \qquad \left(P_\alpha - \frac{\partial\gamma}{\partial q^\alpha}\right) dq^\alpha - \left({}^{\backprime}q^\alpha - \frac{\partial\gamma}{\partial {}^{\backprime}P_\alpha}\right) d{}^{\backprime}P_\alpha$$

$$+ \left(L - P_\alpha y^\alpha - {}^{\backprime}L + {}^{\backprime}P_\alpha {}^{\backprime}y^\alpha - \frac{\partial\gamma}{\partial t}\right) dt \;=\; 0 \;.$$

Since $(t,\; q^\beta,\; {}^{\backprime}P_\beta)$ are independent, (47.16) is satisfied if and only if

$$(47.17) \qquad P_\alpha(t;\; q^\beta;\; y^\beta) \;=\; \partial\gamma(t;\; q^\beta;\; {}^{\backprime}P_\beta)/\partial q^\alpha$$

$$(47.18) \qquad {}^{\backprime}q^\alpha \;=\; \partial\gamma(t;\; q^\beta;\; {}^{\backprime}P_\beta)/\partial {}^{\backprime}P_\alpha$$

$$`L - `P_\alpha `y^\alpha = L - P_\alpha y^\alpha - \partial\gamma(t;\ q^\beta;\ `P_\beta)/\partial t\ . \tag{47.19}$$

We now turn to the problem of obtaining the actual transformations associated with the equations (47.17) through (47.19). We assume that the variational system under consideration comes about from a mechanics problem. In this event, we have

$$\det(\partial^2 L/\partial y^\alpha \partial y^\beta) = \det(\partial P_\alpha/\partial y^\beta) \neq 0 \tag{47.20}$$

from $P_\alpha = \partial L/\partial y^\alpha$. We may thus solve $P_\alpha = \partial L(t;\ q^\beta;\ y^\beta)/\partial y^\alpha$ for the y's so as to obtain

$$y^\alpha = Y^\alpha(t;\ q^\beta;\ P_\beta)\ . \tag{47.21}$$

An elimination of P_β between (47.21) and (47.17) thus yields

$$y^\alpha = Y^\alpha(t;\ q^\beta; \partial\gamma(t;\ q^\beta;\ `P_\beta)/\partial q^\alpha) = \bar{Y}^\alpha(t;\ q^\beta;\ `P_\beta)\ . \tag{47.22}$$

Let us assume that $\gamma(t;\ q^\beta;\ `P_\beta)$ satisfies the condition

$$\det(\partial^2\gamma/\partial q^\alpha \partial `P_\beta) \neq 0\ . \tag{47.23}$$

This condition allows us to solve (47.18) for q^α in terms of $(t,\ `q^\beta,\ `P_\beta)$:

$$q^\alpha = Q^\alpha(t;\ `q^\beta;\ `P_\beta)\ . \tag{47.24}$$

When this is substituted into (47.22), we then obtain

$$y^\alpha = \bar{Y}^\alpha(t;\ Q^\beta(t;\ `q^\mu;\ `P_\mu); `P_\beta) = \underline{Y}^\alpha(t;\ `q^\beta;\ `P_\beta)\ .$$

We thus obtain the following system of parametric equations for the determination of the function $\overset{-1}{\rho}{}^\alpha_1$, $\overset{-1}{\rho}{}^\alpha_2$:

$$\begin{gathered} q^\alpha = Q^\alpha(t;\ `q^\beta;\ `P_\beta)\ , \quad y^\alpha = \underline{Y}^\alpha(t;\ `q^\beta;\ `P_\beta)\ , \\ `P_\alpha = \partial`L(t;\ `q^\beta;\ `y^\beta)/\partial`y^\alpha\ . \end{gathered} \tag{47.25}$$

It follows from these results that the equations for the transformation ρ^{-1} can be determined once we have the function $`L(t; `q^\beta; `y^\beta)$. The function $`L$, however, must be such that there is a choice of the function $\gamma(t; q^\beta; `P_\beta)$ such that (47.19) is satisfied. This implied relation between $`L$ and γ is what allows us to solve problems by transformation-theoretic methods, as we show in the next section.

48. TRANSFORMATION-THEORETIC SOLUTIONS - THE LAGRANGIAN MECHANICS ANALOG OF HAMILTON-JACOBI THEORY

The results established in the previous section give us a Lagrange transformation to a new variational system whenever the functions $`L(t; `q^\beta; `y^\beta)$ and $\gamma(t; q^\beta; `P_\beta)$ are selected in such a manner that (47.19) is satisfied. Now, there is little point to constructing such transformations unless they lead to simplifications in the new system that outweigh the trouble of constructing the transformations; the ideal situation being one in which the new system is so simple that it can be integrated in close form. We know of such systems, however, from the results established in Chapter V; namely, free systems that are characterized by $`L_F = \frac{1}{2}`y^\alpha m_{\alpha\beta}`y^\beta$. Now, there is little point for settling for less than the simplest free system. Accordingly, we take

$$(48.1) \qquad `L = \frac{1}{2}`y^\alpha \delta_{\alpha\beta}`y^\beta , \qquad \delta_{\alpha\beta} = \begin{cases} 1 & \alpha=\beta \\ 0 & \alpha\neq\beta \end{cases} ,$$

and inquire into whether $\gamma(t; q^\beta; `P_\beta)$ can be chosen so as to yield satisfaction of the one remaining equation (47.19).

The first thing we note is that (48.1) and $`P_\alpha = \partial`L/\partial y^\alpha$ yield

$$(48.2) \qquad `P_\alpha = \delta_{\alpha\beta}`y^\beta , \qquad `y^\alpha = \delta^{\alpha\beta}`P_\beta$$

and hence we have

$$(48.3) \qquad \grave{}L - \grave{}P_\alpha \grave{}y^\alpha = -\frac{1}{2}\grave{}P_\alpha \delta^{\alpha\beta}\grave{}P_\beta = -\frac{1}{2}\sum_{\mu=1}^{3N}(\grave{}P_\mu)^2 .$$

We also have

$$(48.4) \qquad P_\alpha = \partial\gamma/\partial q^\alpha$$

from (47.17), and

$$(48.5) \qquad y^\alpha = Y^\alpha(t;\ q^\beta;\ \partial\gamma/\partial q^\beta)$$

from (47.21) and (47.17). A substitution of (48.3) through (48.5) into (47.19) thus yields the following condition in order that (48.1) shall hold:

$$(48.6) \qquad -\frac{1}{2}\grave{}P_\alpha \delta^{\alpha\beta}\grave{}P_\beta = L(t;\ q^\beta;\ Y^\beta(t;\ q^\mu;\ \partial\gamma/\partial q^\mu)) - \frac{\partial\gamma}{\partial q^\alpha} Y^\alpha(t;\ q^\beta;\ \partial\gamma/\partial q^\beta) - \frac{\partial\gamma}{\partial t} .$$

However, all of the variables that occur in (48.6) consist of the list $(t;\ q^\beta;\ \grave{}P_\beta)$ and γ is a function of exactly these variables. The equation (48.6) can thus be looked upon as a partial differential equation for the determination of the generating function $\gamma(t;\ q^\beta;\ \grave{}P_\beta)$. Interpreted in this way, *the partial differential equation* (48.6) *is the Lagrangian mechanics analog of the Hamilton-Jacobi equation in Hamiltonian mechanics.* If $\gamma(t; q^\beta; \grave{}P_\beta)$ is determined by solving (48.6), then γ is referred to as a *principal function.*

In order to make this analog somewhat clearer, we introduce $S(t;\ q^\beta;\ \grave{}P_\beta)$ by

$$(48.7) \qquad \gamma(t;\ q^\beta;\ \grave{}P_\beta) = S(t;\ q^\beta;\ \grave{}P_\beta) + \frac{1}{2}t\grave{}P_\alpha \delta^{\alpha\beta}\grave{}P_\beta .$$

A direct substitution of this into (48.6) gives

$$(48.8) \quad L(t;\ q^\beta;\ Y^\beta(t;\ q^\mu,\ \partial S/\partial q^\mu)) - \frac{\partial S}{\partial q^\alpha} Y^\alpha(t;\ q^\beta;\ \partial S/\partial q^\beta) - \frac{\partial S}{\partial t} = 0 .$$

Thus, since $y^\alpha = Y^\alpha(t; q^\beta; P_\beta)$ is obtained by solving $P_\alpha = \partial L(t; q^\beta; y^\beta)/\partial y^\alpha$ for the y's, the analogy is direct, for any solution of (48.8) is a solution of the Hamilton-Jacobi partial differential equation that would obtain if the variational system were transcribed into Hamiltonian form. This is further corroborated by noting that (16.26) shows that $L - P_\alpha y^\alpha = L - \frac{\partial L}{\partial y^\alpha} y^\alpha = -E$ and E is the total energy function of the variational system (i.e., the Hamiltonian written in terms of $(t; q^\beta; y^\beta)$). We note that the substitution (48.7) combines with (47.17) and (47.18) to yield

$$(48.9) \qquad P_\alpha(t; q^\beta; y^\beta) = \partial S(t; q^\beta; {}^\backprime P_\beta)/\partial q^\alpha ,$$

$$(48.10) \qquad {}^\backprime q^\alpha = t\delta^{\alpha\mu}\, {}^\backprime P_\mu + \partial S(t; q^\beta; {}^\backprime P_\beta)/\partial\, {}^\backprime P_\alpha .$$

Further, since the variables ${}^\backprime P_\beta$ have been used to replace the ${}^\backprime y^\beta$ by (48.2), the specific transformation equations, (47.25) become

$$(48.11) \qquad q^\alpha = Q^\alpha(t; {}^\backprime q^\beta; {}^\backprime P_\beta) , \quad y^\alpha = \underline{Y}^\alpha(t; {}^\backprime q^\beta; {}^\backprime P_\beta) .$$

The important thing to note at this point is that the transformation equations (48.11) actually give a complete solution of the original variational problems. In order to see this, we first note that (48.1) gives

$$(48.12) \qquad \partial\, {}^\backprime L/\partial\, {}^\backprime q^\beta = 0 .$$

Thus, if ${}^\backprime\Phi: \mathbb{R} \to {}^\backprime K \mid {}^\backprime\Phi(t) = t,\ {}^\backprime\Phi({}^\backprime q^\alpha) = {}^\backprime\phi^\alpha(t),\ {}^\backprime\Phi({}^\backprime y^\alpha) = D\,{}^\backprime\phi^\alpha(t)$ defines a Newtonian motion of the new system, (48.2) shows that

$$(48.13) \qquad D\,{}^\backprime P_\alpha = D\delta_{\alpha\beta} D\,{}^\backprime\phi^\beta(t) = 0 .$$

Accordingly,

$$(48.14) \qquad {}^\backprime P_\alpha = C_\alpha , \quad {}^\backprime y^\alpha = \delta^{\alpha\mu} C_\mu , \quad {}^\backprime\phi^\alpha(t) = tC_\beta\delta^{\beta\alpha} + c^\alpha ,$$

where the C's and the c's are constants. When these are substituted into (48.11) we obtain the solution

(48.15) $$\Phi(q^\alpha) = Q^\alpha(t;\ tC_\mu\delta^{\mu\beta}+c^\beta;\ C_\beta) = \phi^\alpha(t) ,$$

(48.16) $$\Phi(y^\alpha) = \underline{Y}^\alpha(t;\ tC_\mu\delta^{\mu\beta}+c^\beta;\ C_\beta) = D\phi^\alpha(t) .$$

There are several important observations that need to be made concerning obtaining solutions by the method given above. First, the function $S(t;\ q^\beta;\ {}^{\backprime}P_\beta)$, that is to be obtained by solving (48.8), is a function of the 6N+1 variables t, q^β, ${}^{\backprime}P_\beta$ while (48.8) does not contain either the variables ${}^{\backprime}P_\beta$ or the derivatives $\partial S/\partial{}^{\backprime}P_\beta$. In addition the determinental condition (47.23) and (48.7) combine to give the requirement $\det(\partial^2 S/\partial q^\alpha \partial{}^{\backprime}P_\beta) \neq 0$. Accordingly, the ${}^{\backprime}P$'s have to be identified with the 3N nonadditive integration constants that occur in the *general solution* of (48.8) when considered as a partial differential equation of the first order with the 3N+1 independent variables t, q^α. Thus, as in Hamilton-Jacobi theory†, we must obtain the general solution of (48.8) rather than just a particular solution. We note in passing that the general solution of (48.8) will contain 3N nonadditive constants since (48.8) is linear in $\partial S/\partial t$ but quadratic in $\partial S/\partial q^\alpha$ because L depends on the y's through the kinetic energy function that is quadratic. Second, it is essential that the 3N nonadditive integration constants that obtain in the general solution of (48.8) be identified with the same 3N constants ${}^{\backprime}P_\alpha$ that occur in the term $\frac{1}{2}\, t\, {}^{\backprime}P_\alpha \delta^{\alpha\beta}\, {}^{\backprime}P_\beta$ on the right-hand side of (48.7), for otherwise the principal function Υ would turn out to be a function of 9N+1 quantities and the whole procedure dissolves in confusion. This requirement is clearly implied by the explicit occurrence of the arguments in writing (48.7), but could be overlooked by the reader accustomed to the classic Hamiltonian formulation of the problem.

†Goldstein (Ref. 12); Whittaker (Ref. 13); Pars (Ref. 17).

There is one further result in the context of solutions that is both useful and informative. Since $\gamma = \gamma(t;\, q^\beta;\, {}^\backprime P_\beta)$, we can compute $d\Gamma/dt$ along any solution curve, it being understood that $q^\beta = \phi^\beta(t)$, ${}^\backprime P_\beta = P_\beta = C_\beta$ along any such solution curve and $\Gamma = \gamma(t;\, \phi^\beta(t);\, C_\beta)$:

$$(48.17) \qquad \frac{d\Gamma}{dt}(t;\, \phi^\beta(t);\, C_\beta) = \frac{\partial\Gamma}{\partial t} + \frac{\partial\Gamma}{\partial\phi^\beta} D\phi^\beta .$$

However, (47.17) gives

$$\frac{\partial\Gamma}{\partial\phi^\beta} = \Phi^*\left(\frac{\partial\gamma}{\partial q^\beta}\right) = \Phi^* P_\beta = P_\beta$$

and hence (48.17) becomes

$$(48.18) \qquad \frac{d\Gamma}{dt} = \frac{\partial\Gamma}{\partial t} + P_\alpha D\phi^\alpha = \Phi^*\left(\frac{\partial\gamma}{\partial t} + P_\alpha y^\alpha\right) .$$

Now, because $\partial\gamma/\partial q^\alpha = P_\alpha$, $y^\alpha = Y^\alpha(t;\, q^\beta;\, \partial\gamma/\partial q^\beta)$, (48.18) can be written as

$$(48.19) \qquad \frac{d\Gamma}{dt} = \Phi^*\left(\frac{\partial\gamma}{\partial t} + \frac{\partial\gamma}{\partial q^\alpha} Y^\alpha\right) ,$$

and hence (48.6) can be used to obtain

$$(48.20) \qquad \frac{d\Gamma}{dt} = \Phi^*(L + \frac{1}{2}\, {}^\backprime P_\alpha \delta^{\alpha\beta}\, {}^\backprime P_\beta) = L(t;\, \phi^\alpha(t);\, D\phi^\alpha(t)) + \frac{1}{2} C_\alpha \delta^{\alpha\beta} C_\beta .$$

An integration then gives

$$(48.21) \qquad \Gamma(t;\, \phi^\alpha(t);\, C_\alpha) - \Gamma(t_0;\, \phi^\alpha(t_0);\, C_\alpha) = \frac{1}{2}(t-t_0) C_\alpha \delta^{\alpha\beta} C_\beta + \int_{t_0}^{t} L(\lambda;\, \phi^\alpha(\lambda);\, D\phi^\alpha(\lambda))d\lambda .$$

The principal function of a dynamical system of variational form is determined along any orbit of the dynamical system by the indefinite action integral $\int_{t_0}^{t} L(\lambda;\, \phi^\alpha(\lambda);\, D\phi^\alpha(\lambda))d\lambda$, *the value of the principal function for* $t = t_0$ *, and the constants* $C_\alpha = {}^`P_\alpha$ *that obtain from the transformation of the given system to one in uniform rectilinear motion.*

49. A USEFUL ALTERNATIVE - SOLUTION BY MEANS OF A PARTICULAR INTEGRAL

The necessity of obtaining the general solution of the partial differential equation (48.8) is quite often a heavy burden, even for simple problems. Fortunately, there is an alternative in the Lagrangian formulation that requires that one obtain only a particular solution of a certain partial differential equation that is such as to satisfy a single determinental condition. This comes about by going back to the first class of Lagrange transformations obtained in Section 47. This class of Lagrange transformations uses t , q^β , ${}^`q^\beta$ as the independent variables so that the generating function is given by

$$(49.1) \qquad \eta = u(t;\ q^\beta;\ {}^`q^\beta) .$$

Under these conditions, the equations that define the Lagrange transformation are given by

$$(49.2) \qquad P_\alpha = \partial u/\partial q^\alpha ,$$

$$(49.3) \qquad {}^`P_\alpha = -\ \partial u/\partial {}^`q^\alpha ,$$

$$(49.4) \qquad {}^`L - {}^`P_\alpha {}^`y^\alpha = L - P_\alpha y^\alpha - \partial u/\partial t .$$

We must, however, satisfy the determinental condition

$$(49.5) \qquad \det(\partial^2 u/\partial q^\alpha \partial {}^`q^\beta) \neq 0$$

in order that (49.3) can be solved for the q's so as to obtain

$$(49.6) \qquad q^\alpha = \overset{-1}{\rho}{}^\alpha_1(t;\ {}^`q^\beta;\ {}^`y^\beta) .$$

Further, since L is representative of a mechanical system in variational form, $\det(\partial^2 L/\partial y^\alpha \partial y^\beta) \neq 0$, and we can solve $P_\alpha = \partial L(t; q^\beta; y^\beta)/\partial y^\beta$ for the y's in terms of the variables t, q^β, P_β so as to obtain $y^\alpha = Y^\alpha(t; q^\beta; P_\beta)$. Accordingly, use of (49.2) gives

$$(49.7) \qquad y^\alpha = Y^\alpha(t; q^\beta; \partial u/\partial q^\beta) .$$

With these results at hand, we now proceed as in Section 48 to demand that the new system be a system undergoing uniform rectilinear motion; that is, we take

$$(49.8) \qquad \grave{}L = \frac{1}{2}\, \grave{}y^\alpha \delta_{\alpha\beta}\, \grave{}y^\beta .$$

Under these circumstances, we have

$$(49.9) \qquad \grave{}P_\beta = \delta_{\beta\alpha}\, \grave{}y^\alpha , \qquad \grave{}y^\alpha = \delta^{\alpha\beta}\, \grave{}P_\beta ,$$

$$(49.10) \qquad \grave{}L - \grave{}P_\alpha\, \grave{}y^\alpha = -\frac{1}{2}\, \grave{}y^\alpha \delta_{\alpha\beta}\, \grave{}y^\beta = -\frac{1}{2}\, \grave{}P_\alpha \delta^{\alpha\beta}\, \grave{}P_\beta .$$

A substitution of (49.2), (49.3), (49.7) and (49.10) into (49.4) thus gives us the following equation for the determination of $u(t; q^\beta; \grave{}q^\beta)$:

$$(49.11) \qquad L(t; q^\beta; Y^\beta(t; q^\gamma; \partial u/\partial q^\gamma)) - \frac{\partial u}{\partial q^\alpha} Y^\alpha(t; q^\beta; \partial u/\partial q^\beta) + \frac{1}{2} \frac{\partial u}{\partial\, \grave{}q^\beta} \delta^{\alpha\beta} \frac{\partial u}{\partial\, \grave{}q^\beta} = \frac{\partial u}{\partial t} .$$

Any particular solution of this first order partial differential equation in the 6N+1 independent variables $t, q^\beta, \grave{}q^\beta$ that satisfies the determinental condition (49.5) will be suitable for our purposes. This is quite easily seen, for if $\grave{}\Phi : \mathbb{R} \to \grave{}K \mid \grave{}\Phi(t) = t$ defines a Newtonian motion of the new system, then

$$(49.12) \qquad D\grave{}P_\alpha = \partial\grave{}L/\partial\grave{}\phi^\alpha(t) , \qquad \grave{}P_\alpha = \partial\grave{}L/\partial D\grave{}\phi^\alpha(t) .$$

However, (49.8), (49.9) and (49.12) give

$$D\grave{}P_\alpha = 0 , \quad \grave{}P_\alpha = \delta_{\alpha\beta} D\grave{}\phi^\alpha ,$$

so that we have the immediate quadratures

$$(49.13) \qquad \grave{}P_\alpha = C_\alpha , \quad \grave{}\phi^\alpha(t) = C_\beta \delta^{\beta\alpha} t + c^\alpha .$$

Thus, if $\phi^\alpha(t)$ is the inverse image of $\grave{}\phi^\alpha(t)$ under the map ρ , then (49.3) and (49.13) give

$$(49.14) \qquad C_\alpha = \{\partial u(t;\, \phi^\beta(t);\, \grave{}\phi^\beta(t))/\partial\grave{}\phi^\alpha(t)\}\Big|_{\grave{}\phi^\beta = C_\gamma \delta^{\gamma\beta} t + c^\beta} ,$$

which is the same thing as

$$(49.15) \qquad \phi^\alpha(t) = {}^{-1}\rho_1^\alpha(t;\, C_\gamma \delta^{\gamma\beta} t + c^\beta;\, C_\gamma \delta^{\gamma\beta})$$

that obtains from (49.6), (49.9) and (49.13).

There is an illuminating result that we can now obtain which is similar to (48.21). Let $\{\phi^\alpha(t)\}$ and $\{\grave{}\phi^\alpha(t)\}$ define Newtonian motions for the variational systems characterized by L and $\grave{}L$, where $\grave{}L = L\circ\rho$ and $\{\phi^\alpha\}$ is mapped onto $\{\grave{}\phi^\alpha\}$ by the map ρ that is generated by the procedure given above. Further let ρ be generated by $u(t;\, q^\beta;\, \grave{}q^\beta)$ and set

$$(49.16) \qquad U(t) = u(t;\, \phi^\beta(t);\, \grave{}\phi^\beta(t)) .$$

It then follows that

$$(49.17) \qquad \frac{dU}{dt} = \frac{\partial u}{\partial t} + \frac{\partial u}{\partial\phi^\alpha} D\phi^\alpha + \frac{\partial u}{\partial\grave{}\phi^\alpha} D\grave{}\phi^\alpha .$$

However, (49.2), (49.3) give $\partial u/\partial\phi^\alpha = P_\alpha$, $\partial u/\partial\grave{}\phi^\alpha = -\grave{}P_\alpha$, so that

$$(49.18) \qquad \frac{dU}{dt} = \frac{\partial u}{\partial t} + P_\alpha D\phi^\alpha - \grave{}P_\alpha D\grave{}\phi^\alpha .$$

We now solve for $\partial u(t;\, \phi^\beta(t);\, \grave{}\phi^\beta(t))/\partial t$ from (49.4) when all quantities in this equation are evaluated along the orbit of the dynamical system in K and along its image in $\grave{}K$. This gives

$$(49.19)\qquad \frac{\partial u}{\partial t} = L - P_\alpha D\phi^\alpha - {}^{\backprime}L + {}^{\backprime}P_\alpha D^{\backprime}\phi^\alpha ,$$

so that (49.18) yields

$$(49.20)\qquad \frac{dU}{dt} = L(t;\ \phi^\beta(t);\ D\phi^\beta(t)) - {}^{\backprime}L(t;\ {}^{\backprime}\phi^\beta(t);\ D^{\backprime}\phi^\beta(t)) .$$

A direct integration thus yields

$$(49.21)\qquad U(t;\ \phi^\beta(t);\ {}^{\backprime}\phi^\beta(t)) - U(t_0;\ \phi^\beta(t_0);\ {}^{\backprime}\phi^\beta(t_0))$$

$$= \int_{t_0}^{t} \{L(\tau;\ \phi^\beta(\tau);\ D\phi^\beta(\tau)) - {}^{\backprime}L(\tau;\ {}^{\backprime}\phi^\beta(\tau);\ D^{\backprime}\phi^\beta(\tau))\}d\tau .$$

However, ${}^{\backprime}L = \frac{1}{2} D^{\backprime}\phi^\alpha \delta_{\alpha\beta} D^{\backprime}\phi^\beta = \frac{1}{2}c_\alpha \delta^{\alpha\beta} c_\beta$, ${}^{\backprime}\phi^\alpha(t) = c_\beta \delta^{\beta\alpha} t + c^\alpha$, so that we finally obtain

$$(49.22)\quad U(t;\ \phi^\beta(t);\ c_\gamma\delta^{\gamma\beta}t+c^\beta) - U(t_0;\ \phi^\beta(t_0);\ c_\gamma\delta^{\alpha\beta}t_0+c^\beta)$$

$$= \int_{t_0}^{t} L(\tau;\ \phi^\beta(\tau);\ D\phi^\beta(\tau))d\tau - \frac{1}{2}c_\alpha\delta^{\alpha\beta}c_\beta(t-t_0) .$$

It is also of interest to note that (49.20) shows that L and ${}^{\backprime}L$ differ by a total derivative so that their associated Euler equations are equivalent.

50. E-TRANSFORMATIONS OF NONCONSERVATIVE, NONHOLONOMIC SYSTEMS

We know that a nonconservative, nonholonomic system is completely characterized by its fundamental exterior form

$$(50.1)\qquad F = d(J+j) + N \wedge dt$$

and the constraint forms ω^i , $i=1,\ldots,R$, for the equations that govern such systems are given by

$$(50.2)\qquad \Phi^*\omega^i = 0 ,\quad i=1,\ldots,R$$

and

$$(50.3)\qquad \Phi^*(V\rfloor F) = 0$$

for all $V\varepsilon T(K)$. We also know that any system of differential equations that admits F as a balance form is such as to satisfy (50.3), for this is the content of the theorem established in Section 45. Accordingly, let us introduce the following notation so that we may take over the discussion in Section 46 to the general case:

$$(50.4)\qquad J(L) = P_\alpha dq^\alpha - (P_\alpha y^\alpha - L)dt\ , \qquad P_\alpha = \partial L/\partial y^\alpha$$

$$(50.5)\qquad j(N) = N_\alpha^2 dq^\alpha - N_\alpha^2 y^\alpha dt\ .$$

An invertible map

$$(50.6)\qquad \rho\colon K \to {}^\backprime K|{}^\backprime t = t\ , \qquad {}^\backprime q^\alpha = \rho_1^\alpha(t;\ q^\beta;\ y^\beta)\ ,$$

$$ {}^\backprime y^\alpha = \rho_2^\alpha(t;\ q^\beta;\ y^\beta)$$

is said to define an E-*transformation* (extended Lagrange transformation) if and only if there exists a ${}^\backprime L(L;\rho)$ for each L and a scalar valued function $\mu(\rho)$ such that

$$(50.7)\ \rho^{-1*}\{dJ(L) + dj(N) + N\wedge dt\} = dJ({}^\backprime L) + dj({}^\backprime N) + {}^\backprime N\wedge dt\ ,$$

$$(50.8)\ \rho^{-1*}N = {}^\backprime N + d\mu$$

in which case we set

$$(50.9)\ \rho^{-1*}\omega^i = {}^\backprime\omega^i\ .$$

We note explicitly at this point, that (50.7) is the direct analog of (46.10), stating as it does that a balance form is mapped onto a balance form. The additional condition (50.8) does not arise in variational systems for there is no N to be dealt with. The condition (50.8) seems to be necessary for the more general systems considered here, at least as far as the author can perceive.

There is an obvious simplification that can be achieved by using (50.8) to simplify (50.7). In particular, since $`t=t$, we have

$$\rho^{-1*}N \wedge dt = `N \wedge dt + d\mu \wedge dt = `N \wedge dt + d(\mu dt) ,$$

and hence the following result obtains.

An invertible map

$$(50.10) \qquad \rho : K \to `K|`t = t , \quad `q^{\alpha} = \rho_1^{\alpha}(t;\ q^{\beta};\ y^{\beta}) ,$$

$$`y^{\alpha} = \rho_2^{\alpha}(t;\ q^{\beta};\ y^{\beta})$$

defines an E-*transformation if and only if there exists an* $`L(L;\rho)$ *for each* L *and a scalar valued function* $\mu(\rho)$ *such that*

$$(50.11) \qquad \rho^{-1*}d\{J(L) + j(N)\} = d\{J(`L) + j(`N)\} - d(\mu dt) ,$$

$$(50.12) \qquad \rho^{-1*}N = `N + d\mu ,$$

in which case the constraint forms $`\omega^i$ *are given by*

$$(50.13) \qquad \rho^{-1*}\omega^i = `\omega^i .$$

In what follows, we refer to the function $\mu(\rho)$ associated with an E-transformation ρ as the *remainder* of ρ .

51. PROPERTIES OF E-TRANSFORMATIONS

The properties of E-transformations under composition are important, so we shall dispense with this aspect first.

The collection of E-*transformations forms a group under composition. If* μ_1 *is the remainder associated with* ρ_1 *and* μ_2 *is the remainder associated with* ρ_2 *, then the remainder* μ_{21} *associated with* $\rho_2 \circ \rho_1$ *is given by*

$$(51.1) \qquad \mu_{21} = \mu_2 + \rho_2^{-1*}\mu_1 ,$$

and the remainder associated with the identity transformation is the zero function.

The proof is as follows. First, it follows from (50.11) through (50.13) that the identity map defines an E-transformation and that the remainder of the identity transformation is the zero function. Now, let

$$\rho_1^{-1*}N = \grave{}N + d\mu_1 \,, \qquad \rho_2^{-1*}\grave{}N = \grave{}\grave{}N + d\mu_2 \,.$$

Since $((\rho_2\circ\rho_1)^{-1})^* = \rho_2^{-1*}\circ\rho_1^{-1*}$, we obtain

$$\begin{aligned}((\rho_2\circ\rho_1)^{-1})^*N &= \rho_2^{-1*}(\grave{}N+d\mu_1) = \rho_2^{-1*}\grave{}N + \rho_2^{-1*}d\mu_1 \\ &= \grave{}\grave{}N + d\mu_2 + d(\rho_2^{-1*}\mu_1) = \grave{}\grave{}N + d(\mu_2 + \rho_2^{-1*}\mu_1) \\ &= \grave{}\grave{}N + d\mu_{21} \,.\end{aligned}$$

This establishes (51.1) and the fact that (50.12) is valid for the composition $\rho_2\circ\rho_1$. Similarly, (50.11) and (51.1) yield

$$\begin{aligned}((\rho_2\circ\rho_1)^{-1})^*d\{J(L)+j(N)\} &= \rho_2^{-1*}[d\{J(\grave{}L)+j(\grave{}N)\}-d(\mu_1 dt)] \\ &= d\{J(\grave{}\grave{}L)+j(\grave{}\grave{}N)\} - d(\mu_2 dt) - d(\rho_2^{-1*}\mu_1 dt) \\ &= d\{J(\grave{}\grave{}L)+j(\grave{}\grave{}N)\} - d\{(\mu_2+\rho_2^{-1*}\mu_1)dt\} \\ &= d\{J(\grave{}\grave{}L)+j(\grave{}\grave{}N)\} - d(\mu_{21}dt) \,,\end{aligned}$$

and the result is established.

An invertible map $\rho:K\to\grave{}K|\grave{}t = t$ *defines an E-transformation if and only if there exists an* $\grave{}L(L;\rho)$ *for each* L *and two scalar-valued functions* $\mu(\rho)$, $\eta(\rho)$ *such that*

$$(51.2) \qquad \rho^{-1*}\{J(L)+j(N)\} = J(\grave{}L) + j(\grave{}N) - \mu dt + d\eta \,,$$

(51.3) $$\rho^{-1*}N = {}^{\backprime}N + d\mu .$$

The condition (51.3) is the same as the condition (50.12). Since $\Phi^* d = d\Phi^*$, the condition (50.11) becomes

$$d[\rho^{-1*}\{J(L)+j(N)\}] = d\{J({}^{\backprime}L)+j({}^{\backprime}N)\} - d(\mu dt) ,$$

and (51.2) follows from the Poincaré lemma. The converse if obvious and the result is established.

The subject of E-transformations being a new advent in mechanics, the extent to which they can be used to simplify the equations of motion of a nonconservative, nonholonomic system is not known. The following analysis of a particular class of E-transformations is thus given in the hopes of encouraging the interested reader to investigate general dynamical systems from the transformation-theoretic point of view.

Since ρ is assumed invertible, we can take t and any $6N$ of the $12N$ variables q^α, y^α, ${}^{\backprime}q^\alpha$, ${}^{\backprime}y^\alpha$ as independent variables. The simplest choise is to take $(t,\ q^\beta,\ {}^{\backprime}q^\beta)$, and assume that all other variables are expressed in terms of these by the equations that define ρ and ρ^{-1} . If we set

(51.4) $$\mu = u(t;\ q^\beta;\ {}^{\backprime}q^\beta) + A_\alpha(t;\ q^\beta;\ {}^{\backprime}q^\beta)y^\alpha + B_\alpha(t;\ q^\beta;\ {}^{\backprime}q^\beta){}^{\backprime}y^\alpha ,$$

and recall that

$$N = N^0 dt + N^1_\alpha dq^\alpha + N^2_\alpha dy^\alpha ,$$

$${}^{\backprime}N = {}^{\backprime}N^0 dt + {}^{\backprime}N^1_\alpha d{}^{\backprime}q^\alpha + {}^{\backprime}N^2_\alpha d{}^{\backprime}y^\alpha ,$$

then (51.3) gives

$$N^0 dt + N^1_\alpha dq^\alpha + N^2_\alpha dy^\alpha = {}^{\backprime}N^0 dt + {}^{\backprime}N^1_\alpha d{}^{\backprime}q^\alpha + {}^{\backprime}N^2_\alpha d{}^{\backprime}y^\alpha + du$$

$$+ A_\alpha dy^\alpha + y^\alpha dA_\alpha + B_\alpha d{}^{\backprime}y^\alpha + {}^{\backprime}y^\alpha dB_\alpha .$$

This gives us

$$(51.5)\qquad N^2_\alpha = A_\alpha\ , \quad {}^{\backprime}N^2_\alpha = -\,B_\alpha$$

and

$$(51.6)\quad N^0dt + N^1_\alpha dq^\alpha = {}^{\backprime}N^0dt + {}^{\backprime}N^1_\alpha d{}^{\backprime}q^\alpha + \frac{\partial u}{\partial t} + \frac{\partial u}{\partial q^\alpha}dq^\alpha + \frac{\partial u}{\partial {}^{\backprime}q^\alpha}d{}^{\backprime}q^\alpha$$
$$+ y^\beta\left(\frac{\partial A_\beta}{\partial t}dt + \frac{\partial A_\beta}{\partial q^\alpha}dq^\alpha + \frac{\partial A_\beta}{\partial {}^{\backprime}q^\alpha}d{}^{\backprime}q^\alpha\right)$$
$$+ {}^{\backprime}y^\beta\left(\frac{\partial B_\beta}{\partial t}dt + \frac{\partial B_\beta}{\partial q^\alpha}dq^\alpha + \frac{\partial B_\beta}{\partial {}^{\backprime}q^\alpha}d{}^{\backprime}q^\alpha\right).$$

Since t, q^β, and ${}^{\backprime}q^\beta$ are independent, (51.6) can be satisfied if and only if

$$(51.7)\qquad {}^{\backprime}N^0 = N^0 - y^\beta\frac{\partial A_\beta}{\partial t} - {}^{\backprime}y^\beta\frac{\partial B_\beta}{\partial t} - \frac{\partial u}{\partial t}\ ,$$

$$(51.8)\qquad N^1_\alpha = y^\beta\frac{\partial A_\beta}{\partial q^\alpha} + {}^{\backprime}y^\beta\frac{\partial B_\beta}{\partial q^\alpha} + \frac{\partial u}{\partial q^\alpha}\ ,$$

$$(51.9)\qquad {}^{\backprime}N^1_\alpha = -\,y^\beta\frac{\partial A_\beta}{\partial {}^{\backprime}q^\alpha} - {}^{\backprime}y^\beta\frac{\partial B_\beta}{\partial {}^{\backprime}q^\alpha} - \frac{\partial u}{\partial {}^{\backprime}q^\alpha}\ .$$

Thus, (51.5) together with (51.7)-(51.9) must hold in order that (51.3) be satisfied. Together, they constitute a system of equations for the determination of the $6N+1$ functions ${}^{\backprime}N^0$, ${}^{\backprime}N^1_\alpha$, ${}^{\backprime}N^2_\alpha$, $\alpha=1,\dots,3N$ in terms of the $6N+1$ functions u, A_α, B_α, $\alpha=1,\dots,3N$ of the $6N+1$ variables t, q^α, ${}^{\backprime}q^\alpha$, $\alpha=1,\dots,3N$. It thus remains to satisfy the requirements (51.2). A substitution of (51.4) into (51.2) and use of (51.5), (50.4) and (50.5) give us

$$(51.10)\qquad (P_\alpha+N^2_\alpha)dq^\alpha - P_\alpha y^\alpha dt + Ldt$$
$$= ({}^{\backprime}P_\alpha+{}^{\backprime}N^2_\alpha)d{}^{\backprime}q^\alpha - {}^{\backprime}P_\alpha{}^{\backprime}y^\alpha dt + {}^{\backprime}Ldt - udt + d\eta\ .$$

However, $\eta = \eta(t;\ q^\beta;\ {}^{\backprime}q^\beta)$, and hence the independence of t ,

q^β, $\grave{}q^\beta$ shows that (51.10) can be satisfied only if

$$(51.11)\qquad P_\alpha + N^2_\alpha = \partial\eta/\partial q^\alpha ,$$

$$(51.12)\qquad \grave{}P_\alpha + \grave{}N^2_\alpha = -\,\partial\eta/\partial\grave{}q^\alpha ,$$

$$(51.13)\qquad \grave{}L - \grave{}P_\alpha \grave{}y^\alpha = L - P_\alpha y^\alpha + u - \partial\eta/\partial t .$$

Finally, substituting from (51.5) for N^2_α and $\grave{}N^2_\alpha$, (51.11) and (51.12) become

$$(51.14)\qquad P_\alpha = \partial\eta/\partial q^\alpha - A_\alpha ,$$

$$(51.15)\qquad \grave{}P_\alpha = -\,\partial\eta/\partial\grave{}q^\alpha + B_\alpha .$$

We note that (51.13) is identical in form with (47.10) (with (47.19) with the substitution (47.15)) to within the extra term u. In fact, (51.13) may be viewed as the Lagrangian analog of the Hamilton-Jacobi equation for nonconservative, nonholonomic systems if we take $\grave{}L_F = \frac{1}{2}\,\grave{}y^\alpha \delta_{\alpha\beta}\,\grave{}y^\beta$. This will lead to the "conservative" part of the system becoming a free system. This yields the significant simplification of eliminating all interactions that arise from conservative force fields. The nonconservative, nonholonomic aspects of the problem may thus be thought of as contributing only the single extra term u to the Hamilton-Jacobi equation. However, this equation must be combined with all of the other equations and terms involving the A_α's, B_α's and u that are listed above. The prospects of achieving simplifications by use of the class of transformation conditions derived here appears encouraging. Of course, whether this be the case or not will depend heavily on the rank, class and other particular properties of the non C-H form N that is characteristic of nonconservative, nonholonomic systems under study.

CHAPTER IX

FIELD MECHANICS FOR SYSTEMS WITH LAWS OF BALANCE

The purpose of this chapter is to show that the methods developed in the previous chapters are not restricted to systems with only one independent variable, namely time, but rather have validity and utility in general field mechanics as well. The vehicle for this demonstration will be a collection of fields (functions of several independent variables) that satisfies a system of laws of balance. Such systems constitute a natural generalization of Newtonian systems with nonconservative forces, for they both share the property of the nonexistence of a homogeneous† variational principle.

52. KINEMATIC SPACE AND RELATED TOPICS

Chapter I dealt with the question of the appropriate domain space for mechanical systems consisting of a finite number of particles; i.e., systems with only one independent variable. Field mechanics, on the other hand, deals with quantities that depend on several independent variables, so we start anew. The number of independent variables used in this chapter will be designated by $n>1$; the usual case being that in which $n=4$.

†A system has a homogeneous variational principle if $\delta\int L = 0$. If the system can be represented as $\delta\int L = \int Q_\alpha \, \delta q^\alpha$, then it is said to admit an inhomogeneous variational principle. Such is the case when the forces that act on a dynamical system do not arise from a potential; that is, the Q_α's are the generalized forces of the system that must be specified *a priori*.

Let B_n be an n-dimensional, closed and connected point set whose points are labeled by n-tuples of real numbers $(x^1, x^2, \ldots, x^n) \equiv (x^i)$ with respect to the coordinate cover (x) of the covering space E_n = n-dimensional number space. The n-form of volume in E_n with respect to the cover (x) is denoted by

$$\omega \quad = \quad dx^1 \wedge dx^2 \wedge \ldots \wedge dx^n \ .$$

We assume that there are N fields $\{\phi^1(x^i), \ldots, \phi^N(x^i)\} \equiv \{\phi^\alpha(x^i)\}$ on B_n that constitute the fields of study. These are conveniently described in terms of a *Graph Space* G of $(n+N)$ dimensions with a global coordinate cover $(x^1, \ldots, x^n; q^1, \ldots, q^N) \equiv (x^i; q^\alpha)$. The fields $\{\phi^\alpha(x^i)\}$ are then realized by a map

$$(52.1) \qquad \phi\colon B_n \to G \,|\, \phi(x^i) \;=\; x^i \ , \qquad \phi(q^\alpha) \;=\; \phi^\alpha(x^i) \ .$$

We now use exactly the same procedure as used in Section 4 in order to introduce *Kinematic Space* K of dimension $n+N+nN$ with the global coordinate cover $(x^1, \ldots, x^n;\ q^1, \ldots, q^N;\ y^1_1, \ldots, y^N_n) \equiv (x^i;\ q^\alpha;\ y^\alpha_i)$. The essential requirement that distinguishes the properties of Kinematic space K is that any map ϕ, given by (52.1), shall lift to a map Φ of B_n into K in such a way that we obtain the derivatives of the $\phi^\alpha(x^i)$ in the place of the y's; that is,

$$(52.2) \qquad \Phi\colon B_n \to K \,|\, \Phi(x^i) \;=\; x^i \ ; \qquad \Phi(q^\alpha) \;=\; \phi^\alpha(x^j) \ ;$$

$$\Phi(y^\alpha_i) \;=\; D_i\phi^\alpha(x^j) \;\equiv\; \partial\phi^\alpha(x^j)/\partial x^i \ .$$

This requirement is easily made universal by the demand that

$$(52.3) \qquad \Psi^*(dq^\alpha - y^\alpha_i\, dx^i) \;=\; 0 \ , \quad \alpha=1,\ldots,N$$

for all maps $\Psi\colon B_n \to K$ such that $\Psi^*(\omega) \neq 0$. To see this, suppose that

$$\Psi(x^i) \;=\; \psi^i(t^j) \ , \quad \Psi(q^\alpha) \;=\; \psi^\alpha(t^j) \ , \quad \forall (t^j) \in B_n \ ,$$

then (52.3) yields $0 = \Psi^*(dq^\alpha - y_i^\alpha\, dx^j) = \left(\frac{\partial\psi^\alpha}{\partial t^j} - \Psi^*(y_i^\alpha)\frac{\partial\psi^i}{\partial t^j}\right)dt^j$, and we obtain $\Psi^*(y_i^\alpha)\partial\psi^i/\partial t^j = \partial\psi^\alpha/\partial t^j$. However, $\Psi^*(\omega) = \det(\partial\psi^i/\partial t^j)\, dt^1 \wedge \ldots \wedge dt^n \neq 0$ and hence the matrix $((\partial\psi^i/\partial t^j)) = ((\partial x^i/\partial t^j))$ is invertible. The above calculation thus yields $\Psi^*(y_i^\alpha) = (\partial\psi^\alpha/\partial t^j)(\partial t^j/\partial x^i) = \partial\psi^\alpha(t^j(x^k))/\partial x^i$ and the result is established.

It is useful at this point to introduce certain notational conveniences. We use D_i to denote the operation of total coordinate differentiation with respect to x^i . Thus, for instance,

$$D_i f(x^j;\ \phi^\alpha(x^j)) \;=\; \partial f/\partial x^i + (\partial f/\partial\phi^\alpha)(\partial\phi^\alpha/\partial x^i) \ .$$

We have already introduced the notation ω for the volume n-form of E_n . Since we will not introduce a metric in E_n , we define n specific (n-1)-forms π_i , $i=1,\ldots,n$, by the requirements

$$(52.4) \qquad dx^i \wedge \pi_j \;=\; \delta^i_j\ \omega \ .$$

These equations have the unique solution

$$(52.5) \qquad \pi_i \;=\; (-1)^{(n+1)(i+1)} dx^{i+1} \wedge \ldots \wedge dx^n \wedge dx^1 \wedge \ldots \wedge dx^{i-1} \ ,$$

from which we obtain the following results:

$$(52.6) \qquad d\pi_i \;=\; 0\ , \quad i=1,\ldots,n\ ;$$

$$(52.7) \qquad d(F^i\pi_i) \;=\; D_jF^i\ dx^j \wedge \pi_i \;=\; D_jF^j\ \omega\ , \quad F^i \varepsilon \Lambda^0(E_n)\ ;$$

$$(52.8) \qquad \int_{B_n} d(F^i\pi_i) \;=\; \int_{\partial B_n} F^i\pi_i\ , \quad F^i\varepsilon\Lambda^0(E_n)\ ;$$

$$(52.9) \qquad V \lrcorner\, \omega \;=\; v^i\pi_i$$

where $V = v^i\ \partial/\partial x^i + v^\alpha\ \partial/\partial q^\alpha + v_i^\alpha\ \partial/\partial y_i^\alpha$ is a vector field on K .

It is now a straightforward task (use the same methods as applied in Section 4 together with (52.3) and due allowance for the presence of n independent variables) to obtain the structure of vector fields on K. The same results obtain, however, if we simply require that any vector field

$$V_G = v^i(x^j;\ q^\beta)\partial/\partial x^i + v^\alpha(x^j;\ q^\beta)\partial/\partial q^\alpha$$

on G lift to a corresponding vector field

$$V = v^i(x^j;\ q^\beta)\partial/\partial x^i + v^\alpha(x^j;\ q^\beta)\partial/\partial q^\beta + v_i^\alpha(x^j;\ q^\beta;\ y_j^\beta)\partial/\partial y_i^\alpha$$

on K in such a manner that any finite Lie algebra on G lift to a finite Lie algebra on K with the same constants of structure. It is then a straightforward matter to establish the following results.

The quantity

$$(52.10)\qquad V = v^i(x^j;\ q^\beta)\partial/\partial x^i + v^\alpha(x^j;\ q^\beta)\partial/\partial q^\alpha + v_i^\alpha(x^j;\ q^\beta;\ y_i^\beta)\partial/\partial y_i^\alpha$$

is a vector field on K *if and only if*

$$(52.11)\qquad v_i^\alpha(x^j;\ q^\beta;\ y_j^\beta) = Z_i(v^\alpha - y_j^\alpha v^j)\ ,$$

where Z_i , $i=1,\dots,n$, *are* n *linear operators that are defined by*

$$(52.12)\qquad Z_i \equiv \partial/\partial x^i + y_i^\alpha\ \partial/\partial q^\alpha\ .$$

If $\overset{1}{V}$ *and* $\overset{2}{V}$ *are two elements of* $T(K)$ *(vector fields on* K *) then* $[\overset{1}{V}, \overset{2}{V}] = \pounds_{\overset{1}{V}}\overset{2}{V}\in T(K)$ *and we have*

$$(52.13)\qquad [\overset{1}{V}, \overset{2}{V}] = [\overset{1}{V}_G, \overset{2}{V}_G]^i\ \partial/\partial x^i + [\overset{1}{V}_G, \overset{2}{V}_G]^\alpha\ \partial/\partial q^\alpha + Z_i([\overset{1}{V}_G, \overset{2}{V}_G]^\alpha - y_j^\alpha[\overset{1}{V}_G, \overset{2}{V}_G]^j)\partial/\partial y_i^\alpha\ ,$$

where $[\overset{1}{V}_G, \overset{2}{V}_G] = [\overset{1}{V}_G, \overset{2}{V}_G]^i\ \partial/\partial x^i + [\overset{1}{V}_G, \overset{2}{V}_G]^\alpha\ \partial/\partial q^\alpha$ *is the com-*

mutator of the vector fields $\overset{1}{V}_G$, $\overset{2}{V}_G$ *on* G *that give rise to the vector fields* $\overset{1}{V}$ *and* $\overset{2}{V}$ *on* K .

This result shows that *Kinematic space* K *is the domain of the first extension of all groups on* G *with respect to all mappings* $\Psi: B_n \to G | \Psi(x^i) = x^i$. K is thus distinct from the tangent bundle of G .

We note for future reference the following results. Let $\Psi: B_n \to K | \Psi(x^i) = x^i$, $\Psi(q^\alpha) = \psi^\alpha(x^i)$, then

$$\Psi^* Z_i h(x^j;\, q^\beta) = D_i h(x^j;\, \psi^\beta(x^j)) , \tag{52.14}$$

$$\begin{aligned} \Psi^* v_i^\alpha &= \Psi^* Z_i (v^\alpha - y_j^\alpha v^j) = D_i V^\alpha - (D_j \psi^\alpha)(D_i V^j) \\ &= D_i (V^\alpha - V^j D_j \psi^\alpha) + V^j D_j D_i \psi^\alpha , \end{aligned} \tag{52.15}$$

where we have used the notational convention

$$V^\alpha = v^\alpha(x^j;\, \psi^\alpha(x^j)) , \quad V^i = v^i(x^j;\, \psi^\alpha(x^j)) .$$

A vector field on K is said to be *horizontal* if and only if the coefficients of $\partial/\partial x^i$ vanish for all $i=1,\ldots,n$. We usually denote horizontal vector fields by the generic symbol U so that

$$U = u^\alpha(x^j;\, q^\beta)\partial/\partial q^\alpha + (Z_i u^\alpha)(x^j;\, q^\beta;\, y_j^\beta)\partial/\partial y_i^\alpha . \tag{52.16}$$

Since the orbital equations of such a vector field are given by

$$\frac{dx^i}{d\sigma} = 0 , \quad \frac{dq^\alpha}{d\sigma} = u^\alpha(x^j;\, q^\beta) , \quad \frac{dy_i^\alpha}{d\sigma} = \frac{\partial u^\alpha}{\partial q^\beta} y_i^\beta + \frac{\partial u^\alpha}{\partial x^i} , \tag{52.17}$$

a solution, to within second order terms in σ is given by

$$\begin{aligned} x^i(\sigma) &= x_0^i , \quad q^\alpha(\sigma) = q_0^\alpha + \sigma\, u^\alpha(x_0^j;\, q_0^\beta) + o(\sigma) , \\ y_i^\alpha(\sigma) &= y_{i0}^\alpha + \sigma\left[\frac{\partial u^\alpha(x_0^k;\, q_0^\gamma)}{\partial q_0^\beta} y_{i0}^\beta + \frac{\partial u^\alpha(x_0^k;\, q_0^\gamma)}{\partial x_0^i}\right] + o(\sigma) . \end{aligned} \tag{52.18}$$

Accordingly, if $\Phi: B_n \to K|\Phi(x^i) = x^i$, $\Phi(q^\alpha) = \phi^\alpha(x^j)$, $\Phi(y_i^\alpha) = D_i\phi(x^j)$, then we can compose Φ with the orbits of the horizontal vector field U (i.e., set $x_0^i = x^i$, $q_0^\alpha = \phi^\alpha(x^j)$, etc.) and imbed the map Φ in a one-parameter family of maps

$$(52.19)\quad \Phi_\sigma: B_n \to K|\Phi_\sigma(x^i) = x^i \ , \quad \Phi_\sigma(q^\alpha) = \phi^\alpha(x^j) + \sigma\, u^\alpha(x^j;\, \phi^\beta(x^j)) + o(\sigma) \ ,$$

$$\Phi_\sigma(y_i^\alpha) = D_i\phi^\alpha(x^j) + \sigma\, D_i u^\alpha(x^j;\, \phi^\beta(x^j)) + o(\sigma)$$

such that $\Phi = \Phi_0$. This allows us to construct arbitrary variations by use of the freedom in the choice of the functions $u^\alpha(x^j;\, q^\beta)$. If we write

$$(52.20)\qquad \delta\phi^\alpha(x^j) \;=\; u^\alpha(x^j;\, \phi^\beta(x^j)) \;=\; u^\alpha\circ\Phi \ ,$$

then (52.19) shows that $D_i\delta\phi^\alpha = \delta D_i\phi^\alpha$, variation and differentiation commute, and that

$$(52.21)\quad \Phi_\sigma: B_n \to K|\Phi_\alpha(x^i) = x^i \ , \quad \Phi_\sigma(q^\alpha) = \phi^\alpha(x^j) + \sigma\, \delta\phi^\alpha(x^j) + o(\sigma) \ ,$$

$$\Phi_\sigma(y_i^\alpha) = D_i\phi^\alpha(x^j) + \sigma\, D_i\delta\phi^\alpha(x^j) + o(\sigma) \ .$$

It is also clear that we can achieve the requirement $\delta\phi^\alpha|_{\partial B_n}=0$ by the choice $u^\alpha(x^j;\, q^\beta)|_{\partial B_n}=0$ since $\delta\phi^\alpha|_{\partial B_n} = u^\alpha(x^j;\, \phi^\beta(x^j))|_{\partial B_n}$. Variational processes are thus accessible in exactly the same fashion as given in Chapter I: Φ stationarizes $\int_{B_n}\Phi^*(L\omega)$ if and only if

$$(52.22)\qquad 0 \;=\; \delta\int_{B_n}\Phi^*(L\omega) \;=\; \int_{B_n}\Phi^*\pounds_u(L\omega)$$

for all horizontal vector fields U such that $(u^{\alpha}\circ\Phi)|_{\partial B_n}=0$, where $L\varepsilon\Lambda^0(K)$ is the Lagrangian function. In this event, we obtain the *Euler-Lagrange equations*

$$(52.23)\qquad D_i\,\frac{\partial L}{\partial(D_i\phi^{\alpha})} - \frac{\partial L}{\partial\phi^{\alpha}} = 0\ , \qquad L = L\circ\Phi\ .$$

53. BEHAVIOR UNDER MAPPINGS

The results of the previous section were obtained with respect to the given coordinate cover $(x^i;\ q^{\alpha};\ y^{\alpha}_i)$ of Kinematic Space. Accordingly, the mapping properties of Kinematic Spaces and their vector fields become of particular importance here.

Let

$$(53.1)\qquad \gamma\colon G \to {}^{\backprime}G|{}^{\backprime}x^i = \gamma^i(x^j;\ q^{\beta})\ ;\quad {}^{\backprime}q^{\alpha} = \gamma^{\alpha}(x^j;\ q^{\beta})$$

be an invertible map between Graph Spaces and let

$$(53.2)\qquad V_G = v^i(x^j;\ q^{\beta})\partial/\partial x^i + v^{\alpha}(x^j;\ q^{\beta})\partial/\partial q^{\alpha}$$

be a vector field on G. We then have

$$(53.3)\qquad \gamma_*\colon T(G) \to T({}^{\backprime}G)\,|\,V_G \rightarrowtail {}^{\backprime}V_G = {}^{\backprime}v^i\ \partial/\partial{}^{\backprime}x^i + {}^{\backprime}v^{\alpha}\ \partial/\partial{}^{\backprime}q^{\alpha}\ ,$$

where

$$(53.4)\qquad \begin{aligned} {}^{\backprime}v^i &= \left(\frac{\partial{}^{\backprime}x^i}{\partial x^k}\,v^k + \frac{\partial{}^{\backprime}x^i}{\partial q^{\beta}}\,v^{\beta}\right)\circ\gamma^{-1}\ ,\\ {}^{\backprime}v^{\alpha} &= \left(\frac{\partial{}^{\backprime}q^{\alpha}}{\partial x^k}\,v^k + \frac{\partial{}^{\backprime}q^{\alpha}}{\partial q^{\beta}}\,v^{\beta}\right)\circ\gamma^{-1}\ . \end{aligned}$$

The question to be answered is what happens when V_G is lifted to a vector field on K by $v^{\alpha}_i = Z_i(v^{\alpha} - y^{\alpha}_j v^j)$.

We start with the fact that the orbits of V_G and of $\gamma_* V_G$ are given by

$$(53.5)\qquad \frac{dx^i}{d\sigma} = \upsilon^i(x^j;\ q^\beta)\ ,\qquad \frac{dq^\alpha}{d\sigma} = \upsilon^\alpha(x^j;\ q^\beta)\ ;$$

$$(53.6)\qquad \frac{d\,{}^{`}x^i}{d\sigma} = {}^{`}\upsilon^i({}^{`}x^j;\ {}^{`}q^\beta)\ ,\qquad \frac{d\,{}^{`}q^\alpha}{d\sigma} = {}^{`}\upsilon^\alpha({}^{`}x^j;\ {}^{`}q^\beta)\ .$$

We denote the lift of γ to K by Γ so that $\Gamma\colon K \to {}^{`}K$. The reader is cautioned, however, to note that we do not yet know that ${}^{`}K$ is itself a Kinematic Space. The relations (53.5), (53.6) and the condition $\Psi^*(dq^\alpha - y^\alpha_j dx^j) = 0$ for all Ψ such that $\Psi^*\omega \neq 0$ then imply that

$$(53.7)\qquad \begin{aligned} \upsilon^\alpha(x^j(\sigma);\ q^\beta(\sigma)) &= y^\alpha_j(\sigma)\upsilon^j(x^k(\sigma);\ q^\beta(\sigma))\ ,\\ {}^{`}\upsilon^\alpha({}^{`}x^j(\sigma);\ {}^{`}q^\beta(\sigma)) &= {}^{`}y^\alpha_j(\sigma)\,{}^{`}\upsilon^j(x^k(\sigma);\ q^\beta(\sigma)) \end{aligned}$$

are satisfied on the orbits of the lift of V_G to K and on the orbits of the lift of ${}^{`}V_G$ to ${}^{`}K$, respectively. When (53.4) and (53.6) are substituted into the second of (53.7) and the various terms are collected together, we obtain

$$(53.8)\qquad {}^{`}y^\alpha_i\ \upsilon^k Z_k({}^{`}x^i) = \upsilon^k Z_k({}^{`}q^\alpha)\ .$$

Clearly, the results (53.8) must be required to hold for all possible vector fields on G in order that the resulting transformation properties should be universal, and this can be the case if and only if the following transformation rule holds for the y's :

$$(53.9)\qquad {}^{`}y^\alpha_i\ Z_j({}^{`}x^i) = Z_j({}^{`}q^\alpha)\ .$$

An invertible map $\gamma\colon G \to {}^{`}G \mid {}^{`}x^i = \gamma^i(x^j;\ q^\beta)\ ,\ {}^{`}q^\alpha = \gamma^\alpha(x^j;\ q^\beta)$ *lifts to a map* $\Gamma\colon K \to {}^{`}K \mid {}^{`}x^i = \gamma^i(x^j;\ q^\beta)\ ,\ {}^{`}q^\alpha = \gamma^\alpha(x^j;\ q^\beta)\ ,$ ${}^{`}y^\alpha_i\ Z_j({}^{`}x^i) = Z_j({}^{`}q^\alpha)$.
It is clear that Γ is *regular* only on its *domain of regularity* $R(\Gamma) = \{(x^i;\ q^\alpha;\ y^\alpha_i) \mid \det(Z_j({}^{`}x^i)) \neq 0\}$. In what follows, we only consider maps Γ on $R(\Gamma)$.

We now distinguish certain preferred classes of maps of

Kinematic Space. A map $\Gamma: K \to {}^{`}K$ is said to be *admissible* if and only if ${}^{`}x^i = \gamma^i(x^j; q^\beta) = x^i$; that is, Γ leaves the domain space E_n invariant. An admissible map thus gives the following transformation rules:

$$(53.10) \qquad \Gamma: K \to {}^{`}K | {}^{`}x^i = x^i, \quad {}^{`}q^\alpha = \gamma^\alpha(x^j; q^\beta), \quad {}^{`}y_i^\alpha = Z_i({}^{`}q^\alpha),$$

so that $R(\Gamma) = K$. The y's thus transform linearly under an admissible map:

$$(53.11) \qquad {}^{`}y_i^\alpha = \frac{\partial {}^{`}q^\alpha}{\partial x^i} + y_i^\beta \frac{\partial {}^{`}q^\alpha}{\partial q^\beta} .$$

A map Γ is said to be a *factor* map if and only if

$$(53.12) \qquad \Gamma: K \to {}^{`}K | {}^{`}x^i = x^i, \quad {}^{`}q^\alpha = \gamma^\alpha(q^\beta), \quad {}^{`}y_i^\alpha = Z_i({}^{`}q^\beta) ,$$

in which case $R(\Gamma) = K$. The y's thus transform linearly and homogeneously under a factor map,

$$(53.13) \qquad {}^{`}y_i^\alpha = y_i^\beta \frac{\partial {}^{`}q^\alpha}{\partial q^\beta} ,$$

so that they may be considered as the components of a vector field on the N-dimensional space with coordinate functions (q^α) if we restrict the maps to be factor maps.

A map Γ is said to be a *domain* map if and only if

$$(53.14) \qquad \begin{aligned} &\Gamma: K \to {}^{`}K | {}^{`}x^i = \gamma^i(x^j) , \quad \overset{-1}{\Gamma}{}^{*}\omega \neq 0 , \\ &{}^{`}q^\alpha = G_\beta^\alpha(x, {}^{`}x) q^\beta + \lambda^\alpha(x, {}^{`}x) ; \end{aligned}$$

that is, the independent variables undergo an invertible mapping so that $R(\Gamma) = K$ and the q's transform as linear geometric object fields. In this case, the y's transform by

$$(53.15) \qquad {}^{`}y_i^\alpha \frac{\partial {}^{`}x^i}{\partial x^j} = Z_j(G_\beta^\alpha q^\beta + \lambda^\alpha) = \frac{\partial \lambda^\alpha}{\partial x^j} + \frac{\partial \lambda^\alpha}{\partial {}^{`}x^k} \frac{\partial {}^{`}x^k}{\partial x^j} + \\ + \left(\frac{\partial G_\beta^\alpha}{\partial x^j} + \frac{\partial G_\beta^\alpha}{\partial {}^{`}x^k} \frac{\partial {}^{`}x^k}{\partial x^j} \right) q^\beta + y_j^\beta G_\beta^\alpha .$$

Thus, if the q's transform as scalar fields $(G^\alpha_\beta = \delta^\alpha_\beta,\ \lambda^\alpha = 0)$, we obtain ${}^\backprime y^\alpha_i\, \partial {}^\backprime x^i/\partial x^j = y^\alpha_j$; that is, the y's transform as the components of a covector on E_n for each value of the index α.

We now restrict attention to admissible maps. In this event we have

$$(53.16)\qquad {}^\backprime y^\alpha_i = Z_i({}^\backprime q^\alpha) = \partial {}^\backprime q^\alpha/\partial x^i + y^\beta_i\, \partial {}^\backprime q^\alpha/\partial q^\beta .$$

Now, for any orbit of $V\varepsilon T(K)$, we obtain

$$(53.17)\qquad \frac{d{}^\backprime y^\alpha_i}{d\sigma} = \frac{\partial}{\partial x^j}\left[Z_i({}^\backprime q^\alpha)\right]v^j + \frac{\partial}{\partial q^\gamma}\left[Z_i({}^\backprime q^\alpha)\right]v^\gamma$$

$$+ \frac{\partial}{\partial y^\gamma_k}\left[Z_i({}^\backprime q^\alpha)\right]\, Z_k(v^\gamma - y^\gamma_\ell\, v^\ell)$$

$$\stackrel{\text{def}}{=} {}^\backprime v^\alpha_i({}^\backprime x^j;\ {}^\backprime q^\beta;\ {}^\backprime y^\beta_j) ,$$

where $d{}^\backprime x^i/d\sigma = {}^\backprime v^i$, $d{}^\backprime q^\alpha/d\sigma = {}^\backprime v^\alpha$, $dx^i/d\sigma = v^i$, $dq^\alpha/d\sigma = v^\alpha$. It thus follows, in exactly the same fashion as given in Section 6, that

$$(53.18)\qquad {}^\backprime v^\alpha_i = {}^\backprime Z_i({}^\backprime v^\alpha - {}^\backprime y^\alpha_j\, {}^\backprime v^j) ,$$

where

$$(53.19)\qquad {}^\backprime Z_i = \partial/\partial {}^\backprime x^i + {}^\backprime y^\alpha_i\, \partial/\partial {}^\backprime q^\alpha .$$

This establishes the following result.

An admissible map

$$(53.20)\qquad \Gamma\colon K \to {}^\backprime K \mid {}^\backprime x^i = x^i,\ {}^\backprime q^\alpha = \gamma^\alpha(x^i;\ q^\beta),\ {}^\backprime y^\alpha_i = Z_i({}^\backprime q^\alpha)$$

maps Kinematic Space K *into a Kinematic space* ${}^\backprime K$.

It can likewise be shown that $\Gamma(R(\Gamma))$ *is a Kinematic Space*, so that ${}^\backprime v^\alpha_i = {}^\backprime Z_i({}^\backprime v^\alpha - {}^\backprime y^\alpha_j v^j)$ holds on $\Gamma(R(\Gamma))$.

54. LAWS OF BALANCE AND THEIR FUNDAMENTAL DIFFERENTIAL FORMS

We now proceed directly to the study of fields that satisfy a given system of laws of balance on a given $B_n \subset E_n$. Let $\{W_\alpha(a^j;\ b^\beta;\ c^\beta_j),\ W^i_\alpha(a^j;\ b^\beta;\ c^\beta_j) \mid i,j=1,\ldots,n;\ \alpha,\beta=1,\ldots,N\}$ be $N+nN$ functions of $n+N+nN$ arguments. A collection of N fields $\{\phi^\alpha(x^i) \mid \alpha=1,\ldots,N\}$ is said to be $(W_\alpha;\ W^i_\alpha)$-*balanced* on $B_n \subset E_n$ if and only if the functions $\{\phi^\alpha(x^j)\}$ satisfy the balance equations

$$D_i W^i_\alpha(x^k;\ \phi^\beta(x^k);\ D_j\phi^\beta(x^k)) = W_\alpha(x^k;\ \phi^\beta(x^k);\ D_j\phi^\beta(x^k)) , \quad \alpha = 1,\ldots,N , \tag{54.1}$$

at all points (x^i) of B_n. We note that balanced fields do not necessarily satisfy systems of second order partial differential equations, for nothing stops the possibility of some of the nN functions W^i_α from being identically zero. Balanced fields do include, however, most of the laws of physics, for such laws can usually be manipulated into forms similar to (54.1). Further, Stokes' theorem can be used to rewrite (54.1) in the equivalent form

$$\int_{\partial B_n} W^i_\alpha \pi_i = \int_{B_n} W_\alpha \omega$$

from which we have taken the term "balanced fields." Finally, we note that (54.1) includes those cases in which there is an underlying variational principle, for (54.1) becomes the Euler-Lagrange equations if there exists a function $L(x^i;\ \phi^\alpha(x^i);\ D_j\phi^\alpha(x^i))$ such that $W_\alpha = \partial L/\partial\phi^\alpha$, $W^i_\alpha = \partial L/\partial D_i\phi^\alpha$. This will not be the case in general, however.

Since we assume no underlying variational principle, it seems preferable to proceed directly to the construction of the fundamental differential forms on K that underly the whole theory. This construction parallels the construction given in Chapters V and VI for the case of only one independent variable, and we shall use the same symbolism where possible in order to place the analogous forms in sharp relief. We start by using the given functions (W_α, W_α^i) to construct a 1-form

$$(54.2) \qquad \mathcal{W} = \mathcal{W}_i(x^j;\, q^\beta;\, y_j^\beta)dx^i + \mathcal{W}_\alpha(x^j;\, q^\beta;\, y_j^\beta)dq^\alpha + \mathcal{W}_\alpha^i(x^j;\, q^\beta;\, y_j^\beta)dy_i^\alpha$$

where

$$(54.3) \qquad \Phi^*\mathcal{W}_\alpha = W_\alpha , \qquad \Phi^*\mathcal{W}_\alpha^i = W_\alpha^i$$

and $\Phi: B_n \to K|\Phi(x^i) = x^i$, $\Phi(q^\alpha) = \phi^\alpha(x^i)$ is the map associated with the lift to K of the graph of the field $\{\phi^\alpha(x^i)\}$. We purposely leave the functions $\mathcal{W}_i(x^j;\, q^\beta;\, y_j^\beta)$ in (54.2) undetermined. This freedom in the choice of these functions will be of particular convenience later on. Next, we define the n-form J on K by (see (36.10))

$$(54.4) \qquad J = (dq^\alpha - y_j^\alpha\, dx^j) \wedge \mathcal{W}_\alpha^i\, \pi_i = dq^\alpha \wedge \mathcal{W}_\alpha^i\, \pi_i - y_j^\alpha\, \mathcal{W}_\alpha^j\, \omega ,$$

where the last equality follows from (54.2). We thus have

$$(54.5) \qquad \Phi^* J = 0 \qquad \forall \Phi^* | \Phi^*\omega \neq 0$$

because $\Phi^*(dq^\alpha - y_j^\alpha\, dx^j) = 0$ for all such maps. The (n+1)-form F is then defined on K by

$$(54.6) \qquad F = \mathcal{W} \wedge \omega + dJ = (dq^\alpha - y_j^\alpha dx^j) \wedge (\mathcal{W}_\alpha \omega - d\mathcal{W}_\alpha^i \wedge \pi_i) .$$

We note that F is closed if and only if $\mathcal{W} \wedge \omega$ is closed. This will be of importance when we come to the study of the existence of a variational principle for a balanced field.

The intrinsic utility of the form F is most easily perceived by evaluating $V \rfloor F$ for $V \in T(K)$. A straightforward computation based on (54.6), (52.9), (52.10) and (52.11) yields

$$(54.7)\qquad V\rfloor F = V\rfloor(W\wedge\omega) + V\rfloor dJ$$

$$= W_\alpha v^\alpha\omega - W_\alpha dq^\alpha\wedge(v^i\pi_i)$$

$$- (v^\alpha - y^\alpha_j v^j)\,dW^i_\alpha\wedge\pi_i$$

$$+ (dq^\alpha - y^\alpha_j\,dx^j)\wedge[V\rfloor(dW^i_\alpha\wedge\pi_i)]\ .$$

Now, let Φ be an arbitrary map of B_n into K such that $\Phi(x^i) = x^i$, $\Phi(q^\alpha) = \phi^\alpha(x^j)$, $\Phi(y^\alpha_i) = D_i\phi^\alpha(x^j)$, then (54.7) yields

$$(54.8)\qquad \Phi^*(V\rfloor F) = (W_\alpha - D_i W^i_\alpha)(V^\alpha - V^j D_j\phi^\alpha)\omega\ ,$$

where

$$(54.9)\qquad V^i = v^i\circ\Phi\ ,\quad V^\alpha = v^\alpha\circ\Phi\ ,\quad W_\alpha = W_\alpha\circ\Phi\ ,\quad W^i_\alpha = W^i_\alpha\circ\Phi\ .$$

If $\{\phi^\alpha(x^j)\}$ is $(W_\alpha,\ W^i_\alpha)$-balanced, then (54.8) shows that $\Phi^*(V\rfloor F) = 0$ for all $V\varepsilon T(K)$. Conversely, if $\Phi^*(V\rfloor F) = 0$ for all $V\varepsilon T(K)$, then an integration of the n-form $\Phi^*(V\rfloor F) = 0$ over B_n and use of (54.8) and the fundamental lemma of the calculus of variations shows that $\{\phi^\alpha(x^j)\}$ is $(W_\alpha,\ W^i_\alpha)$-balanced. We have thus established the following result.

The functions $\phi^\alpha(x^j)$, $\alpha=1,\ldots,N$, *that define the map*

$$(54.10)\qquad \Phi\colon B_n\to K|\Phi(x^i) = x^i,\ \ \Phi(q^\alpha) = \phi^\alpha(x^j),\ \ \Phi(y^\alpha_i) = D_i\phi^\alpha(x^j)$$

form a $(W_\alpha,\ W^i_\alpha)$*-balanced field on* B_n *if and only if the* n-*form* $\Phi^*(V\rfloor F)$ *satisfies*

$$(54.11)\qquad \Phi^*(V\rfloor F) = 0$$

for all $V\varepsilon T(K)$.

This theorem, in effect, allows us to study balanced fields in terms of the properties of the n-dimensional surfaces in Kinematic Space that are defined by maps Φ of the type given by (54.10).

We obtain in this way an intrinsic economy and a convenience, for we can shift our considerations directly to Kinematic space through study of the quantity $V\rfloor F$ which is a linear functional on $T(K)$ with values in $\Lambda^n(K)$. Further, since the forms J and F are made up from the 1-form $\mathcal{W}$, we can view the 1-form $\mathcal{W}$ as characteristic of a balanced field $\{\phi^\alpha(x^j)\}$ under action by the map $\Phi^*:\Lambda^k(K) \to \Lambda^k(B_n)$. The above theorem can thus be restated in the following manner.

A map Φ: $B_n \to K|\Phi(x^i) = x^i$ *defines a balanced field on* B_n *that is characterized by the* 1-*form* $\mathcal{W}$ *on* K *if and only if* (54.11) *holds for all* $V \varepsilon T(K)$.

There is one further n-form that we need to introduce. This n-form is defined by

$$\Omega(V) = V\rfloor F + d(V\rfloor J) , \tag{54.12}$$

so that we have

$$\Omega(V_1 + V_2) = \Omega(V_1) + \Omega(V_2) . \tag{54.13}$$

If we use (54.6), (54.12) yields $\Omega(V) = V\rfloor(\mathcal{W}\wedge\omega)+V\rfloor dJ+d(V\rfloor J)$, from which we obtain the more useful form

$$\Omega(V) = \pounds_V J + V\rfloor(\mathcal{W}\wedge\omega) . \tag{54.14}$$

The n-form $\Omega(V)$ is the basic ingredient whereby we obtain a large collection of currents on B_n that are conserved for any given balanced field. This is shown in the next section.

There is one further aspect of balanced fields that we need to consider. It should be clear from the start that the functions (W_α, W^i_α) that are used to represent a given system of field equations as balance equations $D_iW^i_\alpha - W_\alpha = 0$ are not uniquely determined by the given field equations. For instance, let $a^\alpha_\beta(x^j; \phi^\gamma(x^j))$ be an arbitrary collection of n^2 functions such that $\det(a^\alpha_\beta) \neq 0$ for all (x^i) in B_n and all functions $\{\phi^\gamma(x^j)\}$. It then follows that

$$(54.15) \qquad D_i\bar{W}^i_\beta - \bar{W}_\beta = a^\gamma_\beta(D_iW^i_\alpha - W_\alpha) \quad \forall\{\phi^\alpha(x^j)\}$$

defines an equivalence relation between (W_α, W^i_α) and $(\bar{W}_\beta, \bar{W}^i_\beta)$, for any $(W_\alpha; W^i_\alpha)$-balanced field is a $(\bar{W}_\alpha, \bar{W}^i_\alpha)$-balanced field, any $(\bar{W}_\alpha, \bar{W}^i_\alpha)$-balanced field is a (W_α, W^i_α)-balanced field, and the relations (54.15) are symmetric, reflexive and transitive. Since $a^\alpha_\beta(D_iW^i_\alpha - W_\alpha) = D_i(a^\alpha_\beta W^i_\alpha) - a^\alpha_\beta W_\alpha - W^i_\alpha D_i a^\alpha_\beta$, it follows that (54.15) implies the relations

$$(54.16) \qquad \begin{aligned} \bar{W}^i_\beta &= a^\alpha_\beta W^i_\alpha \\ \bar{W}_\beta &= a^\alpha_\beta W_\alpha + W^i_\alpha D_i a^\alpha_\beta . \end{aligned}$$

Remembering that $\Phi^*\omega^i_\alpha = W^i_\alpha$, $\Phi^*\omega_\alpha = W_\alpha$, the relations (54.16) are lifted to K by

$$(54.17) \qquad \begin{aligned} \bar{\omega}^i_\beta &= a^\alpha_\beta \omega^i_\alpha \\ \bar{\omega}_\beta &= a^\alpha_\beta \omega_\alpha + \omega^i_\alpha Z_i a^\alpha_\beta \end{aligned}$$

since $a^\alpha_\beta \varepsilon \Lambda^0(G)$ implies $\Phi^*(Z_i a^\alpha_\beta) = D_i a^\alpha_\beta$. This motivates the following definition.

Two 1-forms

$$\omega = \omega_i dx^i + \omega_\alpha dq^\alpha + \omega^i_\alpha dy^\alpha_i$$

$$\bar{\omega} = \omega_i dx^i + \bar{\omega}_\alpha dq^\alpha + \bar{\omega}^i_\alpha dy^\alpha_i$$

are said to be B-*equivalent* (balance equivalent) if and only if there exist n^2 functions $a^\alpha_\beta(x^j, q^\gamma)$ such that

$$(54.18) \qquad \det(a^\alpha_\beta) \neq 0 ,$$

$$(54.19) \qquad \bar{\omega}_\beta = a^\alpha_\beta \omega_\alpha + \omega^i_\alpha Z_i a^\alpha_\beta ,$$

$$(54.20) \qquad \bar{\omega}^i_\beta = a^\alpha_\beta \omega^i_\alpha .$$

The fundamental pairs of forms $W \wedge \omega$, $\bar{W} \wedge \omega$; $J(W)$, $J(\bar{W})$; $F(W)$, $F(\bar{W})$; $\Omega(W;V)$, $\Omega(\bar{W};V)$ are then said to be B-equivalent. Clearly, the dynamical structure inferred from any two B-equivalent forms become equivalent due to the requirement $\det(a^{\alpha}_{\beta}) \neq 0$.

If we write out $\bar{W}$ and use (54.19), (54.20), we obtain

$$(54.21) \qquad \bar{W} = \bar{W}_i dx^i + (a^{\alpha}_{\beta} W_{\alpha} + W^i_{\alpha} Z_i a^{\alpha}_{\beta}) dq^{\beta} + a^{\alpha}_{\beta} W^i_{\alpha} dy^{\beta}_i .$$

Let us define a new basis for $\Lambda^1(K)$ by

$$(54.22) \qquad \begin{aligned} d\bar{x}^i &= dx^i \\ \xi^{\alpha} &= a^{\alpha}_{\beta}\, dq^{\beta} \\ \xi^{\alpha}_i &= a^{\alpha}_{\beta}\, dy^{\beta}_i + Z_i a^{\alpha}_{\beta}\, dq^{\beta} , \end{aligned}$$

then (54.21) yields

$$(54.23) \qquad \bar{W} = W_i d\bar{x}^i + W_{\alpha} \xi^{\alpha} + W^i_{\alpha} \xi^{\alpha}_i .$$

Accordingly, if W and $\bar{W}$ are B-equivalent then $\bar{W}$ is obtained from W by replacing the natural basis $(dx^i;\ dq^{\alpha};\ dy^{\alpha}_i)$ by the basis $(dx^i;\ \xi^{\alpha};\ \xi^{\alpha}_i)$ and retaining the same coefficients.

Let $I(K)$ denote the class of all 1-forms

$$(54.24) \qquad I = I_i\, dx^i + I_{\alpha}\, dq^{\alpha} + I^i_{\alpha}\, dy^{\alpha}_i$$

on K such that every map $\Phi: B_n \to K \mid \Phi(x^i) = x^i$ is $(\Phi^* I_{\alpha}, \Phi^* I^i_{\alpha})$-balanced. The class $I(K)$ is referred to as the *identity class* since any $\{\phi^{\alpha}(x^j)\}$ satisfies the balance equations $D_i \Phi^* I^i_{\alpha} = \Phi^* I_{\alpha}$ identically throughout B_n. The identity class is of obvious importance in the study of equivalence of balanced fields, so we shall obtain its explicit characterization. Set $I_{\alpha} = \Phi^* I_{\alpha}$, $I^i_{\alpha} = \Phi^* I^i_{\alpha}$, then Φ is $(I_{\alpha}, I^i_{\alpha})$-balanced if and only if

$$(54.25) \qquad 0 = I_{\alpha} - D_i I^i_{\alpha} = I_{\alpha} - \frac{\partial I^i_{\alpha}}{\partial \phi^{\beta}} D_i \phi^{\beta} - \frac{\partial I^i_{\alpha}}{\partial D_j \phi^{\beta}} D_i D_j \phi^{\alpha} .$$

Since the only terms in (54.25) that contain second derivatives of the ϕ's are the last and $D_i D_j \phi^\alpha$ is symmetric in i and j, (54.25) can hold identically in the ϕ's only if

$$(54.26) \qquad \partial I^i_\alpha / \partial D_j \phi^\beta + \partial I^j_\alpha / \partial D_i \phi^\beta = 0 ,$$

in which case (54.25) becomes

$$(54.27) \qquad 0 = I_\alpha - \frac{\partial I^i_\alpha}{\partial x^i} - \frac{\partial I^i_\alpha}{\partial \phi^\beta} D_i \phi^\beta .$$

Now, conditions (54.26) and (54.27) lift to K in a trivial manner to give

$$(54.28) \qquad \partial I^i_\alpha / \partial y^\beta_j + \partial I^j_\alpha / \partial y^\beta_i = 0 ,$$

$$(54.29) \qquad I_\alpha = \partial I^i_\alpha / \partial x^i + y^\beta_i \, \partial I^i_\alpha / \partial q^\beta = Z_i (I^i_\alpha) .$$

We have thus established the following result.

The identity class, $I(K)$, *consists of* 1-*forms given by*

$$(54.30) \qquad I = I_i dx^i + Z_i (I^i_\alpha) dq^\alpha + I^i_\alpha dy^\alpha_i$$

where I_i *are arbitrary elements of* $\Lambda^0(K)$ *and* I^i_α *are any elements of* $\Lambda^0(K)$ *such that*

$$(54.31) \qquad \partial I^i_\alpha / \partial y^\beta_j + \partial I^j_\alpha / \partial y^\beta_i = 0 .$$

If $U \varepsilon T(K)$ is a horizontal vector and $I \varepsilon I(K)$, then (54.30) shows that

$$(54.32) \qquad U \rfloor I = Z_i (u^\alpha I^i_\alpha) ,$$

and hence $\Phi^*(U \rfloor I) = \Phi^* Z_i (u^\alpha I^i_\alpha)$. However, satisfaction of (54.31) implies that $\Phi^* Z_i (u^\alpha I^i_\alpha) = D_i \Phi^* (u^\alpha I^i_\alpha)$, and we obtain

$$(54.33) \qquad \Phi^*(U \rfloor I) = D_i \Phi^* (u^\alpha I^i_\alpha) .$$

This result shows the relation between elements of $I(K)$ and divergences of vector fields on B_n . In fact, (54.33) gives $\Phi^*(U\rfloor(I\wedge\omega)) = D_i\Phi^*(u^\alpha I_\alpha^i)\omega$ from which we conclude that

$$(54.34)\qquad \int_{B_n}\Phi^*(U\rfloor(I\wedge\omega)) = \int_{\partial B_n}\Phi^*(u^\alpha I_\alpha^i)\pi_i$$

for every horizontal $U\varepsilon T(K)$ and every $I\varepsilon I(K)$. Thus, if U generates a variation of Φ that vanishes on ∂B_n , then $\int_{B_n}\Phi^*(U\rfloor I\wedge\omega) = 0$ for every Φ .

The action of B-equivalence on $I(K)$ is given by $I_\alpha^i \to a_\alpha^\beta I_\beta^i$, $I_\alpha \to a_\alpha^\beta I_\beta + Z_i(a_\alpha^\beta)I_\beta^i = a_\alpha^\beta Z_i I_\beta^i + Z_i(a_\alpha^\beta)I_\beta^i = Z_i(a_\alpha^\beta I_\beta^i)$. Further, since $a_\alpha^\beta\varepsilon\Lambda^0(G)$, $\{a_\alpha^\beta I_\beta^i\}$ satisfies (54.29) because $\{I_\beta^i\}$ does. We thus have the following result.

B-*equivalence is a mapping of* $I(K)$ *to* $I(K)$.

The results concerning $I(K)$ allow us to establish a general equivalence relation for balance equations. Two 1-forms $\underset{1}{W}$ and $\underset{2}{W}$ are said to be F-*equivalent* (field equivalent) if and only if there exists an element I of $I(K)$ and a 1-form $\underset{1}{\tilde{W}}$ that is B-equivalent to $\underset{1}{W}$ such that

$$(54.35)\qquad \underset{2}{W} = \underset{1}{\tilde{W}} + I .$$

Thus, $\underset{1}{W}$ *and* $\underset{2}{W}$ *are* F-*equivalent if and only if there exists a nonsingular* $((a_\beta^\alpha))$ *with* $a_\beta^\alpha\varepsilon\Lambda^0(G)$ *and a collection* $\{I_\alpha^i\}$ *of elements of* $\Lambda^0(K)$, *satisfying* (54.31), *such that*

$$(54.36)\qquad \underset{2}{W}_\alpha = Z_i(I_\alpha^i) + a_\alpha^\beta\underset{1}{W}_\beta + \underset{1}{W}{}_\beta^i Z_i(a_\alpha^\beta)$$

$$(54.37)\qquad \underset{2}{W}{}_\alpha^i = I_\alpha^i + a_\alpha^\beta\underset{1}{W}{}_\beta^i .$$

As an example, suppose that $N=1$, $n=2$, $\lambda(q)\neq 0$ and $\underset{1}{W} = \left(\lambda(q)f(q) + y_1y_2\frac{d\lambda(q)}{dq}\right)dq + \lambda(q)y_1dy_2$, so that $\underset{1}{W}_1 = \lambda(q)f(q) + y_1y_2\frac{d\lambda(q)}{dq}$, $\underset{1}{W}{}_1^1 = 0$, $\underset{1}{W}{}_1^2 = \lambda(q)y_1$, $d\underset{1}{W} = dy_1\wedge d(\lambda(q)y_2)$. Thus, $d\underset{1}{W}\neq 0$ even for λ=constant. If we choose $a_1^1 = \frac{1}{\lambda}$, then a B-equivalence generated by a_1^1 yields $\underset{1}{\tilde{W}}_1 = f(q)$, $\underset{1}{\tilde{W}}_1^1 = 0$, $\underset{1}{\tilde{W}}_1^2 = y_1 dy_2$, $d\underset{1}{\tilde{W}} = dy_1 \wedge dy_2$. Now, $I = b(y_2 dy_1 - y_1 dy_2) \varepsilon I(K)$ and we obtain $\underset{2}{W} = f(q)dq + by_2 dy_1 + (1-b)y_1 dy_2$ is F-equivalent to $\underset{1}{W}$ for any constant b , and $d\underset{2}{W} = (1-2b)dy_1 \wedge dy_2$. We can thus achieve $d\underset{2}{W} = 0$ by taking $b = \frac{1}{2}$. This yields $\underset{2}{W} = f(q)dq + \frac{1}{2}(y_2 dy_1 + y_1 dy_2) = d\left[\int_0^q f(\rho)d\rho + \frac{1}{2} y_1 y_2\right]$ which is F-equivalent to the original system, and the field equation is the f-Gordan equation $D_1 D_2 \phi = f(\phi)$ in two dimensions. This shows that there exists one choice of the W's , namely $W_1 = f$, $W_1^1 = \frac{1}{2} y_2$, $W_1^2 = \frac{1}{2} y_1$, for which the f-Gordan equation is characterized by an exact 1-form $W = d\left[\int_0^q f(\rho)d\rho + \frac{1}{2} y_1 y_2\right]$.

If $\Gamma: K \to {}^\backprime K$ is an admissible map, we have ${}^\backprime x^i = x^i$,

$$ {}^\backprime q^\alpha = \gamma^\alpha(x^j;\, q^\beta) , \quad {}^\backprime y_i^\alpha = Z_i \gamma^\alpha = y_i^\beta \frac{\partial \gamma^\alpha}{\partial q^\beta} + \frac{\partial \gamma^\alpha}{\partial x^i} . \tag{54.38} $$

Thus, since Γ^* maps $\Lambda^k({}^\backprime K) \to \Lambda^k(K)$, we obtain

$$ \begin{aligned} W_i dx^i + W_\alpha dq^\alpha + W_\alpha^i dy_i^\alpha &= {}^\backprime W_i dx^i + {}^\backprime W_\beta d\gamma^\beta + {}^\backprime W_\beta^i dZ_i \gamma^\beta \\ &= \left({}^\backprime W_i + \frac{\partial \gamma^\beta}{\partial x^i}\, {}^\backprime W_\beta + Z_j\left(\frac{\partial \gamma^\beta}{\partial x^i}\right) {}^\backprime W_\beta^j\right) dx^i \\ &\quad + \left({}^\backprime W_\beta \frac{\partial \gamma^\beta}{\partial q^\alpha} + W_\beta^i Z_i \left(\frac{\partial \gamma^\beta}{\partial q^\alpha}\right)\right) dq^\alpha + {}^\backprime W_\beta^i \frac{\partial \gamma^\beta}{\partial q^\alpha} dy_i^\alpha . \end{aligned} \tag{54.39} $$

Accordingly, since the functional form of W_i and ${}^\backprime W_i$ are arbitrary, we obtain

$$ W_\alpha = {}^\backprime W_\beta A_\alpha^\beta + {}^\backprime W_\beta^i Z_i A_\alpha^\beta , \tag{54.40} $$

$$ W_\alpha^i = {}^\backprime W_\beta^i A_\alpha^\beta , \tag{54.41} $$

with

$$(54.42) \qquad A^\beta_\alpha = \partial\gamma^\beta(x^j;\, q^\gamma)/\partial q^\alpha \,.$$

Since Γ is invertible, the matrix $((A^\beta_\alpha))$ has an inverse which we denote by $((\overset{-1}{A}{}^\alpha_\beta))$. If we then solve (54.40) and (54.41) for ${}^\backprime W_\alpha$ and ${}^\backprime W^i_\alpha$ we obtain the following result.

An admissible map $\Gamma: K \to {}^\backprime K \,|\, {}^\backprime x^i = x^i$, ${}^\backprime q^\alpha = \gamma^\alpha(x^j;\, q^\beta)$, ${}^\backprime y^\alpha_i = Z_i\gamma^\alpha$ *with* $A^\alpha_\beta = \partial\gamma^\alpha/\partial q^\beta$ *induces a map* Γ^* *of a* $(W_\alpha;\, W^i_\alpha)$*-balanced system on* K *to a* $({}^\backprime W_\alpha,\, {}^\backprime W^i_\alpha)$*-balanced system on* ${}^\backprime K$ *where*

$$(54.43) \qquad {}^\backprime W_\beta = \overset{-1}{A}{}^\alpha_\beta W_\alpha + Z_i(\overset{-1}{A}{}^\alpha_\beta)W^i_\alpha \,,$$

$$(54.44) \qquad {}^\backprime W^i_\beta = \overset{-1}{A}{}^\alpha_\beta W^i_\beta \,.$$

A comparison of (54.43), (54.44) with (54.19), (54.20) shows that we can identify $({}^\backprime W_\beta,\, {}^\backprime W^i_\beta)$ on ${}^\backprime K$ with $(\bar W_\beta,\, \bar W^i_\beta)$ on K provided we take $a^\alpha_\beta = A^\alpha_\beta$. Accordingly, *the image of a* $(W_\alpha; W^i_\alpha)$*-balanced system under an admissible map can always be realized in the original Kinematic Space* K *by a* B*-equivalent system generated by* $a^\alpha_\beta = \overset{-1}{A}{}^\alpha_\beta$. However, we see that $\overset{-1}{A}{}^\alpha_\beta = \partial q^\alpha(x^j;\, {}^\backprime q^\gamma)/\partial {}^\backprime q^\beta$, while the identification of admissible maps with B-equivalence replaces the ${}^\backprime q$'s by the q's . This has the effect of giving $a^\alpha_\beta = \partial h^\alpha(x^j;\, q^\lambda)/\partial q^\beta$, and hence we have $\partial a^\alpha_\beta/\partial q^\gamma = \partial a^\alpha_\gamma/\partial q^\beta$. It thus follows that *all images of a given balanced system under admissible maps can be realized by all* B*-equivalent systems with* $\partial a^\alpha_\beta/\partial q^\gamma = \partial a^\alpha_\gamma/\partial q^\beta$, $\det(a^\alpha_\beta) \neq 0$.

55. NOETHERIAN VECTOR FIELDS AND THEIR ASSOCIATED CURRENTS

A vector field $V \in T(K)$ is said to be a *Noetherian vector field* of the balanced fields characterized by W if and only if

$$(55.1) \qquad d\Omega(\lambda V) = 0 \quad \forall \lambda \in \mathbb{R} \,.$$

The collection of all Noetherian vector fields of the balanced fields characterized by $\mathcal{W}$ is denoted by $N(\mathcal{W})$. The reason for referring to such vector fields as Noetherian vector fields is that they give rise to results for balanced fields that are similar to the Noether theorems for fields that obey a variational principle. In fact, we shall show in the next section that Noetherian vector fields provide significant generalizations of the theorems of E Noether when there is an underlying variational principle.

If $V \varepsilon N(\mathcal{W})$, then $\Omega(\lambda V)$ is a closed n-form on K by (55.1). If B_n is a star-like region of E_n, then the domain of definition of $\Omega(\lambda V)$ is star-like since this domain has the structure $B_n \times E_N \times E_{nN}$. Since the Poincaré lemma holds for such domains, we conclude that $V \varepsilon N(\mathcal{W})$ if and only if there exists an (n-1)-form M such that $\Omega(\lambda V) = dM$. However, $\Omega(\lambda V)$ is homogeneous of degree one in λV, as shown in (54.13), and hence $\Omega(\lambda V) = dM$ can hold for all constants λ if and only if there exists an n-form $\mathcal{B}$ on K such that $M = \lambda V \rfloor \mathcal{B}$. We have thus established the following result.

If V is a Noetherian vector field of the balanced field characterized by $\mathcal{W}$ then V satisfies

$$(55.2) \qquad \Omega(V) = d(V \rfloor \mathcal{B})$$

for some n-*form* $\mathcal{B}$ *on* K.

This theorem can also be turned around the other way, for we clearly have $V \varepsilon N(\mathcal{W})$ for any given $\mathcal{B}$ provided (55.2) has a non-vacuous solution set for that given $\mathcal{B}$. It is therefore useful to introduce the notation $N(\mathcal{W};\mathcal{B})$ for the collection of all Noetherian vector fields that satisfy (55.2) with fixed $\mathcal{B} \varepsilon \Lambda^n(K)$. We than have

$$N(\mathcal{W}) = \bigcup_{\mathcal{B} \varepsilon \Lambda^n(K)} N(\mathcal{W};\mathcal{B})$$

and the following result.

$V \in N(W;B)$ if and only if V satisfies $\Omega(V) = d(V \rfloor B)$.

There is an alternative formulation of the conditions for $V \in N(W;B)$ that obtains from use of (54.14):

$$\Omega(V) = \pounds_V J + V \rfloor (W \wedge \omega) = d(V \rfloor B) = d(V \rfloor B) + V \rfloor dB - V \rfloor dB$$

$$= \pounds_V B - V \rfloor dB \; .$$

This together with (55.2) yields the following result.

$V \in N(W;B)$ if and only if V satisfies

$$(55.3) \qquad \pounds_V (J-B) = - V \rfloor (dB + W \wedge \omega) \; .$$

It follows immediately from (55.3) that *the elements of $N(W;B)$ form a vector space over the real number field.* However $\pounds_{\lambda V}(J-B) = \lambda \, \pounds_V (J-B) + d\lambda \wedge (V \rfloor (J-B))$, and hence $V \in N(W;B)$ implies $\lambda V \in N(W;B)$ if and only if $\lambda \in \Lambda^0(K)$ is such that $d\lambda \wedge (V \rfloor (J-B)) = 0$. Thus, $N(W;B)$ does not form a module over $\Lambda^0(K)$, in general. It is also clear from these results that $N(W)$ does not, in general, form a vector space over the real number field, as is immediately obvious from what happens to (55.3) in going from one B to another. The remaining question is whether the vector space $N(W;B)$ forms a Lie algebra. Let V_1 and V_2 belong to $N(W;B)$, then $[V_1, V_2] = \pounds_{V_1} V_2 = - \pounds_{V_2} V_1 \in T(K)$. It then follows directly from (A8.21) that

$$- \pounds_{[V_1,V_2]}(J-B) = 2[V_1, V_2] \rfloor (dB + W \wedge \omega) + V_2 \rfloor \pounds_{V_1} (dB + W \wedge \omega)$$

$$- V_1 \rfloor \pounds_{V_2} (dB + W \wedge \omega) \; ,$$

and hence $[V_1, V_2]$ satisfies (55.3) if $dB + W \wedge \omega = 0$. Collecting the above results, we establish the following theorem.

The Noetherian vector fields $N(W;B)$ form a vector space over the real number field. This vector space becomes a Lie algebra with product $[V_1, V_2] = \pounds_{V_1} V_2$ if

$$(55.4) \qquad d\mathcal{B} + \mathcal{W} \wedge \omega = 0 .$$

This result is characteristic of balanced fields in that their Noetherian vector fields fail to form a Lie algebra except in the case in which (55.4) is satisfied. In this respect, there is a significant difference between balanced fields and fields that admit a variational principle, as we shall see in the next section.

The intrinsic utility of Noetherian vector fields of balanced fields comes about as follows. Consider a map

$$\Phi: B_n \to K | \Phi(x^i) = x^i, \quad \Phi(q^\alpha) = \phi^\alpha(x^j), \quad \Phi(y_i^\alpha) = D_i\phi^\alpha(x^j) ,$$

then Φ^* pulls $\Lambda^k(K)$ back to $\Lambda^k(B_n)$. If $V \varepsilon N(\mathcal{W};\mathcal{B})$, then (54.12) holds. Allowing Φ^* to act on both sides of (54.12) gives

$$(55.5) \qquad \Phi^*\Omega(V) = \Phi^* d(V \rfloor \mathcal{B}) = d\Phi^*(V \rfloor \mathcal{B}) .$$

Now, (54.12) gives

$$(55.6) \qquad \Phi^*\Omega(V) = \Phi^*(V \rfloor F) + \Phi^* d(V \rfloor F) = \Phi^*(V \rfloor F) + d\Phi^*(V \rfloor J) .$$

Use of (54.8) to evaluate $\Phi^*(V \rfloor F)$ gives us $\{W_\alpha - D_i W_\alpha^i\}(v^\alpha - v^j D_j \phi^\alpha)\omega$, while (54.4) yields

$$(55.7) \qquad \Phi^*(V \rfloor J) = (v^\alpha - v^j D_j \phi^\alpha) W_\alpha^i \pi_i .$$

Accordingly, (55.6) yields

$$(55.8) \qquad \Phi^*\Omega(V) = (W^\alpha - D_i W_\alpha^i)(v^\alpha - v^j D_j \phi^\alpha)\omega + d\left[(v^\alpha - v^j D_j \phi^\alpha) W_\alpha^i \pi_i\right] .$$

Since $\mathcal{B} \varepsilon \Lambda^n(K)$, $V \rfloor \mathcal{B} \varepsilon \Lambda^{n-1}(K)$ and hence $\Phi^*(V \rfloor \mathcal{B}) \varepsilon \Lambda^{n-1}(B_n)$. However, π_i , $i=1,\ldots,n$ constitute a basis for $\Lambda^{n-1}(B_n)$, from which we conclude that there exist functions B^i on B_n such that

$$(55.9) \qquad \Phi^*(V \rfloor \mathcal{B}) = B^i \pi_i .$$

For example, if $\mathcal{B} = R\omega$, $R\varepsilon\Lambda^0(K)$, then $V\rfloor\mathcal{B} = Rv^i\pi_i$ by (52.9), and (55.9) yields $\Phi^*(V\rfloor\mathcal{B}) = RV^i\pi_i$; that is, $B^i = RV^i$. For the case where n=2 and $\mathcal{B} = d(R^i\pi_i)$, $R^i\varepsilon\Lambda^0(G)$, we obtain $V\rfloor\mathcal{B} = \frac{\partial R^i}{\partial x^i} v^j\pi_j + \frac{\partial R^i}{\partial q^\beta}(v^\beta\pi_i + (V\rfloor\pi_i)dq^\beta)$. Hence

$$\Phi^*(V\rfloor\mathcal{B}) = \left(\frac{\partial R^i}{\partial x^i} V^j + \frac{\partial R^j}{\partial \phi^\beta} V^\beta\right)\pi_j + \left(\frac{\partial R^1}{\partial \phi^\beta} V^2 - \frac{\partial R^2}{\partial \phi^\beta} V^1\right)(D_2\phi^\beta\pi_1 - D_1\phi^\beta\pi_2) ,$$

and we have

$$B^j = \frac{\partial R^i}{\partial x^i} V^j + \frac{\partial R^j}{\partial \phi^\beta} V^\beta + \left(\frac{\partial R^1}{\partial \phi^\beta} V^2 - \frac{\partial R^2}{\partial \phi^\beta} V^1\right)(D_2\phi^\beta\delta^j_1 - D_1\phi^\beta\delta^j_2) .$$

Clearly, in the general case, the forms of the functions B^i will be determined by the n-form $\mathcal{B}$ and will be linear in the functions V^i, V^α and their first total derivatives. We now substitute (55.8) and (55.9) into (55.5). This gives

$$(55.10) \qquad 0 = (W_\alpha - D_iW^i_\alpha)(V^\alpha - V^jD_j\phi^\alpha)\omega + d\left[(W^i_\alpha V^\alpha - W^i_\alpha(D_j\phi^\alpha)V^j - B^i)\pi_i\right] .$$

However, the coefficient of π_i in the above expression is a function defined on B_n, in which case (52.7) can be used. Thus, we finally have

$$(55.11) \qquad 0 = (W_\alpha - D_iW^i_\alpha)(V^\alpha - V^jD_j\phi^\alpha)\omega + D_i\left[W^i_\alpha V^\alpha - W^i_\alpha(D_j\phi^\alpha)V^j - B^i\right]\omega ,$$

from which the following result obtains.

Let $V\varepsilon N(W;\mathcal{B})$ *for given* $W\varepsilon\Lambda^1(K)$, $\mathcal{B}\varepsilon\Lambda^n(K)$, *then the functions* $\{\phi^\alpha(x^j)\}$ *that define the map*

$$(55.12) \quad \Phi: B_n \to K|\Phi(x^i) = x^i, \quad \Phi(q^\alpha) = \phi^\alpha(x^j), \quad \Phi(y^\alpha_i) = D_i\phi^\alpha(x^j)$$

satisfy the identity

$$(55.13)\qquad 0 = (W_\alpha - D_i W^i_\alpha)(V^\alpha - V^j D_j \phi^\alpha) + D_i(W^i_\alpha V^\alpha - W^i_\alpha (D_j \phi^\alpha) V^j - B^i) \; ,$$

where

$$(55.14)\qquad \Phi^*(V \lrcorner B) = B^i \pi_i$$

and $W_\alpha = \mathcal{W}_\alpha \circ \Phi$, $W^i_\alpha = \mathcal{W}^i_\alpha \circ \Phi$, $V^i = v^i \circ \Phi$, $V^\alpha = v^\alpha \circ \Phi$. *If the functions* $\phi^\alpha(x^j)$ *form a balanced field on* B_n *characterized by* $\mathcal{W}$, *then we obtain the conservation law*

$$(55.15)\qquad D_i(W^i_\alpha V^\alpha - W^i_\alpha (D_j \phi^\alpha) V^j - B^i) = 0 \; .$$

Thus every balanced field admits the conserved current

$$(55.16)\qquad K^i = W^i_\alpha V^\alpha - W^i_\alpha (D^j \phi^\alpha) V^i - B^i$$

and there are as many conserved currents as there are linearly independent (with constant coefficients) elements of $N(\mathcal{W};B)$ *for all* $B \in \Lambda^n(K)$.

There is an obvious generalization of the above results that obtain from the fact that the field equations are invariant under F-transformations.

Let $((a^\alpha_\beta))$ *be a nonsingular matrix of elements of* $\Lambda^0(G)$, *let* $\{I^i_\alpha\}$ *be a collection of elements of* $\Lambda^0(K)$ *that satisfy* (54.31), *and define* $\tilde{\mathcal{W}}$ *by*

$$(55.17)\qquad \tilde{\mathcal{W}} = \tilde{\mathcal{W}}_i dx^i + (Z_i(I^i_\alpha) + a^\beta_\alpha \mathcal{W}_\beta + \mathcal{W}^i_\beta Z_i(a^\beta_\alpha))dq^\alpha + (I^i_\alpha + a^\beta_\alpha \mathcal{W}^i_\beta) dy^\alpha_i$$

so that $\tilde{\mathcal{W}}$ *is an* F-*transformation image of* $\mathcal{W}$. *Let* $V \in N(\tilde{\mathcal{W}};B)$ *for given* $B \in \Lambda^n(K)$. *Then the functions* $\{\phi^\alpha(x^j)\}$ *that define the map* (55.12) *satisfy the identity*

$$(55.18) \qquad 0 = a_\alpha^\beta (W_\beta + D_i W_\beta^i)(V^\alpha - V^j D_j \phi^\alpha) + D_i [(I_\alpha^i + a_\alpha^\beta W_\beta^i)(V^\alpha - V^j D_j \phi^\alpha) - B^i]$$

where the quantities B^i *are defined by* (55.14). *If the functions* $\phi^\alpha(x^j)$ *form a balanced field on* B_n *characterized by* W *, then we obtain the conservation law*

$$(55.19) \qquad D_i [(I_\alpha^i + a_\alpha^\beta W_\beta^i)(V^\alpha - V^j D_j \phi^\alpha) - B^i] = 0 .$$

Thus every balanced field admits the conserved current

$$(55.20) \qquad K^i = (I_\alpha^i + a_\alpha^\beta W_\beta^i)(V^\alpha - V^j D_j \phi^\alpha) - B^i$$

and there are as many conserved currents as there are linearly independent elements of $N(\tilde{W};B)$ *for all* $B \varepsilon \Lambda^n(K)$ *, all non-singular matrices* $((a_\alpha^\beta))$ *of elements of* $\Lambda^0(G)$ *and all collections* $\{I_\alpha^i\}$ *of elements of* $\Lambda^0(K)$ *that satisfy* (54.31). This latter collection of conserved currents is obviously significantly larger than that given by (55.16) for all $B \varepsilon \Lambda^0(K)$.

One would tend to believe that there should be a preferred class of choices of the n-form B for general balanced systems, even though such a choice is not evident from the results established this far. A partial substantiation of this belief can be realized by using the homotopy operator H introduced in Section A10 of the Appendix to obtain a unique canonical representation of the (n+1)-form $W \wedge \omega$. Let us assume that the region B_n of E_n is star-shaped with respect to one of its points P_0 , and that the coordinate cover of B_n has been so chosen that the point P_0 is at the origin of the coordinate system. We then define the vector field $\bar{X}$ on K by

$$(55.21) \qquad \bar{X} = x^i \frac{\partial}{\partial x^i} + q^\alpha \frac{\partial}{\partial q^\alpha} + y_i^\alpha \frac{\partial}{\partial y_i^\alpha} ,$$

which can be thought of as a "radius" vector field in K for the given coordinate cover. Thus, since $W \wedge \omega \in \Lambda^{n+1}(K)$, we obtain the unique representation

$$(55.22) \qquad W \wedge \omega = d\chi + Q ,$$

where $\chi \in \Lambda^n(K)$, $Q \in \Lambda^{n+1}(K)$ are antiexact forms,

$$(55.23) \qquad \chi = H(W \wedge \omega) = \int_0^1 \bar{X} \lrcorner (\bar{W} \wedge \omega)\lambda^n d\lambda$$

$$= \left(\int_0^1 \bar{X} \lrcorner \bar{W}\lambda^n d\lambda\right)\omega - \left(\int_0^1 \bar{W}\lambda^n d\lambda\right) \wedge x^i \pi_i ,$$

$$(55.24) \qquad Q = H(dW \wedge \omega) = \int_0^1 \bar{X} \lrcorner (d\bar{W} \wedge \omega)\lambda^{n+1} d\lambda ,$$

and $\bar{W} = W_i(\lambda x^i;\lambda q^\alpha;\lambda y_i^\alpha)dx^i + W_\alpha(\lambda x^i;\lambda q^\alpha;\lambda y_i^\alpha)dq^\alpha + W_\beta^j(\lambda x^i;\lambda q^\alpha;\lambda y_i^\alpha)dy_j^\alpha$. We now substitute (55.22) into (55.3) to obtain $\pounds_V(J-B) = -V \lrcorner (dB+d\chi+Q)$, from which we obtain the following result.

$V \in N(d\chi+Q;B)$ *if and only if* V *satisfies*

$$(55.25) \qquad \pounds_V(J-B) = -V \lrcorner (d(B+\chi)+Q) .$$

The obvious simplification now obtains if we choose B so that $d(B+\chi)' = 0$; that is, $B = d\eta - \chi$ where $\eta \in \Lambda^{n-1}(K)$ is arbitrary. This gives

$$(55.26) \qquad \pounds_V(J+\chi-d\eta) = -V \lrcorner Q$$

for any $V \in N(d\chi+Q;d\eta-\chi)$, and in particular, we see that

$$(55.27) \qquad \pounds_V(J+\chi-d\eta) = 0$$

for all $V \in N(d\chi+Q;d\eta-\chi)$ such that $V \lrcorner Q = 0$. This class of V's transports the n-form $J+\chi+d\eta$ without deformation.

Returning to (55.26), exterior differentiation yields $d\pounds_V(J+\chi+d\eta) = \pounds_V d(J+\chi) = -d(V \lrcorner Q) = -\pounds_V Q + V \lrcorner dQ = -\pounds_V Q + V \lrcorner d(W \wedge \omega)$, since (55.24) implies that $dQ = d(W \wedge \omega)$. Accordingly, we see that $V \in N(d\chi+Q;d\eta-\chi)$ implies $\pounds_V\{d\chi+Q+dJ\} = \pounds_V(W \wedge \omega+dJ) = \pounds_V F =$

$= V \rfloor dQ = V \rfloor d(W \wedge \omega) = V \rfloor d(W \wedge \omega + dJ) = V \rfloor dF$, and hence $d(V \rfloor F) = 0$.

If $V \in N(d\chi + Q; d\eta - \chi)$ *, then* $V \rfloor F$ *is an exact form and* V *transports* F *without deformation if* $V \rfloor dF = 0$.

The above results, although useful in specific problems, is not altogether satisfactory. This follows from the fact that the vector field $\bar{X}$ defined by (55.21) is not a tangent vector field on K , and hence χ and Q do not behave nicely when K undergoes a mapping. This problem is easily overcome, however. Define the tangent vector field $\hat{X}$ on K by

$$(55.28) \qquad \hat{X} = q^\alpha \frac{\partial}{\partial q^\alpha} + Z_i(q^\alpha)\frac{\partial}{\partial y_i^\alpha} = q^\alpha \frac{\partial}{\partial q^\alpha} + y_i^\alpha \frac{\partial}{\partial y_i^\alpha} .$$

This serves to define the zero form $\hat{\chi}$ and the 1-form $\hat{Q}$ by

$$(55.29) \qquad \hat{\chi} = \int_0^1 \hat{X} \rfloor \hat{W} d\lambda , \quad \hat{Q} = \int_0^1 \lambda \hat{X} \rfloor (d\hat{W})|_x d\lambda ,$$

where $\hat{W} = W_\beta(x^i;\lambda q^\alpha;\lambda y_i^\alpha)dq^\beta + W_\beta^j(x^i;\lambda q^\alpha;\lambda y_i^\alpha)dy_j^\beta$ and $(d\hat{W})|_x$ denotes the fact that x^i are held constant in computing $d\hat{W}$. It then follows directly that

$$(55.30) \qquad W \wedge \omega = d\hat{\chi} \wedge \omega + \hat{Q} \wedge \omega$$

and that $dW \wedge \omega = d\hat{Q} \wedge \omega$. Accordingly, we can replace χ by $\hat{\chi}\omega$ and Q by $\hat{Q} \wedge \omega$ in (55.25)-(55.27) and obtain

$$(55.31) \qquad \mathfrak{L}_V(J-B) = -V \rfloor (d(B+\hat{\chi}\omega) + \hat{Q} \wedge \omega) ,$$

and, for $B = d\eta - \hat{\chi}\omega$, $B^i = \hat{\chi}V^i + \Phi^* d\eta$,

$$(55.32) \qquad \mathfrak{L}_V(J+\hat{\chi}\omega - d\eta) = -V \rfloor (\hat{Q} \wedge \omega) .$$

Thus, for those V such that $V \rfloor (\hat{Q} \wedge \omega) = 0$, we obtain

$$(55.33) \qquad \mathfrak{L}_V(J+\hat{\chi}\omega - d\eta) = 0 .$$

The exact same reasoning as used previously shows that $\mathfrak{L}_V F = V \rfloor dF$

if (55.32) and $V \rfloor (\hat{Q} \wedge \omega) = 0$; that is $d(V \rfloor F) = 0$ if $V \rfloor dF = V \rfloor d(W \wedge \omega) = V \rfloor (d\hat{Q} \wedge \omega) = 0$ and $B = d\eta - \hat{\chi}\omega$.

56. VARIATIONAL SYSTEMS

A $(W_\alpha; W^i_\alpha)$-balanced field is said to be *variational* if and only if its balance equations, $D_i W^i_\alpha = W_\alpha$, are the Euler-Lagrange equations for some Lagrangian function $L(x^i; \phi^\alpha(x^i); D_j\phi^\alpha(x^i))$; that is

$$(56.1) \qquad W_\alpha = \partial L/\partial\phi^\alpha , \quad W^i_\alpha = \partial L/\partial(D_i\phi^\alpha) .$$

Since this situation is that which most often obtains in modern field theory, it is natural to inquire as to what restrictions must be placed on a balanced system so that it be variational.

It was noted in Section 54 that the (n+1)-form F is closed if and only if the (n+1)-form $W \wedge \omega$ is closed, in which case we have

$$(56.2) \qquad d(W \wedge \omega) = dW \wedge \omega = 0 .$$

Since the Poincaré lemma holds in K , (56.2) is satisfied if and only if there exists an n-form ρ on K such that

$$(56.3) \qquad W \wedge \omega = d\rho .$$

However, $W \wedge \omega$ is a simple (n+1)-form since ω is a simple n-form. Accordingly, since $W \wedge \omega$ is closed, it is an exterior "gradient product" and (56.3) can be satisfied if and only if there exists an $L \varepsilon \Lambda^0(K)$ such that

$$(56.4) \qquad \rho = L\omega + d\eta$$

where $\eta \varepsilon \Lambda^{n-1}(K)$ is arbitrary. Under these conditions, (56.3) becomes $W \wedge \omega = d(L\omega) = dL \wedge \omega$, so that we obtain

(56.5) $(W-dL)\wedge\omega = 0$.

This equation then yields the following expression for W :

(56.6) $W = dL + \sigma_i\, dx^i$,

for some $\sigma_i \varepsilon \Lambda^0(K)$, $i=1,\ldots,n$. A combination of (56.6) with (54.2) now yields

$$W_i = \frac{\partial L}{\partial x^i} + \sigma_i\ , \quad W_\alpha = \frac{\partial L}{\partial q^\alpha}\ , \quad W^i_\alpha = \frac{\partial L}{\partial y^\alpha_i}$$

so that a composition with the pull-back, Φ^* , of a map $\Phi:B_n \to K|$ $\Phi(x^i) = x^i$, $\Phi(q^\alpha) = \phi^\alpha(x^i)$ gives (56.1). The result $W_i = \partial L/\partial x^i + \sigma_i$ could have been anticipated, for we selected the functions W_i in quite an arbitrary manner in constructing the 1-form W from the given functions $(W_\alpha;\ W^i_\alpha)$. Thus, the functions σ_i occur in just the manner needed in order that we may achieve $W_i - a_i = \partial L/\partial x^i$; that is, (56.6) yields $dW = d\sigma_i \wedge dx^i$. These considerations establish the following result.

A balanced field characterized by the 1-*form* W *is variational if and only if the* (n+1)-*form* F *is closed, and this condition is equivalent to the equivalent conditions*

(56.7) $dW = d\sigma_i \wedge dx^i$, $W = dL + \sigma_i\, dx^i$

for some $\sigma_i \varepsilon \Lambda^0(K)$, $i=1,\ldots,n$. *If these conditions are satisfied, then there exists an* $L\varepsilon\Lambda^0(K)$ *such that*

(56.8) $W\wedge\omega = d\rho$, $\rho = L\omega + d\eta$

for arbitrary $\eta\varepsilon\Lambda^{n-1}(K)$.

The function $L\varepsilon\Lambda^0(K)$ that occurs in (56.8) is referred to as the *Lagrangian function* for the variational system under study.

The next obvious thing to do is to obtain the representations of the forms J, F and $\Omega(V)$ for variational systems. We start by noting that (56.8) implies $W_\alpha = \partial L/\partial q^\alpha$ and $W^i_\alpha = \partial L/\partial y^\alpha_i$, so that we may introduce the notation

$$(56.9)\qquad P^i_\alpha = \partial L/\partial y^\alpha_i$$

as a guide to similarities between this chapter and the previous ones. A straightforward substitution of the results given above into (54.4) and (54.5) yields

$$(56.10)\qquad J = dq^\alpha \wedge P^i_\alpha \pi_i - y^\alpha_i P^i_\alpha \omega ,$$

$$(56.11)\qquad F = W \wedge \omega + dJ = d(\rho + J) = d(L\omega + J)$$

$$= \left(\frac{\partial L}{\partial q^\alpha} dq^\alpha - y^\alpha_i dP^i_\alpha\right) \wedge \omega - dq^\alpha \wedge dP^i_\alpha \wedge \pi_i .$$

Particular note should be taken of the fact that J and F are independent of the choice of the $\eta \varepsilon \Lambda^{n-1}(K)$ that occurs in (56.8), and that

$$(56.12)\qquad \Phi^* J = 0 , \quad \Phi^* F = \Phi^* d(L\omega + J) = d\Phi^*(L\omega + J) - d\Phi^* L\omega = 0$$

for any $\Phi: B_n \to K | \Phi^*(x^i) = x^i$. It is now a straightforward matter to use (56.11) and (54.12) in order to obtain

$$(56.13)\qquad V \rfloor F = V \rfloor d(\rho + J) = \pounds_V(\rho + J) - d[V \rfloor (\rho + J)]$$

$$= \pounds_V(L\omega + J) - d[V \rfloor (L\omega + J)] ,$$

$$(56.14)\qquad \Omega(V) = V \rfloor F + d(V \rfloor J) = \pounds_V(\rho + J) - d(V \rfloor \rho)$$

$$= \pounds_V(L\omega + J) - d(LV \rfloor \omega) .$$

Hence $V \rfloor F$ and $\Omega(V)$ are also independent of the choice of the $\eta \varepsilon \Lambda^{n-1}(K)$ that occurs in (56.8).

The connection between variational systems and variational principles can now be obtained directly from (56.14). We first restrict V to be a horizontal vector field U, so that $u^i=0$ and $U \rfloor \omega = u^i \pi_i = 0$. In this case, (56.14) yields

$$(56.15) \qquad \Omega(U) = \pounds_U(L\omega + J)$$

for all horizontal vector fields U on K. If Φ is any map of B_n into K such that $\Phi(x^i) = x^i$, $\Phi(q^\alpha) = \phi^\alpha(x^i)$, then

$$\int_{B_n} \Phi^* \Omega(U) = \int_{B_n} \Phi^* \pounds_U(L\omega+J) = \delta \int_{B_n} \Phi^*(L\omega+J)$$

$$= \delta \int_{B_n} \Phi^*(L\omega)$$

because $\Phi^* J = 0$. Accordingly, the results established at the end of Section 52 show that Φ *stationarizes the action integral*

$$(56.16) \qquad A[\Phi] = \int_{B_n} \Phi^*(L\omega) = \int_{B_n} \mathrm{L}\omega, \quad \mathrm{L} = L\circ\Phi$$

for all variations $\delta\phi^\alpha = u^\alpha \circ \Phi$ *that vanish on* ∂B_n *if and only if* $\int_{B_n} \Phi^* \Omega(U) = 0$ *for all* U *such that* $(u^\alpha \circ \Phi)|_{\partial B_n} = 0$. Another way of stating this is as follows. *If the functions* $\phi^\alpha(x^i)$, *that define the map* Φ, *form a* $\left(\frac{\partial L}{\partial \phi^\alpha}, P^i_\alpha\right)$*-balanced field, then* $\int_{B_n} \Phi^* \Omega(U) = 0$ *for all horizontal vector fields* U *such that* $(u^\alpha \circ \Phi)|_{\partial B_n} = 0$, *and* $A[\Phi]$ *assumes a stationary value.* The problem of solving a variational system (i.e., of finding functions $\phi^\alpha(x^i)$ such that $D_i P^i_\alpha = \partial L/\partial \phi^\alpha$) is thus equivalent to the variational problem of stationarizing the action integral (56.16), and, more importantly, is equivalent to annihilating the integral $\int_{B_n} \Phi^* \Omega(U)$ for all horizontal U such that $(u^\alpha \circ \Phi)|_{\partial B_n} = 0$.

We now apply the results obtained in Section 55 to variational systems. Since $W = dL + \sigma_i dx^i$, we are concerned with Noetherian vector fields that belong to $N(dL + \sigma_i dx^i)$. Such vector fields are comprised from the subsets $N(dL + \sigma_i dx^i; B)$ that are solutions of (55.3):

$$(56.17)\qquad \pounds_V(J-B) = -V\rfloor(dB+d(L\omega+d\eta)) = -V\rfloor d(B+L\omega)\ .$$

Again, we note that the terms involving $\eta\epsilon\Lambda^{n-1}(K)$ drop out. Now, the elements of $N(dL+\sigma_i dx^i;B)$ form a Lie algebra if $d(B+L\omega) = 0$, in which case we obtain $B+L\omega = d\mu$ for some $\mu\epsilon\Lambda^{n-1}(K)$. We thus obtain the following important result.

$N(dL+\sigma_i dx^i;\ d\mu-L\omega)$ forms a Lie algebra for each $\mu\epsilon\Lambda^{n-1}(K)$, in which case the vector fields generating this Lie algebra satisfy

$$(56.18)\qquad \pounds_V(J+L\omega-d\mu) = 0\ .$$

The elements of $N(dL+\sigma_i dx^i;\ d\mu-L\omega)$ thus transport the n-form $J+L\omega-d\mu$ in K without deformation.

This situation is ironic. We saw that $\rho = L\omega+d\eta$, $W\wedge\omega = d\rho$ characterizes variational systems, where $\eta\epsilon\Lambda^{n-1}(K)$ is arbitrary, but the arbitrary (n-1)-form η drops out from all of the resulting equations. On the other hand, the demand that B be chosen so that $N(dL+\sigma_i dx^i;B)$ form a Lie algebra reintroduces an arbitrary (n-1)-form μ by $B = d\mu-L\omega$. In fact, if we identity μ with $-\eta$, then $B = -d\eta-L\omega = -\rho$ and (55.18) becomes $\pounds_V(J+\rho) = 0$; that is $N(dL+\sigma_i dx^i;-\rho)$ *is a Lie algebra whose elements transport the n-form $J+\rho$ without deformation.* We also note in passing that (56.10) yields the familiar evaluation

$$(56.19)\qquad J + L\omega - d\mu = dq^\alpha\wedge P^i_\alpha\pi_i - (P^i_\alpha y^\alpha_i - L)\omega - d\mu\ .$$

Without further ado, we simple restate the previous theorem that relates Noetherian vector fields and conserved currents, as it obtains in the case of variational systems (see the theorem at the end of Section 55).

Let $V\epsilon N(dL+\sigma_i dx^i;B)$, for given $B\epsilon\Lambda^n(K)$, then the functions $\phi^\alpha(x^j)$ that define the map

$$(56.20)\ \Phi\colon B_n \to K|\Phi(x^i) = x^i,\quad \Phi(q^\alpha) = \phi^\alpha(x^i),\quad \Phi(y_j) = D_j\phi^\alpha(x^i)$$

satisfy the identity

$$(56.21)\qquad 0 = \left(\frac{\partial L}{\partial\phi^\alpha} - D_i\frac{\partial L}{\partial(D_i\phi^\alpha)}\right)(V^\alpha - V^j D_j\phi^\alpha)$$
$$+ D_i\left(\frac{\partial L}{\partial(D_i\phi^\alpha)}V^\alpha - V^j\frac{\partial L}{\partial(D_i\phi^\alpha)}D_j\phi^\alpha - B^i\right),$$

where

$$(56.22)\qquad \Phi^*(V\rfloor B) = B^i\pi_i,\quad L = L\circ\Phi,\quad V^i = v^i\circ\Phi,\quad V^\alpha = v^\alpha\circ\Phi .$$

If the functions $\phi^\alpha(x^i)$ *form a balanced field on* B_n *that is characterized by* $dL + \sigma_i dx^i$ *, then we obtain the conservation laws*

$$(56.23)\qquad D_i\left(\frac{\partial L}{\partial(D_i\phi^\alpha)}V^\alpha - V^j\frac{\partial L}{\partial(D_i\phi^\alpha)}D_j\phi^\alpha - B^i\right) = 0$$

for each element of $N(dL+\sigma_i dx^i;B)$. *Thus, every solution of the Euler-Lagrange equations* $D_i P^i_\alpha = \partial L/\partial\phi^\alpha$, $P^i_\alpha = \partial L/\partial(D_i\phi^\alpha)$ *admits the conserved current*

$$(56.24)\qquad K^i = \frac{\partial L}{\partial(D_i\phi^\alpha)}V^\alpha - V^j\frac{\partial L}{\partial(D_i\phi^\alpha)}D_j\phi^\alpha - B^i$$

and there are as many independent conserved currents as there are linearly independent (with constant coefficient) *elements of* $N(dL+\sigma_i dx^i;B)$ *for all* $B\epsilon\Lambda^n(K)$.

The collections of conserved currents of particular interest are those that are associated with the $N(dL+\sigma_i dx^i;B)$ that form Lie algebras. This is the case when $B = d\mu - L\omega$, in which case the quantities B^i are given by the definition $B^i\pi_i = \Phi^*(V\rfloor B) = \Phi^*(V\rfloor d\mu - LV\rfloor\omega)$. We know, however, that $\Phi^*(LV\rfloor\omega) = \Phi^*(Lv^i\pi_i) = LV^i\pi_i$. Further, since $V\rfloor d\mu\epsilon\Lambda^{n-1}(K)$, $\Phi^*(V\rfloor d\mu)\epsilon\Lambda^{n-1}(E_n)$, $\Phi^*(v^i) = V^i$, $\Phi^*(v^\alpha) = V^\alpha$, $\Phi^*(v^\alpha_i) = D_i(V^\alpha - V^j D_j\phi^\alpha) + V^i D_j D_i\phi^\alpha$, we can write $\Phi^*(V\rfloor d\mu) = (M^i_k V^k + M^i_\alpha V^\alpha + M^{ki}_\alpha D_k(V^\alpha - V^j D_j\phi^\alpha))\pi_i$, where the functions M^i_k, M^i_α and M^{ki}_α are uniquely determined by the structure of $d\mu$. A combination of these results yields

$$B^i = -LV^i + M^i_k V^k + M^i_\alpha V^\alpha + M^{ki}_\alpha D_k(V^\alpha - V^j D_j \phi^\alpha)$$

with $B^i = -LV^i$ for the choice $d\mu=0$. When this is substituted into (56.24), we obtain the conserved current

$$(56.25) \qquad K^i = \left(\frac{\partial L}{\partial(D_i\phi^\alpha)} - M^i_\alpha\right)V^\alpha - \left(\frac{\partial L}{\partial(D_i\phi^\alpha)} D_j\phi^\alpha - L\delta^i_j + M^i_j\right)V^j$$
$$- M^{ki}_\alpha D_k(V^\alpha - V^j D_j\phi^\alpha)$$

which reduces to the well-known form

$$(56.26) \qquad K^i = \frac{\partial L}{\partial(D_i\phi^\alpha)} V^\alpha - \left(\frac{\partial L}{\partial(D_i\phi^\alpha)} D_j\phi^\alpha - L\delta^i_j\right)V^j$$

when $d\mu=0$. The momentum-energy "tensor"

$$T^i_j = \frac{\partial L}{\partial(D_i\phi^\alpha)} D_j\phi^\alpha - L\delta^i_j$$

thus immerges in a natural way whenever $N(dL+\sigma_i dx^i;B)$ forms a Lie algebra.

The relation between Noetherian vector fields, conserved currents, and global symmetries of the system is easily obtained. Let Ψ_t be a 1-parameter family of mappings of K to K that is defined by $\Psi_t(x^i) = \psi^i(t;\, x^j;\, q^\beta)$, $\Psi_t(q^\alpha) = \psi^\alpha(t;\, x^j;\, q^\beta)$, $\Psi_t(y^\alpha_i) = \psi^\alpha_i(t;\, x^j;\, q^\beta;\, y^\beta_j)$ and satisfies the conditions Ψ_0 = identity,

$$(56.27) \qquad \psi^\alpha_i Z_j(\psi^i) = Z_j(\psi^\alpha) .$$

The restriction (56.27) follows from (53.9) in order that the y's transform in a manner consistent with the structure of Kinematic Space. If this family of maps generates a *global symmetry* of $J+L\omega-d\mu$, then we have

$$(56.28) \qquad \Psi^*_t(J+L\omega-d\mu) = J + L\omega - d\mu .$$

Accordingly, the vector field ξ that is tangent to Ψ_t at $t=0$

transports $J+L\omega-d\mu$ without deformation. It then follows that

$$(56.29)\qquad \pounds_{\xi}(J+L\omega-d\mu) = 0$$

and hence $\xi \epsilon N(dL+\sigma_i dx^i;\ d\mu-L\omega)$ if $\xi \epsilon T(K)$. We thus need to calculate the vector field ξ and show that it belongs to $T(K)$. By definition,

$$(56.30)\qquad \xi = \left.\frac{\partial\phi^i}{\partial t}\right|_{t=0}\frac{\partial}{\partial x^i} + \left.\frac{\partial\psi^\alpha}{\partial t}\right|_{t=0}\frac{\partial}{\partial q^\alpha} + \left.\frac{\partial\psi_i^\alpha}{\partial t}\right|_{t=0}\frac{\partial}{\partial y_i^\alpha}$$

so that we need only verify the relations

$$\left.\frac{\partial\psi_i^\alpha}{\partial t}\right|_{t=0} = Z_i\left(\left.\frac{\partial\psi^\alpha}{\partial t}\right|_{t=0} - y_j^\alpha \left.\frac{\partial\psi^j}{\partial t}\right|_{t=0}\right)$$

in order to show that ξ is a vector field on K. This, however, follows immediately from differentiation of (56.27) with respect to t, evaluation at $t=0$, and noting that $Z_j(\psi^i)|_{t=0}=\delta_j^i$, $\psi_i^\alpha|_{t=0}=y_i^\alpha$ because ψ_0 = identity. We thus conclude that *the tangent vector field to a global symmetry map of* $J+L\omega-d\mu$ *belongs to* $N(dL+\sigma_i dx^i;\ d\mu-L\omega)$ *and yields a conserved current for any solution of the field equations* $D_i P_\alpha^i = \partial L/\partial\phi^\alpha$, $P_\alpha^i = \partial L/\partial(D_i\phi^\alpha)$.

We have seen that a given balanced system is variational if and only if $d(W\wedge\omega) = 0$. Suppose, therefore, that the given system is such that $d(W\wedge\omega) \neq 0$. It is then reasonable to ask whether there is a F-equivalent system that is variational; i.e. whether there exists a non-singular matrix $((a_\beta^\alpha(x^j;\ q^\gamma)))$ and a collection $\{I_\alpha^i\}$ of elements of $\Lambda^0(K)$, satisfying (54.31), such that $d(\bar{W}\wedge\omega) = 0$ for W and $\bar{W}$ being related by (54.36), (54.37). This is the case if and only if

$$(56.31)\quad d(a_\beta^\alpha W_\alpha + Z_i I_\beta^i + W_\alpha^i Z_i a_\beta^\alpha)\wedge dq^\beta\wedge\omega + d(a_\beta^\alpha W_\alpha^i + I_\beta^i)\wedge dy_i^\beta\wedge\omega = 0 .$$

If a_β^α and I_α^i can be found that satisfy these conditions, then previous results show there exists an $L\epsilon\Lambda^0(K)$ such that

$$(56.32)\qquad \bar{W}_\beta = \frac{\partial L}{\partial q^\beta},\quad \bar{W}_\beta^i = \frac{\partial L}{\partial y_i^\beta} .$$

When (54.36), (54.37) are used again, the relations (56.32) give

$$(56.33)\qquad \begin{aligned} a^{\alpha}_{\beta}W_{\alpha} + W^{i}_{\alpha}Z_{i}a^{\alpha}_{\beta} + Z_{i}I^{i}_{\beta} &= \partial L/\partial q^{\beta} , \\ a^{\alpha}_{\beta}W^{i}_{\alpha} + I^{i}_{\beta} &= \partial L/\partial y^{\beta}_{i} . \end{aligned}$$

Let $\overset{-1}{a}{}^{\beta}_{\gamma}$ denote the components of the inverse of the nonsingular matrix $((a^{\alpha}_{\beta}))$. The system of relations (56.33) then yields the relations

$$(56.34)\qquad W^{i}_{\alpha} = \overset{-1}{a}{}^{\beta}_{\alpha}\left(\frac{\partial L}{\partial y^{\beta}_{i}} - I^{i}_{\beta}\right) ,$$

$$(56.35)\qquad \begin{aligned} W_{\alpha} &= \overset{-1}{a}{}^{\beta}_{\alpha}\left(\frac{\partial L}{\partial q^{\beta}} - Z_{i}I^{i}_{\beta}\right) - \overset{-1}{a}{}^{\mu}_{\alpha}\,\overset{-1}{a}{}^{\gamma}_{\rho}Z_{i}(a^{\rho}_{\mu})\left(\frac{\partial L}{\partial y^{\gamma}_{i}} - I^{i}_{\gamma}\right) \\ &= \overset{-1}{a}{}^{\beta}_{\alpha}\left(\frac{\partial L}{\partial q^{\beta}} - Z_{i}I^{i}_{\beta}\right) + Z_{i}(\overset{-1}{a}{}^{\beta}_{\alpha})\left(\frac{\partial L}{\partial y^{\beta}_{i}} - I^{i}_{\beta}\right) . \end{aligned}$$

These considerations thus give the following result.

If there exists a nonsingular matrix $((a^{\alpha}_{\beta}))$ *of functions of* $(x^{j}; q^{\gamma})$ *and a collection* $\{I^{i}_{\alpha}\}$ *of elements of* $\Lambda^{0}(K)$ *such that*

$$(56.36)\qquad d\bar{W}\wedge\omega = 0 ,$$

$$(56.37)\qquad \partial I^{i}_{\alpha}/\partial y^{\beta}_{j} + \partial I^{j}_{\alpha}/\partial y^{\beta}_{i} = 0 ,$$

where

$$(56.38)\qquad \bar{W}_{i} = W_{i},\quad \bar{W}_{\beta} = a^{\alpha}_{\beta}W_{\alpha} + W^{i}_{\alpha}Z_{i}(a^{\alpha}_{\beta}) + Z_{i}I^{i}_{\beta},\quad \bar{W}^{i}_{\beta} = a^{\alpha}_{\beta}W^{i}_{\alpha} + I^{i}_{\beta} ,$$

then the given $(W_{\alpha}, W^{i}_{\alpha})$*-balanced system is* F*-equivalent to a variational system. There thus exists an* $L\in\Lambda^{0}(K)$ *such that* $\bar{W}\wedge\omega = dL\wedge\omega$ *and* (56.34), (56.35) *hold.* An immediate consequence of this theorem is the following useful result.

If a given system is F*-equivalent to a variational system with fundamental* 1*-form* $\bar{W}$ *, then reformulation of the problem in* *terms of the equivalent form* $\bar{W}$ *gives a variational system.* This result gives a complete answer to the question of the imbedding of a given system of field equations in a variational principle. For instance, we saw at the end of Section 54 that the f-Gordan equation $D_1 D_2 = f(\phi)$ obtains from $W = d\left[\int_0^{\phi} f(\mu)\,d\mu\right] + y_1 dy_2$, which is not closed. However, adding $I = \frac{1}{2}(y_2 dy_1 - y_1 dy_2) \varepsilon I(K)$ to W yields $\tilde{W} = W + I = d\left[\int_0^{\phi} f(\mu)\,d\mu\right] + \frac{1}{2}(y_1 dy_2 + y_2 dy_1) = d\left[\int_0^{\phi} f(\mu)\,d\mu + \frac{1}{2} y_1 y_2\right]$ is closed with the Lagrangian $L = \int_0^{\phi} f(\mu)\,d\mu + \frac{1}{2} y_1 y_2$.

57. E-TRANSFORMATIONS

The only transformations that have been used thus far in this chapter are regular transformations of K that come from the lifting of invertible transformations of Graph Space; that is, transformations of the form

$$\Gamma: K \to {}^{\backprime}K | {}^{\backprime}x^i = \gamma^i(x^j; q^\beta), \quad {}^{\backprime}q^\alpha = \gamma^\alpha(x^j; q^\beta), \quad {}^{\backprime}y_i^\alpha Z_j \gamma^i = Z_j \gamma^\alpha$$

with $\det(Z_j \gamma^i) \neq 0$. We now consider more general classes of transformations that are modeled on those constructed in Section 50 for nonconservative, nonholonomic systems with one independent variable. In writing the equations that define this class of transformations, we use the notation $F(W)$ and $J(W)$ to designate the (n+1)-form and the n-form that are constructed from the 1-form W as specified in Section 55.

An invertible map

$$(57.1) \qquad \rho: K \to {}^{\backprime}K | {}^{\backprime}x^i = x^i , \quad {}^{\backprime}q^\alpha = \rho^\alpha(x^j; q^\beta; y_j^\beta) ,$$

$${}^{\backprime}y_i^\alpha = \rho_i^\alpha(x^j; q^\beta; y_j^\beta)$$

is said to constitute an E-*transformation* if and only if there exists a 1-form ${}^{\backprime}W(\rho, W)$ for each 1-form W such that

(57.2) $$\rho^{-1*}F(W) = F(\grave{}W) .$$

Since $(\rho_2 \circ \rho_1)^{-1*}F(W) = \rho_2^{-1*} \circ \rho_1^{-1} F(W) = \rho^{-1*}F(\grave{}W) = F(\grave{}\grave{}W)$, obtain the following immediate conclusion.

E-transformations form a group under composition.

Let us suppose that the systems characterized by W and $\grave{}W$ are related by an E-transformation ρ. If ρ is such that $\grave{}K = \rho(K)$ is a Kinematic Space, then ρ_* maps horizontal vector fields $U = u^\alpha\, \partial/\partial q^\alpha + Z_i u^\alpha\, \partial/\partial y_i^\alpha$ on K onto horizontal vector fields $\grave{}U = \grave{}u^\alpha\, \partial/\partial\grave{}q^\alpha + \grave{}Z_i \grave{}u^\alpha\, \partial/\partial\grave{}y_i^\alpha$ on $\grave{}K$ because $\grave{}x^i = x^i$. It thus follows that $\grave{}U \rfloor F(\grave{}W) = \grave{}U \rfloor \rho^{-1*}F(W) = \rho_* U \rfloor \rho^{-1*}F(W) = \rho^{-1*}\{U \rfloor F(W)\}$. Further, since $\grave{}\Phi = \rho \circ \Phi$ for any map $\Phi: B_n \to K$, we have $\grave{}\Phi^* = (\rho \circ \Phi)^* = \Phi^* \circ \rho^*$, and hence $\grave{}\Phi^* \grave{}U \rfloor F(\grave{}W) = \grave{}\Phi^* \rho^{-1*}\{U \rfloor F(W)\} = \Phi^* U \rfloor F(W)$. We know, however, that $\Phi^* U \rfloor F(W) = (D_i W_\alpha^i - W_\alpha)\delta\phi^\alpha \omega$, $\grave{}\Phi^* \grave{}U \rfloor F(\grave{}W) = (D_i \grave{}W_\alpha^i - \grave{}W_\alpha)\delta\grave{}\phi^\alpha \omega$, where $\delta\phi^\alpha = u^\alpha \circ \Phi$, $\delta\grave{}\phi^\alpha = \grave{}u^\alpha \circ \grave{}\Phi$. It thus follows that

$$(D_i \grave{}W_\alpha^i - \grave{}W_\alpha)\delta\grave{}\phi^\alpha \omega = (D_i W_\alpha^i - W_\alpha)\delta\phi^\alpha \omega$$

holds for every map $\Phi: B_n \to K$ and every horizontal vector field U. An integration over B_n thus yields

$$\int_{B_n} (D_i \grave{}W_\alpha^i - \grave{}W_\alpha)\delta\grave{}\phi^\alpha \omega = \int_{B_n} (D_i W_\alpha^i - W_\alpha)\delta\phi^\alpha \omega .$$

From this we see that if $\grave{}\Phi = \rho \circ \Phi$ gives a solution of the system $D_i \grave{}W_\alpha^i = \grave{}W_\alpha$ characterized by $\grave{}W$ then the arbitrary nature of the functions $\delta\phi^\alpha$ allows us to use the fundamental lemma of the calculus of variations to conclude that Φ gives a solution of the system $D_i W_\alpha^i = W_\alpha$. The same argument shows that the converse is also true and we have established the following result.

If an E-transformation ρ maps Kinematic Space K to another Kinematic Space $\grave{}K$, then ρ maps all solutions of any balanced system on K onto solutions of the image balanced system on $\grave{}K$ and ρ^{-1} maps all solutions of any image balanced system on $\grave{}K$

onto solutions of the corresponding preimage balanced system on K . E-transformations are thus very useful constructs provided we make sure that they map Kinematic Space to another Kinematic Space.

From now on, we shall write equations like (57.2) in the equivalent form $F(W) = F({}^{\backprime}W)$, where it is understood that the equations of transformation are used to write both sides of the equality in terms of the same system of arguments. Since $F(W) = W \wedge \omega + dJ(W)$, (57.2) yields

$$(57.3) \qquad d\{J(W) - J({}^{\backprime}W)\} = ({}^{\backprime}W - W) \wedge \omega ,$$

from which we conclude that $({}^{\backprime}W-W) \wedge \omega$ is closed. Thus, since $({}^{\backprime}W-W) \wedge \omega$ is also simple, (57.3) can hold only if there exists a $\mu \varepsilon \Lambda^0(K, {}^{\backprime}K)$ such that

$$(57.4) \qquad W \wedge \omega = {}^{\backprime}W \wedge \omega + d\mu \wedge \omega .$$

The function μ that occurs in (57.4) is referred to as the *remainder function* of the E-transformation. Under satisfaction of the consistency condition (57.4), (57.3) can be satisfied if and only if there exists an $\eta \varepsilon \Lambda^{n-1}(K, {}^{\backprime}K)$, called the *generating form* of the E-transformation, such that

$$(57.5) \qquad J({}^{\backprime}W) = J(W) + \mu\omega + d\eta .$$

We thus have the following results.

An invertible map ρ *, of the form given by* (57.1), *is an E-transformation if and only if there exists a* $\mu \varepsilon \Lambda^0(K, {}^{\backprime}K)$ *and an* $\eta \varepsilon \Lambda^{n-1}(K, {}^{\backprime}K)$ *such that*

$$(57.6) \qquad J({}^{\backprime}W) = J(W) + \mu\omega + d\eta ,$$

$$(57.7) \qquad W \wedge \omega = {}^{\backprime}W \wedge \omega + d(\mu\omega) ,$$

where it is understood that the equations of transformation are used to express both sides of the above equalities in terms of the same arguments.

An E-transformation is said to be *special* if the generating $(n-1)$-form η is given by

$$(57.8)\qquad \eta = N^i\pi_i\ ,\qquad N^i\varepsilon\Lambda^0(K,{}^\backprime K)\ ,\qquad i=1,\ldots,n\ ,$$

in which case we refer to the functions N^i as *generating functions* of the E-transformation.

In analogy with the case of only one independent variable we make the substitution

$$(57.9)\qquad N^i = {}^\backprime W^i_\alpha {}^\backprime q^\alpha + \gamma^i(x^j;\ q^\beta;\ {}^\backprime W^j_\beta)$$

and consider all quantities to be functions of the arguments x^j, q^β, ${}^\backprime W^i_\beta$. Since $J(W) = (dq^\alpha - y^\alpha_i\, dx^j)\wedge W^i_\alpha\pi_i$, a substitution of (57.9) into (57.6) yields

$$\begin{aligned} d{}^\backprime q^\alpha\wedge{}^\backprime W^i_\alpha\pi_i - {}^\backprime y^\alpha_i {}^\backprime W^i_\alpha\omega &= dq^\alpha\wedge W^i_\alpha\pi_i - y^\alpha_i W^i_\alpha\omega + \mu\omega \\ &+ d{}^\backprime q^\alpha\wedge{}^\backprime W^i_\alpha\pi_i + {}^\backprime q^\alpha d{}^\backprime W^i_\alpha\pi_i + \frac{\partial\gamma^i}{\partial x^i}\,\omega + \frac{\partial\gamma^i}{\partial q}\, dq^\alpha\wedge\pi_i \\ &+ \frac{\partial\gamma^i}{\partial{}^\backprime W^j_\alpha}\, d{}^\backprime W^j_\alpha\wedge\pi_i\ . \end{aligned}$$

Noting that this equality involves only the differentials of the independent arguments x^i, q^α, ${}^\backprime W^i_\alpha$, we obtain the conditions

$$(57.10)\qquad \mu = -\,{}^\backprime y^\alpha_i {}^\backprime W^i_\alpha + y^\alpha_i W^i_\alpha - \frac{\partial\gamma^i}{\partial x^i}\ ,$$

$$(57.11)\qquad W^i_\alpha = -\,\frac{\partial\gamma^i}{\partial q^\alpha}\ ,$$

$$(57.12)\qquad {}^\backprime q^\alpha\delta^i_j = -\,\frac{\partial\gamma^i}{\partial{}^\backprime W^j_\alpha}\ .$$

However, (57.12) can hold for all values of the indices i and j if and only if $\gamma^i = -\,{}^\backprime W^i_\alpha h^\alpha(x^j;\ q^\beta)$. Thus, only a linear dependence on ${}^\backprime W^i_\alpha$ is allowed. A substitution of this back into (57.10)-(57.12) shows that we must have

$$(57.13)\qquad \mu = \left(-\grave{y}_i^\beta + y_i^\alpha \frac{\partial h^\beta}{\partial q^\alpha} + \frac{\partial h^\beta}{\partial x^i}\right)\grave{\omega}_\beta^i ,$$

$$(57.14)\qquad \omega_\alpha^i = \grave{\omega}_\beta^i \frac{\partial h^\beta}{\partial q^\alpha} ,$$

$$(57.15)\qquad \grave{q}^\alpha = h^\alpha(x^j;\, q^\beta) .$$

Now that we have determined the remainder function of μ, we can substitute it into (57.7) to obtain

$$(\omega_\alpha dq^\alpha + \omega_\alpha^i dy_i^\alpha)\wedge\omega = (\grave{\omega}_\alpha d\grave{q}^\alpha + \grave{\omega}_\alpha^i d\grave{y}_i^\alpha)\wedge\omega$$
$$+ d\left\{\left(-\grave{y}_i^\beta + y_i^\alpha \frac{\partial h^\beta}{\partial q^\alpha} + \frac{\partial h^\beta}{\partial x^i}\right)\grave{\omega}_\beta^i\right\}\wedge\omega .$$

Substituting the evaluations given by (57.14) and (57.15) into this relation gives

$$\left(\omega_\alpha dq^\alpha + \omega_\beta^i \frac{\partial h^\beta}{\partial q^\alpha} dy_i^\alpha\right)\wedge\omega = \left\{\grave{\omega}_\beta \frac{\partial h^\beta}{\partial q^\alpha} dq^\alpha + \grave{\omega}_\alpha^i d\grave{y}_i^\alpha\right.$$
$$\left. + d\left(-\grave{y}_i^\beta + y_i^\alpha \frac{\partial h^\beta}{\partial q^\alpha} + \frac{\partial h^\beta}{\partial x^i}\right)\grave{\omega}_\beta^i + \left(-\grave{y}_i^\beta + y_i^\alpha \frac{\partial h^\beta}{\partial q^\alpha} + \frac{\partial h^\beta}{\partial x^i}\right)d\grave{\omega}_\beta^i\right\}\wedge\omega ,$$

from which we obtain

$$(57.16)\qquad \left\{\omega_\alpha - \grave{\omega}_\beta \frac{\partial h^\beta}{\partial q^\alpha} - \grave{\omega}_\beta^i\left(\frac{\partial^2 h^\beta}{\partial x^i \partial q^\alpha} + y_i^\gamma \frac{\partial^2 h^\beta}{\partial q^\gamma \partial q^\alpha}\right)\right\} dq^\alpha \wedge\omega$$
$$= \left(-\grave{y}_i^\beta + y_i^\alpha \frac{\partial h^\beta}{\partial q^\alpha} + \frac{\partial h^\beta}{\partial x^i}\right) d\grave{\omega}_\beta^i \wedge\omega .$$

Accordingly, since x^i, q^α and $\grave{\omega}_\alpha^i$ are independent, (57.16) can hold if and only if

$$(57.17)\qquad \omega_\alpha = \grave{\omega}_\beta \frac{\partial h^\beta}{\partial q^\alpha} + \grave{\omega}_\beta^i\left(\frac{\partial^2 h^\beta}{\partial x^i \partial q^\alpha} + y_i^\gamma \frac{\partial^2 h^\beta}{\partial q^\gamma \partial q^\alpha}\right) ,$$

$$(57.18)\qquad \grave{y}_i^\alpha = y_i^\beta \frac{\partial h^\alpha}{\partial q^\beta} + \frac{\partial h^\alpha}{\partial x^i} .$$

We note as a matter of interest at this point that (57.15) and $\gamma^i = -{}^\backprime W^i_\alpha h^\alpha = -{}^\backprime W^i_\alpha {}^\backprime q^\alpha$ show that $\eta=0$ and that satisfaction of (57.18) gives $\mu=0$ by (57.13). Further, (57.15) shows that the requirement that ρ be invertible is satisfied only if $\det \cdot (\partial h^\alpha/\partial q^\beta) \neq 0$. The following basic result is thus obtained.

An invertible map $\rho: K \to {}^\backprime K | {}^\backprime x^i = x^i$ *is a special* E-*transformation with generating* (n-1)-*form* $\eta = {}^\backprime W^i_\alpha {}^\backprime q^\alpha + \gamma^i(x^j; q^\beta; {}^\backprime W^j_\beta)$ *and independent variables* x^i, q^β, ${}^\backprime W^i_\beta$ *if and only if* $\gamma^i = -{}^\backprime W^i_\alpha h^\alpha(x^j; q^\beta)$,

$$(57.19) \qquad {}^\backprime q^\alpha = h^\alpha(x^j;\, q^\beta)\,, \quad \det(\partial h^\alpha/\partial q^\beta) \neq 0\,,$$

$$(57.20) \qquad {}^\backprime y^\alpha_i = \frac{\partial h^\alpha}{\partial q^\beta}\, y^\beta_i + \frac{\partial h^\alpha}{\partial x^i}\,,$$

$$(57.21) \qquad W^i_\alpha = \frac{\partial h^\beta}{\partial q^\beta}\, {}^\backprime W^i_\beta\,,$$

$$(57.22) \qquad W_\alpha = \frac{\partial h^\beta}{\partial q^\alpha}\, {}^\backprime W_\beta + \left(\frac{\partial^2 h^\beta}{\partial x^i \partial q^\alpha} + y_i\, \frac{\partial^2 h^\beta}{\partial q^\gamma \partial q^\alpha}\right) {}^\backprime W^i_\beta\,,$$

where the last equality is equivalent to the condition

$$(57.23) \qquad W \wedge \omega = {}^\backprime W \wedge \omega\,.$$

Such transformations are thus admissible maps form K *to* ${}^\backprime K$, ${}^\backprime K$ *is a Kinematic Space,* ρ *maps all solutions of the balanced system characterized by* W *onto solutions of the balanced system characterized by* ${}^\backprime W$*, and conversely.*

There is a clear and sharp contrast between the results obtained above for balanced systems with several independent variables and the results obtained in Chapter VIII for systems with only one independent variable. This contrast comes about because of the requirement that (57.12) hold for all values of the indices i and j so that $\gamma^i = -{}^\backprime W^i_\alpha h^\alpha(x^j;\, q^\beta)$. This demands that there be only linear dependence on the variables ${}^\backprime W^i_\alpha$ in the generating n-form η and (57.9) and (57.15) then combine to give $\eta=0$ for the actual E-transformation. This fact, in turn leads to (57.16) from which we obtain the conditions (57.19), (57.20) that show that ρ is an admissible map. It should also be clear at this point that the condition (57.12) imply satisfaction of the necessary integrability conditions whereby the y's and the $`y$'s become partial derivatives of the $\phi$'s and the $`\phi$'s under maps Φ and $`\Phi = \rho\circ\Phi$. Since this is an important point, it is useful to give the details. Consider the collections of 1-forms

$$(57.24)\qquad \gamma^\alpha = dq^\alpha - y_i^\alpha dx^i , \qquad `\gamma^\alpha = d`q^\alpha - `y_i^\alpha dx^i .$$

If we multiply the first of these by $(\partial h^\beta/\partial q^\alpha)$ and use (57.20), we obtain $\frac{\partial h^\beta}{\partial q^\alpha}\gamma^\alpha = \frac{\partial h^\beta}{\partial q^\alpha}(dq^\alpha - y_i^\alpha dx^i) = \frac{\partial h^\beta}{\partial q^\alpha}dq^\alpha - \left(`y_i^\beta - \frac{\partial h^\beta}{\partial x^i}\right)dx^i$ $= \frac{\partial h^\beta}{\partial q^\alpha}dq^\alpha + \frac{\partial h^\beta}{\partial x^i}dx^i + `\gamma^\beta - d`q^\beta = dh^\beta + `\gamma^\beta - d`q^\beta$; that is

$$(57.25)\qquad \frac{\partial h^\beta}{\partial q^\alpha}\gamma^\alpha = `\gamma^\beta + d(h^\beta - `q^\beta) .$$

Thus, satisfaction of (57.19) yields

$$(57.26)\qquad `\gamma^\beta = \gamma^\alpha\, \partial h^\beta/\partial q^\alpha ,$$

and $\det(\partial h^\beta/\partial q^\alpha)$ shows that $`\Phi = \rho\circ\Phi$ annihilates $`\gamma^\beta$ if and only if Φ annihilates γ^α.

A useful insight into the structure of the E-transformations obtained above can be gained from looking at the case in which the original system is variational. For simplicity, let us assume that $W = dL$. The relation (57.23) leads directly to the conclusion that the new system is likewise variational with $`W = d`L$. Thus, (57.23) gives

$$d(L - `L)\wedge\omega = 0$$

with the trivial first integral

$$(57.27) \qquad L = {}^{\backprime}L + f(x^i) .$$

It would thus appear that we obtain something quite different from the Hamilton-Jacobi equation in the present instance. This is appearance only. If we go back and substitute (57.10) into (57.7), we obtain

$$(57.28) \qquad dL \wedge \omega = d{}^{\backprime}L \wedge \omega + d\left(-{}^{\backprime}y_i^{\alpha}{}^{\backprime}W_{\alpha}^i + y_i^{\alpha}W_{\alpha}^i - {}^{\backprime}W_{\alpha}^i \frac{\partial h^{\alpha}}{\partial x^i}\right)$$

with the trivial first integral

$$(57.29) \qquad L - y_i^{\alpha}W_{\alpha}^i = {}^{\backprime}L - {}^{\backprime}y_i^{\alpha}{}^{\backprime}W_{\alpha}^i - {}^{\backprime}W_{\alpha}^i \frac{\partial h^{\alpha}}{\partial x^i} .$$

This now has the form of the Hamilton-Jacobi equation. Accordingly, when the original system is variational, we do indeed obtain an equation similar to the Hamilton-Jacobi equation; it is just that the additional side conditions are different from those that are encountered in the case of only one independent variable and yield the equivalent simple relation (57.27).

Another class of special E-transformations is generated by choosing $N^i = N^i(x^j;\, q^{\beta};\, {}^{\backprime}q^{\beta})$. Under these circumstances, (57.6) gives

$$d{}^{\backprime}q^{\alpha} \wedge {}^{\backprime}W_{\alpha}^i \pi_i - {}^{\backprime}y_i^{\alpha}{}^{\backprime}W_{\alpha}^i \omega = dq^{\alpha} \wedge W_{\alpha}^i \pi_i - y_i^{\alpha}W_{\alpha}^i \omega + \mu\omega$$
$$+ \frac{\partial N^i}{\partial x^i}\omega + \frac{\partial N^i}{\partial q^{\alpha}} dq^{\alpha} \wedge \pi_i + \frac{\partial N^i}{\partial {}^{\backprime}q^{\alpha}} d{}^{\backprime}q^{\alpha} \wedge \pi_i$$

and this can be satisfied if and only if

$$(57.30) \qquad \mu = -{}^{\backprime}y_i^{\alpha}{}^{\backprime}W_{\alpha}^i + y_i^{\alpha}W_{\alpha}^i - \frac{\partial N^i}{\partial x^i} ,$$

$$(57.31) \qquad W_{\alpha}^i = -\frac{\partial N^i}{\partial q^{\alpha}} , \qquad {}^{\backprime}W_{\alpha}^i = \frac{\partial N^i}{\partial {}^{\backprime}q^{\alpha}} .$$

Now that we have determined μ by (57.30), we substitute this into (57.7) to obtain

$$\mathcal{W}_\alpha dq^\alpha \wedge \omega + \mathcal{W}^i_\alpha dy^\alpha_i \wedge \omega = {}^\backprime\mathcal{W}_\alpha d{}^\backprime q^\alpha \wedge \omega + {}^\backprime\mathcal{W}^i_\alpha d{}^\backprime y^\alpha_i \wedge \omega$$
$$+ d\left(-{}^\backprime y^\alpha_i {}^\backprime\mathcal{W}^i_\alpha + y^\alpha_i \mathcal{W}^i_\alpha - \frac{\partial N^i}{\partial x^i}\right)\wedge\omega ,$$

and this reduces to

$$(\mathcal{W}_\alpha dq^\alpha - {}^\backprime\mathcal{W}_\alpha d{}^\backprime q^\alpha)\wedge\omega = \left(-{}^\backprime y^\alpha_i d{}^\backprime\mathcal{W}^i_\alpha - d\left(\frac{\partial N^i}{\partial x^i}\right) + y^\alpha_i d\mathcal{W}^i_\alpha\right)\wedge\omega .$$

A substitution from (57.31) for $\mathcal{W}^i_\alpha$ and ${}^\backprime\mathcal{W}^i_\alpha$ then shows that this relation can be satisfied if and only if

$$\mathcal{W}_\alpha = -{}^\backprime y^\beta_i \frac{\partial^2 N^i}{\partial{}^\backprime q^\beta \partial q^\alpha} - y^\beta_i \frac{\partial^2 N^i}{\partial q^\beta \partial q^\alpha} - \frac{\partial^2 N^i}{\partial x^i \partial q^\alpha} , \tag{57.32}$$

$${}^\backprime\mathcal{W}_\alpha = {}^\backprime y^\beta_i \frac{\partial^2 N^i}{\partial{}^\backprime q^\alpha \partial{}^\backprime q^\alpha} + y^\beta_i \frac{\partial^2 N^i}{\partial q^\beta \partial{}^\backprime q^\alpha} + \frac{\partial^2 N^i}{\partial x^i \partial{}^\backprime q^\alpha} . \tag{57.33}$$

We may summarize these results as follows.

An invertible map $\rho: K \to {}^\backprime K \mid {}^\backprime x^i = x^i$ *is a special* E-*transformation with generating* (n-1)-*form* $\eta = N^i(x^j; q^\beta; {}^\backprime q^\beta)\pi_i$ *only if the following relations hold:*

$$\mathcal{W}_\alpha = -{}^\backprime y^\beta_i \frac{\partial^2 N^i}{\partial{}^\backprime q^\beta \partial q^\alpha} - y^\beta_i \frac{\partial^2 N^i}{\partial q^\beta \partial q^\alpha} - \frac{\partial^2 N^i}{\partial x^i \partial q^\alpha} , \tag{57.34}$$

$$\mathcal{W}^i_\alpha = -\frac{\partial N^i}{\partial q^\alpha} , \tag{57.35}$$

$${}^\backprime\mathcal{W}_\alpha = {}^\backprime y^\beta_i \frac{\partial^2 N^i}{\partial{}^\backprime q^\beta \partial{}^\backprime q^\alpha} + y^\beta_i \frac{\partial^2 N^i}{\partial q^\beta \partial{}^\backprime q^\alpha} + \frac{\partial^2 N^i}{\partial x^i \partial{}^\backprime q^\alpha} , \tag{57.36}$$

$${}^\backprime\mathcal{W}^i_\alpha = \frac{\partial N^i}{\partial{}^\backprime q^\alpha} . \tag{57.37}$$

If these conditions are met, then

$$\begin{aligned}\mathcal{W}\wedge\omega &= {}^\backprime\mathcal{W}\wedge\omega + d\left(-{}^\backprime y^\alpha_i {}^\backprime\mathcal{W}^i_\alpha + y^\alpha_i \mathcal{W}^i_\alpha - \frac{\partial N^i}{\partial x^i}\right)\wedge\omega \\ &= {}^\backprime\mathcal{W}\wedge\omega + d\left(-{}^\backprime y^\alpha_i \frac{\partial N^i}{\partial{}^\backprime q^\alpha} - y^\alpha_i \frac{\partial N^i}{\partial q^\alpha} - \frac{\partial N^i}{\partial x^i}\right)\wedge\omega .\end{aligned} \tag{57.38}$$

A very useful insight into the structure of these relations can be gained from looking at the case in which the original system is variational. For simplicity, let us assume that $W = dL$ so that $W \wedge \omega = d(L\omega)$. (57.38) then shows that ${}^{\backprime}W \wedge \omega$ is closed and we conclude that the new system is also variational since $(W - d\mu) \wedge \omega$ is simple. We can therefore take ${}^{\backprime}W = d{}^{\backprime}L$ with no loss of generality. Under these circumstances, (57.38) yields

$$d\left(L - y_i^\alpha P_\alpha^i - {}^{\backprime}L + {}^{\backprime}y_i^\alpha {}^{\backprime}P_\alpha^i + \frac{\partial N^i}{\partial x^i}\right) \wedge \omega = 0 ,$$

where we have set $P_\alpha^i = \partial L / \partial y_i^\alpha$, ${}^{\backprime}P_\alpha^i = \partial {}^{\backprime}L / \partial {}^{\backprime}y_i^\alpha$. A trivial first integral of this equation is given by

$$(57.39) \qquad L - y_i^\alpha P_\alpha^i + \frac{\partial N^i}{\partial x^i} = {}^{\backprime}L - {}^{\backprime}y_i^\alpha {}^{\backprime}P_\alpha^i .$$

If we assume that the old and the new variational systems are regular in the sense that $\det(\partial P_\alpha^i / \partial y_j^\beta) \neq 0$, $\det(\partial {}^{\backprime}P_\alpha^i / \partial {}^{\backprime}y_j^\beta) \neq 0$, then we can solve $P_\alpha^i = \partial L / \partial y_i^\alpha$ and ${}^{\backprime}P_\alpha^i = \partial {}^{\backprime}L / \partial {}^{\backprime}y_i^\alpha$ so as to obtain $y_i^\alpha = Y_i^\alpha(x^j; q^\beta; P_\beta^j)$, ${}^{\backprime}y_i^\alpha = {}^{\backprime}Y_i^\alpha(x^j; q^\beta; {}^{\backprime}P_\beta^j)$, respectively. Accordingly, when (57.35) and (57.37) are used, (57.39) becomes

$$L\left(x^j; q^\beta; Y_j^\beta\left(x^k; q^\gamma; -\frac{\partial N^k}{\partial q^\gamma}\right)\right) - Y_i^\alpha\left(x^k; q^\gamma; -\frac{\partial N^k}{\partial q^\gamma}\right)\frac{\partial N^i}{\partial q^\alpha} + \frac{\partial N^i}{\partial x^i}$$

$$= {}^{\backprime}L\left(x^k; {}^{\backprime}q^\beta; {}^{\backprime}Y_j^\beta\left(x^k; {}^{\backprime}q^\gamma; \frac{\partial N^i}{\partial {}^{\backprime}q^\alpha}\right)\right) - {}^{\backprime}Y_i^\alpha\left(x^k; {}^{\backprime}q^\gamma; \frac{\partial N^k}{\partial {}^{\backprime}q^\alpha}\right)\frac{\partial N^i}{\partial {}^{\backprime}q^\alpha} ,$$

and this is just the Hamilton-Jacobi equation in Lagrangian form. We thus conclude that *the Hamilton-Jacobi equation is a first integral of the system* (57.34)-(57.38) *when the original system is variational, in which case the new system is also variational.*

In contrast to the previous case, we do not know that E-transformations with generating functions $N^i = N^i(x^j; q^\beta; {}^{\backprime}q^\beta)$ map K to another Kinematic Space ${}^{\backprime}K$. The basic theorem established at the beginning of this section can not be used to conclude that such E-transformations map solutions of the balanced system

$\mathcal{W}$ onto solutions of the image balanced system $`\mathcal{W}$, and conversely. Further, we have not actually obtained the mapping equations, as is easily seen by an examination of the equations (57.34) through (57.38). Thus, we are still one step shy of a complete characterization of E-transformations with generating functions $N^i = N^i(x^j; q^\beta; `q^\beta)$. Both problems will now be solved together.

We start by considering the 1-forms

$$\gamma^\alpha = dq^\alpha - y_i^\alpha dx^i , \quad `\gamma^\alpha = d`q^\alpha - `y_i^\alpha dx^i . \tag{57.40}$$

Suppose that we can find a collection of functions $k^\alpha = k^\alpha(x^j; q^\beta; `q^\beta)$ such that

$$\det(\partial k^\alpha / \partial `q^\beta) = 0 . \tag{57.41}$$

It then follows from (57.40) that

$$\frac{\partial k^\alpha}{\partial `q^\beta} `\gamma^\beta + \frac{\partial k^\alpha}{\partial q^\beta} \gamma^\beta = \frac{\partial k^\alpha}{\partial `q^\beta} d`q^\beta + \frac{\partial k^\alpha}{\partial q^\beta} dq^\beta - \left(\frac{\partial k^\alpha}{\partial `q^\beta} `y_i^\beta + \frac{\partial k^\alpha}{\partial q^\beta} y_i^\beta \right) dx^i . \tag{57.42}$$

Hence, if we posit the relations

$$\frac{\partial k^\alpha}{\partial `q^\beta} `y_i^\beta + \frac{\partial k^\alpha}{\partial q^\beta} y_i^\beta + \frac{\partial k^\alpha}{\partial x^i} = 0 \tag{57.43}$$

between the $`y$'s and the y's , (57.42) yields

$$\frac{\partial k^\alpha}{\partial `q^\beta} `\gamma^\beta + \frac{\partial k^\alpha}{\partial q^\beta} \gamma^\beta = dk^\alpha(x^j; q^\beta; `q^\beta) . \tag{57.44}$$

In view of the condition (57.41), it follows that satisfaction of the independent conditions (57.43) and

$$k^\alpha(x^j; q^\beta; `q^\beta) = c^\alpha , \tag{57.45}$$

for constants c^α , is sufficient in order to obtain explicit transformation equations and to guarantee that the resulting

E-transformation maps K to another Kinematic Space $\grave{K}$. This latter result follows directly from (57.44) and (57.45) since these relations imply that any map $\Phi: B_n \to K$ that annihilates the 1-forms γ^α yields a map $\grave{\Phi} = \rho\circ\Phi$ that annihilates $\grave{\gamma}^\alpha$. We have thus established the following result.

A special E-*transformation* ρ *with generating functions* $N^i = N^i(x^j;\, q^\beta;\, \grave{q}^\beta)$ *maps* K *to another Kinematic Space* $\grave{K}$ *provided there exist functions* $k^\alpha = k^\alpha(x^j;\, q^\beta;\, \grave{q}^\beta)$ *such that*

$$(57.46)\qquad \det(\partial k^\alpha/\partial \grave{q}^\beta) \neq 0\ ,$$

$$(57.47)\qquad \frac{\partial k^\alpha}{\partial \grave{q}^\beta}\,\grave{y}_i^\beta + \frac{\partial k^\alpha}{\partial q^\beta}\,y_i^\beta + \frac{\partial k^\alpha}{\partial x^i} = 0\ .$$

If these conditions are met, then the equations that define the E-*transformation* ρ *are given by* (57.47) *and*

$$(57.48)\qquad k^\alpha(x^j;\, q^\beta;\, \grave{q}^\beta) = c^\alpha$$

for constants c^α.

It is evident from the above that we obtain an E-transformation ρ with these properties for each choice of the constants c^α. Further, if we happen to have $k^\alpha = \grave{q}^\alpha - h^\alpha(x^j;\, q^\beta)$, then (57.47) yields $\grave{y}_i^\alpha = \frac{\partial h^\alpha}{\partial q^\beta}\,y_i^\beta + \frac{\partial h^\alpha}{\partial x^i}$ and (57.48) gives $\grave{q}^\alpha = h^\alpha(x^j;\, q^\beta) + c^\alpha$. These relations agree with the relations (57.19)-(57.20) of the previous case. Accordingly, we may conclude that the present case contains the previous one as a special case in which a separation of the variables $\grave{q}^\alpha$ and the variables q^α is achieved.

A further insight can be gained by considering the system of equations (57.34)-(57.38) and (57.46)-(57.48) as a system of equations on the *Union Space* $\tilde{K} = K \cup \grave{K}$ with coordinate functions $(x^i;\, q^\alpha,\, \grave{q}^\alpha;\, y_i^\alpha,\, \grave{y}_i^\alpha)$. It is clear from the structure of $\tilde{K}$ that the operators on functions on $\tilde{K}$ that play the same role as the operators Z_i on functions on K are given by

$$(57.49) \qquad \tilde{Z}_i = {}^{\backprime}y^\alpha_i \, \partial/\partial{}^{\backprime}q^\alpha + y^\alpha_i \, \partial/\partial q^\alpha + \partial/\partial x^i \; .$$

Use of this operator reduces (57.34), (57.36) and (57.47) to

$$(57.50) \qquad \mathcal{W}_\alpha = - \tilde{Z}_i(\partial N^i/\partial q^\alpha) \; ,$$

$$(57.51) \qquad {}^{\backprime}\mathcal{W}_\alpha = \tilde{Z}_i(\partial N^i/\partial{}^{\backprime}q^\alpha) \; ,$$

and

$$(57.52) \qquad \tilde{Z}_i k^\alpha = 0 \; ,$$

respectively. Accordingly, when (57.50) and (57.51) are substituted into (57.38), we obtain

$$(57.53) \qquad ({}^{\backprime}\mathcal{W}-\mathcal{W}) \wedge \omega = \left\{ \tilde{Z}_i\left(\frac{\partial N^i}{\partial q^\alpha}\right)dq^\alpha + \tilde{Z}_i\left(\frac{\partial N^i}{\partial{}^{\backprime}q^\alpha}\right)d{}^{\backprime}q^\alpha + \frac{\partial N^i}{\partial q^\alpha}\, dy^\alpha_i + \frac{\partial N^i}{\partial{}^{\backprime}q^\alpha}\, d{}^{\backprime}y^\alpha_i \right\} \wedge \omega \; .$$

Conversly, (57.53) implies satisfaction of (57.50), (57.51) and (57.35), (57.37). Further, the (n+1)-form on the right-hand side of (57.53) is the exterior product of the exact element $d({}^{\backprime}y^\alpha_i \, \partial N^i/\partial{}^{\backprime}q^\alpha + y^\alpha_i \, \partial N^i/\partial q^\alpha + \partial N^i/\partial x^i) = d\tilde{Z}_i N^i$ of the identity class $I(K)$ with the n-form ω, as follows from the fact that $N^i = N^i(x^j;\, q^\beta;\, {}^{\backprime}q^\beta)$ is defined on the *Union Graph Space* $\tilde{G}=G\cup{}^{\backprime}G$ with coordinate functions $(x^j;\, q^\alpha;\, {}^{\backprime}q^\alpha)$. The entire content of the above results can thus be restated as follows.

An invertible map $\rho: K \to {}^{\backprime}K | {}^{\backprime}x^i = x^i$ *is a special* E-*transformation with generating functions* $N^i = N^i(x^j;\, q^\alpha;\, {}^{\backprime}q^\alpha)$ *if and only if*

$$(57.54) \qquad ({}^{\backprime}\mathcal{W}-\mathcal{W}) \wedge \omega = d(\tilde{Z}_i N^i) \wedge \omega$$

holds on the Union Space $\tilde{K} = K\cup{}^{\backprime}K$. *If there exist functions* $k^\alpha = k^\alpha(x^j;\, q^\beta;\, {}^{\backprime}q^\beta)$ *defined on the Union Graph Space* $\tilde{G} = G\cup{}^{\backprime}G$

such that

$$(57.55)\qquad \tilde{Z}_i k^\alpha = 0\ , \quad \det(\partial k^\alpha/\partial\grave{}q^\beta) \neq 0\ ,$$

then ρ *maps* K *to another Kinematic Space* $\grave{}K$ *and the equations that define* ρ *are given by* (57.55) *and* $k^\alpha(x^j;\, q^\beta;\, \grave{}q^\beta) = c^\alpha$.

58. F-TRANSFORMATIONS

The results established at the end of the last section suggest an immediate generalization of special E-transformations; namely, replace $\grave{}W - W$ by a B-equivalent 1-form on the Union Space $\tilde{K} = K \cup \grave{}K$ and replace the exact element $d\tilde{Z}_i N^i$ of $I(\tilde{K})$ by an arbitrary element of $I(\tilde{K})$. However, it was shown in Section 54 that B-equivalence induces a map of $I(\tilde{K})$ to $I(\tilde{K})$. Accordingly, it is sufficient to replace $dZ_i N^i$ by an arbitrary element $I(\tilde{K})$. If $\tilde{I}$ denotes an arbitrary element of $I(K)$, we know from the previous results concerning the structure of such elements that

$$(58.1)\qquad \tilde{I}\wedge\omega = I_1\wedge\omega + I_2\wedge\omega$$

where

$$(58.2)\qquad \begin{aligned} I_1\wedge\omega &= \{\tilde{Z}_i I^i_{1\alpha} dq^\alpha + I^i_{1\alpha} dy^\alpha_i\}\wedge\omega \\ I_2\wedge\omega &= \{\tilde{Z}_i I^i_{2\alpha} d\grave{}q^\alpha + I^i_{2\alpha} d\grave{}y^\alpha_i\}\wedge\omega \end{aligned}$$

and the functions $(I^i_{1\alpha},\, I^i_{2\alpha})\in\Lambda^0(K)$ satisfy the relations

$$(58.3)\qquad \begin{aligned} \frac{\partial I^i_{1\alpha}}{\partial y^\beta_j} &= -\frac{\partial I^j_{1\alpha}}{\partial y^\beta_i}\ , & \frac{\partial I^i_{1\alpha}}{\partial\grave{}y^\beta_j} &= -\frac{\partial I^i_{1\alpha}}{\partial\grave{}y^\beta_i}\ , \\ \frac{\partial I^i_{2\alpha}}{\partial y^\beta_j} &= -\frac{\partial I^j_{2\alpha}}{\partial y^\beta_i}\ , & \frac{\partial I^i_{2\alpha}}{\partial\grave{}y^\beta_j} &= -\frac{\partial I^j_{2\alpha}}{\partial\grave{}y^\beta_i}\ . \end{aligned}$$

An invertible map $\rho\colon K \to \grave{}K \mid \grave{}x^i = x^i$ is said to be an F-*transformation* (field transformation) from an initial 1-form W

to a target 1-form $`W$ if and only if there exists an element $I = I_1+I_2$ of $I(\tilde{K})$ such that

$$(58.4) \qquad \rho^{-1*}F(W-I_1) = F(W+I_2) .$$

Since I_1 is a 1-form on K with coefficients in $\Lambda^0(\tilde{K})$ and I_2 is a 1-form on $`K$ with coefficients in $\Lambda^0(\tilde{K})$, as shown by (58.2), all of the symbols occurring in (58.4) are well defined. Clearly, F-transformations reduce to E-transformations for $\tilde{I}=0$. For $I \neq 0$, $(\rho_2 \circ \rho_1)^{-1*}F(W-I_1) = \rho_2^{-1*} \circ \rho_1^{-1} F(W-I_1) = \rho_2^{-1*}F(`W+I_2) = \rho_2^{-1*}F(`W) + \rho_2^{-1*}F(I_2) = F(``W) + \rho_2^{-1*}F(I_2) \neq F(``W+I_2)$. We thus conclude that F-*transformations do not form a group under composition unless* $\tilde{I}=0$. This result is not surprising since an F-transformation maps a given initial 1-form onto a target 1-form; that is, an F-transformation is explicitly related to the initial and target 1-forms and the $\tilde{I} \in I(\tilde{K})$ that achieves this relation. In other words we do not map from the class of 1-forms onto the class of 1-forms, as is the case with E-transformations.

F-transformations do have an intrinsic utility even though they do not possess the group property under composition. In order to see this, we suppose that the F-transformation ρ from the initial 1-form W to the target 1-form $`W$ is such that $K' = \rho(K)$ is another Kinematic Space, in which case $\tilde{K} = K \cup `K$ is a Union Kinematic Space. Under this assumption ρ_* maps any horizontal vector field U on K onto a horizontal vector field $`U = \rho_* U$ on $`K$ since $`x = x$. Accordingly, since $I_1$ and $I_2$ are 1-forms on $K$ and $`K$ respectively with coefficients in $\Lambda^0(\tilde{K})$, we obtain $`\Phi^{*}`U \lrcorner F(`W+I_2) = \Phi^* U \lrcorner F(W-I_1)$ for $`\Phi = \rho \circ \Phi$. Since $F(\ )$ is linear, we thus obtain $`\Phi^{*}`U \lrcorner F(`W) + `\Phi^{*}`U \lrcorner F(I_2) + \Phi^* U \lrcorner F(I_1) = \Phi^* U \lrcorner F(W)$. However, $`U+U = \tilde{U}$ is a horizontal vector field on the Union Kinematic Space $\tilde{K}$ and we can define a map $\tilde{\Phi}: B_n \to \tilde{K}$ as the union of the maps Φ and $`\Phi$. This then gives

$$`\Phi^{*}`U \lrcorner F(`W) + \tilde{\Phi}^* \tilde{U} \lrcorner F(\tilde{I}) = \Phi^* U \lrcorner F(W)$$

because $\tilde{I} = I_1+I_2$ and F in linear. Thus, since $\tilde{I}\varepsilon I(\tilde{K})$, $\tilde{\Phi}^*\tilde{U}\lrcorner F(\tilde{I}) = 0$ for all $\tilde{\Phi}$ and all horizontal $\tilde{U}$, we obtain

$$`\Phi^*`U\lrcorner F(`W) = \Phi^*U\lrcorner F(W) .$$

The same argument as that used in the previous section establishes the following result.

If an F-*transformation* ρ, *from an initial* 1-*form* W *to a target* 1-*form* $`W$, *maps Kinematic Space* $K$ *to another Kinematic Space* $`K$, *then* ρ *maps all solutions of the initial balanced system characterized by* W *onto solutions of the target balanced system characterized by* $`W$, *and conversely.*

From now on, we shall write equations like (58.4) in the equivalent form $F(W-I_1) = F(`W+I_2)$, where it is understood that the equations of transformation are used to write both sides of the equality in terms of the same system of arguments. It should now be clear that all of the results of the previous section can be taken over directly by the correspondence $W\rightarrowtail W-I_1$, $`W\rightarrowtail `W+I_2$. This gives the following results.

An invertible map $\rho: K \to `K|`x^i = x^i$ *is an* F-*transformation from an initial* 1-*form* W *to a target* 1-*form* $`W$ *if and only if there exists an* $\tilde{I} = I_1+I_2\varepsilon I(\tilde{K})$, *a remainder function* $\mu\varepsilon\Lambda^0(\tilde{K})$, *and a generating form* $\eta\varepsilon\Lambda^{n-1}(\tilde{K})$ *such that*

$$J(`W+I_2) = J(W-I_1) + \mu\omega + d\eta , \tag{58.5}$$

$$(W-I_1)\wedge\omega = (`W+I_2)\wedge\omega + d(\mu\omega) , \tag{58.6}$$

where it is assumed that the equations of transformation are used to express both sides of the above equalities in terms of the same arguments.

An F-transformation is said to be *special* if the generating form η is given by

$$\eta = N^i\pi_i , \quad N^i\varepsilon\Lambda^0(\tilde{K}) , \quad i=1,\ldots,n , \tag{58.7}$$

in which case the functions N^i are referred to as *generating functions* of the F-transformation.

An invertible map $\rho\colon K \to {}^{\backprime}K | {}^{\backprime}x^i = x^i$ *is a special* F-*transformation with generating form* $\eta = ({}^{\backprime}W^i_\alpha + I^i_{2\alpha}){}^{\backprime}q^\alpha + \gamma^i(x^j;\, q^\beta;\, {}^{\backprime}W^j_\beta + I^j_{2\beta})$ *from the initial* W *to the target* ${}^{\backprime}W$ *if and only if* $\gamma^i = -({}^{\backprime}W^i_\alpha + I^i_{2\alpha})h^\alpha(x^j;\, q^\beta)$,

$$(58.8) \qquad {}^{\backprime}q^\alpha = h^\alpha(x^j;\, q^\beta) \;, \qquad \det(\partial h^\alpha / \partial q^\beta) \neq 0 \;,$$

$$(58.9) \qquad {}^{\backprime}y^\alpha_i = \frac{\partial h^\alpha}{\partial q^\beta} y^\beta_i + \frac{\partial h^\alpha}{\partial x^i} \;,$$

$$(58.10) \qquad W^i_\alpha - I^i_{1\alpha} = \frac{\partial h^\beta}{\partial q^\alpha}({}^{\backprime}W^i_\beta + I^i_{2\beta})$$

$$(58.11) \qquad W_\alpha - \tilde{Z}_i I^i_{1\alpha} = \frac{\partial h^\beta}{\partial q^\alpha}({}^{\backprime}W_\beta + \tilde{Z}_i I^i_{2\beta}) + \left(\frac{\partial^2 h^\beta}{\partial x^i \partial q^\alpha} + y^\gamma_i \frac{\partial^2 h^\beta}{\partial q^\gamma \partial q^\alpha}\right)({}^{\backprime}W^i_\beta + I^i_{2\beta}) \;,$$

where the last equality is equivalent to the condition

$$(58.12) \qquad ({}^{\backprime}W + I - W) \wedge \omega = 0 \;.$$

Such transformations are thus admissible maps from K *to* ${}^{\backprime}K$, ${}^{\backprime}K$ *is also a Kinematic Space,* ρ *maps all solutions of the balanced system characterized by* W *onto solutions of the balanced system characterized by* ${}^{\backprime}W$, *and conversely. If* W *is variational, then* ${}^{\backprime}W$ *is variational if and only if* I *is variational.*

An invertible map $\rho\colon K \to {}^{\backprime}K | {}^{\backprime}x^i = x^i$ *is a special* F-*transformation with generating* (n-1)-*form* $\eta = N^i(x^j;\, q^\beta;\, {}^{\backprime}q^\beta)\pi_i$ *from the initial* W *to the terminal* ${}^{\backprime}W$ *if and only if there exists an element* $\tilde{I} = I_1 + I_2$ *of* $I(\tilde{K})$ *such that*

$$(58.13) \qquad W_\alpha - \tilde{Z}_i I^i_{1\alpha} = -\tilde{Z}_i \frac{\partial N^i}{\partial q^\alpha} \;,$$

$$(58.14) \qquad W^i_\alpha - I^i_{1\alpha} = -\frac{\partial N^i}{\partial q^\alpha} ,$$

$$(58.15) \qquad {}^{\backprime}W_\alpha + \tilde{Z}_i I^i_{2\alpha} = \tilde{Z}_i \frac{\partial N^i}{\partial {}^{\backprime}q^\alpha} ,$$

$$(58.16) \qquad {}^{\backprime}W^i_\alpha + I^i_{2\alpha} = \frac{\partial N^i}{\partial {}^{\backprime}q^\alpha} .$$

If these conditions are met, then

$$(58.17) \qquad (W-I_1) \wedge \omega = ({}^{\backprime}W+I_2) \wedge \omega - d(\tilde{Z}_i N^i) .$$

This F-*transformation maps* K *to another Kinematic Space* ${}^{\backprime}K$ *if there exists* N *functions* $k^\alpha \varepsilon \Lambda^0(\tilde{G})$ *such that*

$$(58.18) \qquad \det(\partial k^\alpha / \partial {}^{\backprime}q^\beta) \neq 0 ,$$

$$(58.19) \qquad \tilde{Z}_i k^\alpha = 0 .$$

If these additional conditions are met, then the equations defining the F-*transformation are* (58.19) *and*

$$(58.20) \qquad k^\alpha(x^j;\ q^\beta;\ {}^{\backprime}q^\beta) = c^\alpha ,$$

ρ *maps all solutions of the initial balanced system characterized by* W *onto solutions of the terminal balanced system characterized by* ${}^{\backprime}W$, *and conversely.*

An interesting example of F-transformations arises in the study of the nonlinear 1-dimensional wave equation $D_1 D_2 \phi(x^1,x^2) = dF(\phi)/d\phi$. We have already seen that this balanced system can be characterized by the exact 1-form

$$(58.21) \qquad W = d(F(q) + \tfrac{1}{2} y_1 y_2) .$$

If we take W as the initial form and

$$(58.22) \qquad {}^{\backprime}W = d(G({}^{\backprime}q) + \tfrac{1}{2}\, {}^{\backprime}y_1\, {}^{\backprime}t_2)$$

as the terminal form, so that we will have $D_1D_2{}^{`}\phi(x^1,x^2) = dG({}^{`}\phi)/d{}^{`}\phi$, then (58.17) shows that the element $\tilde{I}$ of $I(\tilde{K})$ must also be exact. We therefore consider the case in which

$$(58.23) \qquad \tilde{I} = \frac{1}{2}\, d\{a(q,{}^{`}q)(y_1{}^{`}y_2 - y_2{}^{`}y_1)\} \ .$$

Under these circumstances, (58.13) and (58.15) give

$$(58.24) \qquad \begin{aligned} \omega^1 - I_1^1 &= \frac{1}{2}\,(y_2 - a\,{}^{`}y_2) = -\,\partial N^1/\partial q \ , \\ {}^{`}\omega^1 + I_2^1 &= \frac{1}{2}\,({}^{`}y_2 - a y_2) = \partial N^1/\partial{}^{`}q \ ; \end{aligned}$$

$$(58.25) \qquad \begin{aligned} \omega^2 - I_1^2 &= \frac{1}{2}\,(y_1 + a\,{}^{`}y_1) = -\,\partial N^2/\partial q \ , \\ {}^{`}\omega^2 + I_2^2 &= \frac{1}{2}\,({}^{`}y_1 + a y_1) = \partial N^2/\partial{}^{`}q \ . \end{aligned}$$

We now restrict attention to those systems for which the ranks of the systems (58.24) and (58.25) are minimal. In both systems, this is achieved by the choice $a=1$, in which case consistency of the systems demands that $\partial N^1/\partial{}^{`}q = \partial N^1/\partial q$ , $\partial N^2/\partial{}^{`}q = -\partial N^2/\partial q$. Accordingly, we obtain

$$(58.26) \qquad y_1 + {}^{`}y_1 = -\,2\,s'(q - {}^{`}q) \ , \quad y_2 - {}^{`}y_2 = -\,2\,r'(q + {}^{`}q)$$

with $s' = ds(\xi)/d\xi|_{\xi = q - {}^{`}q}$ , $r' = dr(\xi)/d\xi|_{\xi = q + {}^{`}q}$,

$$(58.27) \qquad N^1 = r(q + {}^{`}q) \ , \quad N^2 = s(q - {}^{`}q) \ ,$$

where the functions r and s are to be determined. Under these conditions, (58.13), (58.15), combined in the form given by (58.17), yield $d\{F(q) - G({}^{`}q) - 2r'(q+{}^{`}q)\, s'(q-{}^{`}q)\} = 0$, so that a direct integration gives us the condition

$$(58.28) \qquad F(q) - G({}^{`}q) = 2\, r'(q + {}^{`}q)\, s'(q - {}^{`}q) \ .$$

If we set $u = q + {}^{`}q$ , $v = q - {}^{`}q$,

$$F\left(\frac{u+v}{2}\right) = f(u+v) - b_1 , \quad G\left(\frac{u-v}{2}\right) = - g(u-v) + b_2$$

(58.29)

$$r'(u) = mh(u) , \qquad s'(v) = \frac{2}{m}\bar{k}(v) ,$$

where b_1, b_2, and m are constants with $m \neq 0$, then (58.28) takes the form of the general d'Alembert equation

$$(58.30) \qquad f(u+v) + g(u-v) = h(u)\,\bar{k}(v) .$$

The nontrivial general solutions of this functional equation are given by†

(58.31)a
$$f(t) = c_1\cos bt + C_1\sin bt + \gamma_1 , \quad g(t) = c_2\cos bt + C_2\sin bt + \gamma_2 ,$$
$$h(t) = c\cos bt + C\sin bt , \qquad \bar{k}(t) = \gamma\cos bt + \Gamma\sin bt ,$$
$$\gamma_2 = -\gamma_1 , \quad 2c_1 = c\gamma - C\Gamma , \quad 2C_1 = C\gamma + c\Gamma ,$$
$$2c_2 = c\gamma + C\Gamma , \quad 2C_2 = C\gamma - c\Gamma ;$$

(58.31)b
$$f(t) = c_1\cosh bt + C_1\sinh bt + \gamma_1 , \quad g(t) = c_2\cosh bt + C_2\sinh bt + \gamma_2 ,$$
$$h(t) = c\cosh bt + C\sinh bt , \qquad \bar{k}(t) = \gamma\cosh bt + \Gamma\sinh bt ,$$
$$\gamma_2 = -\gamma_1 , \quad 2c_1 = c\gamma + C\Gamma , \quad 2C_1 = C\gamma + c\Gamma ,$$
$$2c_2 = c\gamma - C\Gamma , \quad 2C_2 = C\gamma - c\Gamma ;$$

(58.31)c
$$f(t) = c_1 t^2 + C_1 t + \gamma_1 , \quad g(t) = c_2 t^2 + C_2 t + \gamma_2 ,$$
$$h(t) = c + Ct , \qquad \bar{k}(t) = \gamma + \Gamma t ,$$
$$\gamma_1 + \gamma_2 = c\gamma , \quad 2C_1 = C\gamma + c\Gamma , \quad 2C_2 = C\gamma - c\Gamma ,$$
$$c_1 = \frac{C\Gamma}{4} = -c_2 .$$

from which we can determine $F(q)$, $G(\check{q})$, N^1 and N^2 be using (58.29). A combination of (58.26) with the above results shows

---

† J. Aczél: <u>Vorlesungen Über Funktionalgleichungen Und Ihre Anwendungen</u> (Birkhauser Verlag, Basel, 1961), p. 130.

that (58.31)a yields the known auto-Bäcklund transformations of the generalized sine-Gordon equation, (58.31)b yields the auto-Bäcklund transformations of the generalized hyperbolic sine-Gordon equation, while (58.31)c yields the auto-Bäcklund transformations of the generalized Klein-Gordon equation. It would thus appear that the theory of F-transformations may provide a general approach to the study of Bäcklund for partial differential equations. In this context, a sufficient condition for the existence of Bäcklund transformations would appear to be minimality of the generalized rank of the system (58.14), (58.16) and that $W - {}^{\backprime}W$ is closed 1-form on $\tilde{K}$.

Turning to the question of the realization of these transformations, the previous theorem shows that we must satisfy (58.19) for some function $k(q, {}^{\backprime}q)$; that is $\tilde{Z}_i k = 0$, $i=1,2$. With such a function $k(q, {}^{\backprime}q)$, the transformation is realized by $k(\phi, {}^{\backprime}\phi) =$ constant, and maps solutions of the initial system onto solutions of the terminal system. In the present instance, since the equations for the determination of the y's is of minimal rank, the relation (58.26) and $\tilde{Z}_i k = 0$ combine to give

$$
\begin{aligned}
{}^{\backprime}y_1 &= 2s'(q - {}^{\backprime}q)\frac{\partial k}{\partial q}\bigg/\left(\frac{\partial k}{\partial {}^{\backprime}q} - \frac{\partial k}{\partial q}\right) , \\
{}^{\backprime}y_2 &= 2r'(q + {}^{\backprime}q)\frac{\partial k}{\partial q}\bigg/\left(\frac{\partial k}{\partial {}^{\backprime}q} + \frac{\partial k}{\partial q}\right) , \\
y_1 &= -\ 2s'(q - {}^{\backprime}q)\frac{\partial k}{\partial {}^{\backprime}q}\bigg/\left(\frac{\partial k}{\partial {}^{\backprime}q} - \frac{\partial k}{\partial q}\right) , \\
y_2 &= -\ 2r'(q - {}^{\backprime}q)\frac{\partial k}{\partial {}^{\backprime}q}\bigg/\left(\frac{\partial k}{\partial {}^{\backprime}q} + \frac{\partial k}{\partial q}\right) .
\end{aligned}
\tag{58.32}
$$

Of course, these give identical satisfaction of the conditions $0 = \tilde{Z}_i k = y_k \partial k/\partial q + {}^{\backprime}y_i \partial k/\partial {}^{\backprime}q$. In this context, it is of interest to note that $\tilde{Z}_i k = 0$ implies $D_2 D_1 k(\phi, {}^{\backprime}\phi) = F'(\phi)\partial k/\partial\phi + G'({}^{\backprime}\phi)\partial k/\partial {}^{\backprime}\phi$ on using the balance equations $D_1 D_2 \phi = F'(\phi)$, $D_1 D_2 {}^{\backprime}\phi = G'({}^{\backprime}\phi)$.

APPENDIX

A SHORT REVIEW OF THE EXTERIOR CALCULUS

A1. PRELIMINARY CONSIDERATIONS, POINTS AND COORDINATE COVERS

The purpose of this Appendix is to give a brief account of the calculus of exterior forms for the benefit of those readers who are not overly familiar with this discipline. In general, we will simply state the relevant results and refer the reader to the standard treatment of this subject as given in any one of the references listed in the first part of the References. There are several instances where we deviate from the standard presentation of the subject. In these cases, the presentation will include the proofs as well as the statements.

Let U be an open set of an n-dimensional separable Hausdorff space S and let P be a generic point in U . The basic assumption is that U can be placed into a one-to-one differentiable correspondence with an open subset of Euclidean n-dimensional space. This correspondence then induces a coordinate cover of U so that any point P in U is labeled by n coordinate functions $x^1(P),\ldots,x^n(P)$. This is conveniently written $P:(x^1,\ldots,x^n)$. For the study of mechanics, the assumption of an inertial frame of reference says that the underlying space S is itself a Euclidean space. Thus, in applications to mechanics, we take the open set U to be the whole space and the coordinate functions $x^1,\ldots,x^n$ as the coordinates of this Euclidean space. In other words, we assume that the underlying space S is covered with a single global coordinate cover, the Cartesian coordinate cover. We will not use this fact in the remainder of this Appendix. The reason for noting it at this point is that we can dispense with the tedious problem of constructing the transformation functions that relate the coordinate functions of different intersecting open sets U_1, U_2, ... This is, of course, required for the study of a general n-dimensional differentiable manifold, of which Euclidean n-dimensional space is a special case. The results given below will still hold for a general differentiable manifold however, for they hold for each member of its atlas of open sets that constitutes an open cover of S, together with the coordinate functions of each such set. We therefore take it as understood that any point P of S is labeled by its coordinate values $(x^1(P),\ldots, x^n(P))$ of its coordinate functions obtained from the set U that contains P.

A2. VECTOR FIELDS

Let J be an open interval of the real line. The map

$$\sigma\colon J \to S \mid x^\alpha = \sigma^\alpha(t),\ t\varepsilon J \quad \alpha=1,\ldots,n\ , \tag{A2.1}$$

of J into S defines a smooth (C^∞) curve in S if all of the n functions $\{\sigma^\alpha(t)\}$ are smooth (C^∞) on J. Such a curve provides the information that is required in order to compute the tangent vector to the curve at any given point on that curve: simply apply the operation d/dt to each of the functions $\sigma^\alpha(t)$ at a given value of t, say t_1 so as to obtain the n quantities $\{d\sigma^\alpha(t)/dt|_{t=t_1}\}$. This gives an ordered set of n quantities that is somewhat cumbersome to work with. It is therefore preferable to proceed along different lines so that one works with scalar quantities (i.e., quantities that have a 1-dimensional range).

Let $f(x^1,\ldots,x^n) = f(x^\alpha)$ be a smooth function whose domain contains the curve defined by the map σ and whose range is some subset of the real line $\mathbb{R}$. Composition of f with σ gives

(A2.2) $$F(t) = (f\circ\sigma)(t) = f(\sigma^\alpha(t)) .$$

Now, the chain rule of the calculus shows that

(A2.3) $$dF(t)/dt = \sum_{\alpha=1}^{n} \frac{d\sigma^\alpha(t)}{dt} \frac{\partial f}{\partial x^\alpha}\bigg|_{x^\alpha=\sigma^\alpha(t)} = \frac{d\sigma^\alpha(t)}{dt} \frac{\partial f}{\partial x^\alpha}\bigg|_{x=\sigma} ,$$

where the Einstein summation convention is used in writing the second equality. This convention is used throughout. It easily follows from this that we can recover the n-tuple of quantities $d\sigma^\alpha(t)/dt$ from (A2.3) provided we take a sufficiently large collection of functions f and evaluate the corresponding $dF(t)/dt$; simply take $f^j(x^\alpha) = x^j$ to obtain $dF^j(t)/dt = d\sigma^j(t)/dt$. This allows us to replace the notion of a "tangent vector" by a scalar valued operator whose domain consists of all scalar-valued functions on S . Accordingly, we represent a tangent vector at the point x by the operator

(A2.4) $$V = v^\alpha \, \partial/\partial x^\alpha \Big|_x .$$

Let σ_j denote the curve given by $\sigma^\alpha_j(t) = x^\alpha + \delta^\alpha_j t$, ($\delta^\alpha_j = 0$ for $\alpha \neq j$, $\delta^\alpha_j = 1$ for $\alpha = j$). In this event, (A2.3) gives $dF(t)/dt = \partial f/\partial x^j \Big|_{x=\sigma^j}$, and hence $\partial/\partial x^j \Big|_{x=\sigma^j}$, $j=1,\ldots,n$ are linearly independent. Accordingly, (A2.3) shows that $T_x(S)$, the set of all tangent vectors to all curves through a point $\{x^\alpha\}$ in S , is an n-dimensional real vector space with the basis $\partial/\partial x^j$, $j=1,\ldots,n$, called the *tangent space* at the point P with coordinates $\{x^\alpha\}$. This basis is referred to as the *natural basis* for $T_x(S)$, and legitimizes representation of tangent vectors by scalar-valued differential operators of the form (A2.4).

A *vector field* $V(x)$ is a smooth assignment $x \to V(x)\varepsilon T_x(S)$ that we write as a scalar valued operator

(A2.5) $$V(x) = v^\alpha(x^\beta) \, \partial/\partial x^\alpha .$$

The n functions $v^{\alpha}(x)$ are referred to as the *components* of $V(x)$ with respect to the natural basis $\partial/\partial x^{\alpha}$. It is understood in writing (A2.5) that explicit evaluation of $V(x)$ is obtained by allowing V to operate on a sufficiently large class of scalar-valued functions $f(x^{\alpha})$. This, however, is a reasonably small price to pay for replacing n-tuples of ordered functions of position, namely $\{v^{\alpha}(x^{\beta})\}$, by a scalar-valued quantity that can replicate the $v^{\alpha}(x^{\beta})$'s on demand.

Let $V(x)$ be a vector field. An *orbit* of $V(x)$ through the point $\{x_0^{\alpha}\}$ is a smooth curve $\sigma:[-|a|\le t\le |b|] \to S|x^{\alpha} = \sigma^{\alpha}(t)$,

$$\text{(A2.6)} \qquad \frac{d\sigma^{\alpha}(t)}{dt} = v^{\alpha}(\sigma^{\beta}(t)) , \quad \sigma^{\alpha}(0) = x_0^{\alpha} .$$

Since $V(x)$ is smooth, standard existence theorems show that there is an orbit of $V(x)$ that passes through any given point of S. In fact, the following result holds.

Any smooth vector field $V(x)$ *determines and is determined by its orbits.*

Let $U(x) = u^{\alpha}(x)\, \partial/\partial x^{\alpha}$ and $V(x) = v^{\alpha}(x)\, \partial/\partial x^{\alpha}$ be two vector fields. The quantity $[U,V](x)$ that is defined by

$$\text{(A2.7)} \qquad [U,V](x)f = U(Vf) - V(Uf)$$

$$= \left(u^{\beta} \frac{\partial v^{\alpha}}{\partial x^{\beta}} - v^{\beta} \frac{\partial u^{\alpha}}{\partial x^{\beta}}\right)(x) \frac{\partial}{\partial x^{\alpha}} f$$

is a new vector field that is called the Lie bracket or *commutator* of $U(x)$ and $V(x)$. Straightforward computations using (A2.7) show that

$$\text{(A2.8)} \qquad [U,V] = -[V,U] ,$$

$$\text{(A2.9)} \qquad \big[[U,V],W\big] + \big[[V,W],U\big] + \big[[W,U],V\big] = 0 ,$$

$$\text{(A2.10)} \qquad [fU,gV] = fg[U,V] + f(Ug)V - g(Vf)U .$$

These equations show that vector fields form a real *Lie algebra* under the operations of addition, multiplication by numbers, and vector multiplication defined by [,].

A3. EXTERIOR FORMS AND EXTERIOR PRODUCTS

Let $T_x^*(S)$ denote the dual space of the tangent space $T_x(S)$ at the point P with coordinates $\{x^\alpha\}$; that is, the linear space of all maps from $T_x(S)$ into $\mathbb{R}$. If $\omega \varepsilon T_x^*(S)$ and $V = v^\alpha\, \partial/\partial x^\alpha|_x$ $\varepsilon T_x(S)$, we write

$$\text{(A3.1)} \qquad \omega(V) = (w_\alpha v^\alpha)\Big|_x .$$

We use $\Lambda^k(T_x^*)$ to denote the kth exterior power of $T_x^* \equiv \Lambda^1(T_x^*)$ consisting of all real valued maps $\Omega_x(V_1,\ldots,V_k)$ of k variables $V_i \varepsilon T_x$ into $\mathbb{R}$ that are linear in each argument and antisymmetric in each pair of arguments. An *exterior form of degree* k (a k-form) is a smooth assignment $\Omega : x \to \Omega_x \varepsilon \Lambda^k(T_x^*)$, in which case we write $\Omega \varepsilon \Lambda^k(S)$. A *zero form* is just a function on S with range in $\mathbb{R}$. A 1-form is called a *differential form* or a Pfaffian form.

If $f(x^\alpha)$ is a smooth function on S with values in $\mathbb{R}$, then the *differential*, df , of f is defined by

$$\text{(A3.2)} \qquad (df)(V) = Vf = v^\alpha \frac{\partial f}{\partial x^\alpha}$$

for any $V = v^\alpha\, \partial/\partial x^\alpha \varepsilon T(S)$. Since $(df)(V)$ is scalar-valued, we see that $(df)\varepsilon T^*(S) = \Lambda^1(S)$. Now, consider the specific scalar-valued functions $f_i(x^\alpha) = x^i$, $i=1,\ldots,n$. For these functions, (A3.2) gives

$$\text{(A3.3)} \qquad (df_i)(V) = v^\alpha\, \partial x^i/\partial x^\alpha = v^i .$$

Thus, in particular, for $V = \partial/\partial x^j$, we have

$$\text{(A3.4)} \qquad (df_i)(\partial/\partial x^j) = \partial x^i/\partial x^j = \delta^i_j ,$$

from which we conclude that a natural basis for $T^*(S) = \Lambda^1(S)$, that is dual to the basis $\partial/\partial x^\alpha$ of $T(S)$, is given by the n scalar-valued quantities $dx^1, dx^2, \ldots, dx^n$. Accordingly, any $\omega \epsilon \Lambda^1(S)$ can be written as

$$\omega(x) = w_\alpha(x)\, dx^\alpha \tag{A3.5}$$

where the quantities $w_\alpha(x)$ are scalar-valued functions of position that are given by

$$w_i(x) = \omega(x)(\partial/\partial x^i), \quad i=1,\ldots,n. \tag{A3.6}$$

Again, the important point is that the quantity $\omega(x)$, as given by (A3.5), is scalar-valued since both the w's and the dx's are scalar-valued quantities.

There is, however, another way of interpreting (A3.5). If we interpret the n quantities dx^i, $i=1,\ldots,n$ as the tangents to the n independent curves $x_i(t) = x^\alpha + \delta_1^\alpha t$ (i.e., $dx^i = (dx^i/dt)|_x dt$), then the n quantities $w_\alpha(t)$ become the components of the linear functional $\omega(x)$ which, of course, is still scalar-valued. This given another consistent interpretation, provided we remember to change from "dx^i = differential of the scalar-valued function x^i" to "dx^i = symbolic tangent to the curve $x_i^\alpha(t) = x^\alpha + \delta_i^\alpha t$ at $t=0$". These two different interpretations of $w_\alpha(x)\, dx^\alpha$ can cause a surprising amount of difficulty for the casual reader of the exterior calculus, for most presentations, and this one included, freely switch back and forth between the two interpretations in order to achieve certain simplifications in actual calculations.

The above results show, upon noting that $\Lambda^0(S)$ consists of scalar-valued functions on S, that

$$\text{dimension}(\Lambda^0) = 1 \tag{A3.7}$$

and that $\Lambda^1(S)$ is a vector space over the real number field with

$$\text{dimension}(\Lambda^1) = n = \text{dimension}(T). \tag{A3.8}$$

However, (A3.5) shows that the multiplication of any element of Λ^1 by any C^∞ function on S (by any element of Λ^0) gives a new element of Λ^1. Accordingly, Λ^1 is a vector space over the associative algebra of C^∞ functions on S; that is, *Λ^1 is a module over the associative algebra of C^∞ functions on S.*

The ability to multiply elements of Λ^0 by elements of Λ^1 gives the very useful modular property of Λ^1. This same modular property can be obtained for any Λ^k through use of what is known as the exterior product. If $\omega(x)$ is a k-form and $\gamma(x)$ is an ℓ-form, then $(\omega \wedge \gamma)(x)$ is the *exterior product* of $\omega(x)$ and $\gamma(x)$ that is a $(k+\ell)$-form defined by

(A3.9)

$$(\omega \wedge \gamma)(V_1,\ldots,V_{k+\ell})$$

$$= \frac{1}{(k+\ell)!} \sum_{P_{k+\ell}} (\mathrm{sign}\,\sigma)\omega(V_{\sigma(1)},\ldots,V_{\sigma(k)})\gamma(V_{\sigma(k+1)},\ldots,V_{\sigma(k+\ell)}) ,$$

where $\sigma(i)$ is the image of the integer i under the permutation σ belonging to the group $P_{k+\ell}$ of all permutations of $k+\ell$ characters, and the summation in (A3.9) extends over the group $P_{k+\ell}$.

The exterior product possesses the following properties: (i) *$\omega \wedge \gamma$ is linear in ω and in γ;* (ii) *if $\omega \epsilon \Lambda^k$, $\gamma \epsilon \Lambda^\ell$ then*

$$\text{(A3.10)} \qquad \omega \wedge \gamma = (-1)^{k\ell}\, \gamma \wedge \omega ;$$

and (iii)

$$\text{(A3.11)} \qquad \omega \wedge (\gamma \wedge \eta) = (\omega \wedge \gamma) \wedge \eta .$$

Note in particular that (A3.10) shows that

$$\text{(A3.12)} \qquad \omega \epsilon \Lambda^{2r+1} \implies \omega \wedge \omega = 0 ,$$

for $\omega \wedge \omega = (-1)^{(2r+1)^2}\, \omega \wedge \omega = -\,\omega \wedge \omega$. Further, the basis $dx^1, \ldots, dx^n$ of Λ^1 can be used to generate a natural basis for each

Λ^k by the process of exterior multiplication. Thus, a natural basis for Λ^k is given by

$$dx^{\alpha_1} \wedge dx^{\alpha_2} \wedge \ldots \wedge dx^{\alpha_k} , \quad 1 \leq \alpha_1 < \alpha_2 < \ldots < \alpha_k \leq n ,$$

and

(A3.13) $$\text{dimension}(\Lambda^k) = \binom{n}{k} = \frac{n!}{k!(n-k)!} ,$$

(A3.14) $$\text{dimension}(\Lambda^n) = 1 ,$$

(A3.15) $$\omega \varepsilon \Lambda^{n+k} \implies \omega = 0 \quad \text{for any } k>0 .$$

If $\omega(x)\varepsilon\Lambda^k$, we may thus write

(A3.16) $$\omega(x) = \sum_{1\leq\alpha_1<\ldots<\alpha_k\leq n} w_{\alpha_1\ldots\alpha_k}(x)\, dx^{\alpha_1} \wedge \ldots \wedge dx^{\alpha_k} ,$$

where

(A3.17) $$w_{\alpha_1\ldots\alpha_k}(x) = \omega(x)\left(\frac{\partial}{\partial x^{\alpha_1}}, \ldots, \frac{\partial}{\partial x^{\alpha_k}}\right) , \quad 1 \leq \alpha_1 < \ldots < \alpha_k \leq n .$$

If the functions $w_{\alpha_1\ldots\alpha_k}(x)$ are extended to all values of the indices by demanding that they be antisymmetric in every pair of indices, then we can also write

(A3.18) $$\omega(x) = \left(\frac{1}{k!}\right) w_{\alpha_1\ldots\alpha_k}(x)\, dx^{\alpha_1} \wedge \ldots \wedge dx^{\alpha_k} .$$

It thus follows that each $\Lambda^k(S)$ is a module over the associative algebra of C^∞ functions on S.

A4. THE EXTERIOR DERIVATIVE

We have seen that there are two different interpretations that can be attached to the symbols dx^α, namely, tangents to coordinate curves and differentials of scalar-valued coordinate functions. These two interpretations are reconciled by means of the operation of exterior differentiation.

There is a unique map d *from* k-*forms to* (k+1)-*forms, called the exterior derivative, such that*

(i) *if* f *is a* 0-*form, then* df *is the differential of* f ;

(ii) *if* ω *is a* k-*form, then*

(A4.1) $$d(\omega \wedge \gamma) = (d\omega) \wedge \gamma + (-1)^k \omega \wedge (d\gamma) ;$$

(iii) *if* ω *and* γ *are* k-*forms, then*

(A4.2) $$d(\omega+\gamma) = d\omega + d\gamma$$

(iv) *for* $\omega\epsilon\Lambda^k$ *and for any* $0\leq k\leq n$,

(A4.3) $$d(d\omega) = 0 .$$

If $\omega(x)$ is an exterior form such that

(A4.4) $$d\omega = 0 ,$$

then ω is called a *closed* form. If $\omega\epsilon\Lambda^k$ and there exists a $\gamma\epsilon\Lambda^{k-1}$ such that

(A4.5) $$\omega = d\gamma ,$$

then ω is called an *exact* form. Clearly, *every exact form is a closed form.* Also, *the set of all closed elements of* Λ^k *forms a vector subspace* $C^k(S)$ *but not a submodule of* Λ^k , and *the set of all exact elements of* Λ^k *forms a vector subspace* $E^k(S)$ *but not a submodule of* Λ^k (i.e., $C^k(S)$ and $E^k(S)$ are closed under multiplication by numbers but are not closed under multiplication by functions). Thus $E^k(S)$ is a vector subspace of $C^k(S)$ since $\omega = d\gamma$ implies $d\omega = d(d\gamma) = 0$ by (A4.3). A question of fundamental importance is when is a closed form exact; that is, if $d\omega = 0$, can we conclude that there exists a γ such that $\omega = d\gamma$? This question is answered in the affirmative for Euclidean spaces by the Poincaré lemma which we shall come to shortly.

It is also important to note that $d(d\omega) = 0$ provides us with a direct means of obtaining necessary conditions under which a solution of an exterior differential equation can exist. Suppose that $\Gamma \varepsilon \Lambda^1$, $\Omega \varepsilon \Lambda^{k+1}$ are given forms and that we are required to solve

$$\text{(A4.6)} \qquad d\omega = \Gamma \wedge \omega + \Omega$$

for the exterior form $\omega \varepsilon \Lambda^k$. If we take the exterior derivative of both sides of (A4.6) and use (A4.6) to eliminate $d\omega$ from the result, we obtain $(\Gamma \varepsilon \Lambda^1 \Longrightarrow \Gamma \wedge \Gamma = 0)$

$$d(d\omega) = d\Gamma \wedge \omega - \Gamma \wedge d\omega + d\Omega = (d\Gamma - \Gamma \wedge \Gamma) \wedge \omega - \Gamma \wedge \Omega + d\Omega$$

$$= d\Gamma \wedge \omega - \Gamma \wedge \Omega + d\Omega \ .$$

However, we know, by (A4.3) that any ω must satisfy $d(d\omega) = 0$. It thus follows that there can exist an ω satisfying (A4.6) only if

$$\text{(A4.7)} \qquad d\Omega = \Gamma \wedge \Omega - d\Gamma \wedge \omega$$

holds. In other words, (A4.7) constitutes a system of necessary conditions in order that there exists an ω that satisfies (A4.6). Whether or not these conditions are sufficient is altogether another question whose answer will have to wait until we introduce the homotopy operator H in connection with the Poincaré lemma.

The most familiar example of such situations is that encountered in the study of conservative fields of force. Suppose that we are given an n-tuple of functions $f_\alpha(x^\beta)$ that are the components of a field of forces. We can thus construct the differential form

$$\text{(A4.8)} \qquad F = f_\alpha(x^\beta) dx^\alpha \ .$$

Such a system of forces is a conservative system if and only if there exists a scalar-valued potential function $W(x^\beta)$ such that

(A4.9) $\qquad \mathcal{F} = dW$;

that is, $f_\alpha(x^\beta) = \partial W(x^\beta)/\partial x^\alpha$. A conservative system of forces thus gives rise to an exact differential form dW . Now, for given $\mathcal{F}$ (for given $f_\alpha(x^\beta)$), the question arises as to whether (A4.9) can be solved for W ; that is, is the given $\mathcal{F}$ a 1-form of a conservative field of forces. Exterior differentiation of both sides of (A4.9) gives $d\mathcal{F} = d(dW)$. However, $d(dW) = 0$ must hold, from which we obtain the necessary conditions

(A4.10) $\qquad d\mathcal{F} = 0$

in order that $\mathcal{F}$ be a 1-form of a conservative field of forces. In three dimensions, this is nothing but the well-known condition that "curl $\mathcal{F}$" must vanish. However, (A4.10) is not restricted to three-dimensional problems; it gives the correct necessary conditions for the existence of a potential $W(x^\beta)$ in any finite number of dimensions. It should likewise be clear that we can also ask when is $\mathcal{F}$ proportional to the "gradient" of a potential function if $d\mathcal{F} \neq 0$; that is $\mathcal{F} = \lambda(x^\beta)dW(x^\beta)$. Similar arguments show that this can be the case only if $\mathcal{F} \wedge d\mathcal{F} = 0$. The general case, in which $\mathcal{F}$ is the sum of gradients with scalar-valued functions as coefficients, is decided by the Darboux theorem which we will come to shortly.

A5. ALGEBRAIC RESULTS

An exterior k-form Ω is said to be *simple* if and only if there exist k 1-forms $\omega^1,\ldots,\omega^k$ such that

(A5.1) $\qquad \Omega = \omega^1 \wedge \omega^2 \wedge \ldots \wedge \omega^k$.

Since Λ^n is 1-dimensional, every element of Λ^n has the form $f(x^\alpha)dx^1 \wedge \ldots \wedge dx^n = (f(x^\alpha)dx^1) \wedge dx^2 \wedge \ldots \wedge dx^n$. Thus, *every n-form is simple*. It can also be shown that *every* (n-1)-*form is simple*.

If ω is a k-form, then

$$\text{(A5.2)} \qquad \overset{1}{\omega} = \omega(\cdot, V_2, \ldots, V_k)$$

is a 1-form for every collection of k-1 elements $V_2, \ldots, V_k$ of T(S) . We then have the following result. *A k-form ω is simple if and only if*

$$\text{(A5.3)a} \qquad \omega \wedge \omega(\cdot, V_2, \ldots, V_k) = 0$$

for all $V_2, \ldots, V_k \in T(S)$, and this is true if and only if

$$\text{(A5.3)b} \qquad \omega \wedge \omega(\cdot, \cdot, V_3, \ldots, V_k) = 0$$

for all $V_3, \ldots, V_k \in T(S)$.

Nonzero simple forms are important, for they characterize subspaces of $T^*(S) = \Lambda^1(S)$ in an efficient manner. First off, we have the following.

A necessary and sufficient condition that a collection of r 1-forms $\omega^1, \ldots, \omega^r$ be linearly dependent (constant coefficients) is that

$$\text{(A5.4)a} \qquad \omega^1 \wedge \omega^2 \wedge \ldots \wedge \omega^r = 0$$

Conversely, if

$$\text{(A5.4)b} \qquad \omega^1 \wedge \omega^2 \wedge \ldots \wedge \omega^r \neq 0$$

then the 1-forms $\omega^1, \ldots, \omega^r$ span a r-dimensional linear subspace V_r of Λ^1 at each point of S and

$$\text{(A5.5)} \qquad \Omega = \omega^1 \wedge \omega^2 \wedge \ldots \wedge \omega^r$$

is a characteristic simple r-form of the subspace V_r .

This theorem leads to the following characterization of linear subspace properties of Λ^1 .

To each r-dimensional linear subspace V_r of Λ^1 there is a characteristic simple r-form $\Omega \neq 0$ of V_r that is determined *up to a nonzero scalar-valued factor and such that* ω *belongs to* V_r *if and only if* $\omega\wedge\Omega = 0$. *Let* V_{r_1} *and* V_{r_2} *be two subspaces of* Λ^1 *with characteristic simple* r_1*- and* r_2*-forms* Ω_1 *and* Ω_2 . *A necessary and sufficient condition that* V_{r_1} *be contained in* V_{r_2} *is that there exist an* (r_2-r_1)*-form* γ *such that* $\Omega_2=\gamma\wedge\Omega_1$. *A necessary and sufficient condition that* V_{r_1} *and* V_{r_2} *have only the zero* 1*-form in common is* $\Omega_1\wedge\Omega_2 \neq 0$. *If this condition is satisfied then* $\Omega_1\wedge\Omega_2$ *is a characteristic simple* (r_1+r_2)*-form of the subspace* $V_{r_1}+V_{r_2}$.

Suppose that

$$\Omega = \omega^1\wedge\omega^2\wedge\dots\wedge\omega^r \tag{A5.6}$$

is a characteristic simple r-form of a $V_r\subset\Lambda^1$. If a^i_j , $i,j=1,\dots,r$ are the entries of a nonsingular r-by-r matrix, we can define r new 1-forms by

$$\gamma^i = a^i_j\,\omega^j . \tag{A5.7}$$

However, it is an easy calculation to see that

$$\gamma^1\wedge\dots\wedge\gamma^r = \det(a^j_i)\,\omega^1\wedge\dots\wedge\omega^r . \tag{A5.8}$$

Accordingly, we can generate many characteristic simple forms of a given subspace from any one characteristic simple form by the process of constructing the nonsingular transformations (A5.7). We note that

$$\Omega_n = dx^1\wedge\dots\wedge dx^n\ \varepsilon\Lambda^n \tag{A5.9}$$

is a characteristic simple n-form of Λ^1 that is called the *natural volume element* of S for reasons that will become evident when we come to the theory of integration of exterior forms on S . Again, since Λ^n is 1-dimensional, any other characteristic simple form of Λ^n is a multiple of the natural volume element.

The following result, known as Cartan's lemma, is quite useful.

Let $\omega^1,\dots,\omega^k$ *be linearly independent* 1-*forms and suppose that*

(A5.10) $$\omega^1 \wedge \gamma^1 + \dots + \omega^k \wedge \gamma^k = 0 ,$$

then there exist k^2 *quantities* a^i_j , $i,j=1,\dots,k$ *such that*

(A5.11) $$\gamma^i = a^i_j\, \omega^j$$

and

(A5.12) $$a^i_j = a^j_i .$$

Another reason for singling out simple forms is that they possess a very simple evaluation when they act on elements of $T(S)$:

(A5.13) $$(\omega^1 \wedge \dots \wedge \omega^k)(V_1,\dots,V_k) = \frac{1}{k!}\det(\omega^i(V_j)) .$$

It should be evident to the reader that this can be used to discuss linear independence and subspace structure of $T(S)$ in a direct manner similar to the results established above for $\Lambda^1(S)$.

A6. INNER MULTIPLICATION

Since any element of any Λ^k can be generated by forming sums of exterior products of elements of Λ^0 and Λ^1 , we can define an operations on any Λ^k provided we define it on Λ^0 , Λ^1 and give the manner in which it combines with addition and exterior multiplication. We use this ability to define the operation of inner multiplication.

The *inner multiplication*, $V \rfloor \omega$, of an element V of $T(S)$ with an element ω of Λ^k is a map from Λ^k to Λ^{k-1} that is defined by the following conditions:

(i) if $f \in \Lambda^0$, then

(A6.1) $$V \rfloor f = 0 ;$$

(ii) if $\omega \epsilon \Lambda^1$, then

(A6.2) $$V \rfloor \omega = \omega(V) ;$$

(iii) if ω and γ belong to Λ^k, then

(A6.3) $$V \rfloor (\omega+\gamma) = V \rfloor \omega + V \rfloor \gamma ;$$

(iv) if $\omega \epsilon \Lambda^k$, then

(A6.4) $$V \rfloor (\omega \wedge \gamma) = (V \rfloor \omega) \wedge \gamma + (-1)^k \omega \wedge (V \rfloor \gamma) .$$

It follows immediately from this definition that

(A6.5) $$(V_1 \rfloor \omega)(V_2,\dots,V_k) = k\, \omega(V_1,\dots,V_k)$$

for any $\omega \epsilon \Lambda^k$. Furhter, since $\omega(V_1) = w_\alpha v^\alpha$ for $\omega = w_\alpha dx^\alpha$, $V = v^\alpha\, \partial/\partial x^\alpha$, it follows that

$$V \rfloor (w_{\alpha\beta} dx^\alpha \wedge dx^\beta) = v^\alpha w_{\alpha\beta} dx^\beta - v^\beta w_{\alpha\beta} dx^\alpha$$

$$= v^\alpha (w_{\alpha\beta} - w_{\beta\alpha}) dx^\beta ,$$

and so forth for higher degree forms.

The operation of inner multiplication is of obvious use, for it maps in the opposite direction to the map d ; that is $d:\Lambda^k \to \Lambda^{k+1}$ while $V \rfloor : \Lambda^k \to \Lambda^{k-1}$. In fact, the operation of inner multiplication will be instrumental in finding an operator that inverts the operation of exterior differentiation, at least for certain specific subspaces of Λ^k.

A7. BEHAVIOR UNDER MAPS

Let $\Phi : S \to \grave{S}$ be a smooth 1-to-1 map (an invertible map) of the space S to the space $\grave{S}$. If the space $\grave{S}$ is covered by a coordinate cover $(\grave{y}^i)$, then we write

(A7.1) $\Phi\colon S \to {}^{\backprime}S \,|\, {}^{\backprime}y^i = \phi^i(x^\beta)\ , \quad i=1,\dots,{}^{\backprime}n\ .$

Let

(A7.2) $V = v^\alpha(x^\beta)\ \partial/\partial x^\alpha$

be a generic element of $T(S)$, then we define the vector field ${}^{\backprime}V = \Phi_* V$ on ${}^{\backprime}S$ by

(A7.3) ${}^{\backprime}V({}^{\backprime}f) = (\Phi_* V)({}^{\backprime}f) = V({}^{\backprime}f \circ \Phi)\ .$

This means that

$$ {}^{\backprime}v^i \frac{\partial {}^{\backprime}f}{\partial {}^{\backprime}y^i} = v^\beta \frac{\partial}{\partial x^\beta} {}^{\backprime}f(\phi^i(x^\gamma)) = v^\beta \frac{\partial {}^{\backprime}f}{\partial {}^{\backprime}y^j} \frac{\partial \phi^j}{\partial x^\beta}\ , $$

and hence we have

(A7.4) ${}^{\backprime}v^i = v^\beta \dfrac{\partial \phi^i}{\partial x^\beta}\ .$

This shows the reason why we have to require that Φ be invertible, for ${}^{\backprime}v^i = {}^{\backprime}v^i({}^{\backprime}y^j)$, while the right-hand side of (A7.4) is clearly a function of the x's . Thus, in order to solve for the functions ${}^{\backprime}v^i$ of the ${}^{\backprime}y$'s we have to invert the relations ${}^{\backprime}y^i = \phi^i(x^\beta)$ so as to obtain the x's as functions of the ${}^{\backprime}y$'s . This is reflected in the fact that (A7.4) should actually be written as

(A7.4)a ${}^{\backprime}v^i = (v^\beta\ \partial\phi^i/\partial x^\beta) \circ \overset{-1}{\Phi}\ .$

There is a much simpler way of obtaining the map Φ_* . We simply use the chain rule and obtain

$$ v^\alpha \frac{\partial}{\partial x^\alpha} = v^\alpha \frac{\partial {}^{\backprime}y^j}{\partial x^\alpha} \frac{\partial}{\partial {}^{\backprime}y^j} = {}^{\backprime}v^j \frac{\partial}{\partial {}^{\backprime}y^j} \overset{\text{def}}{=} \Phi_*(v^\alpha \frac{\partial}{\partial x^\alpha})\ . $$

This leads immediately to (A7.4). The important thing is to note that Φ_* *maps in the same direction as* Φ , that is from S or quantities defined on S to ${}^{\backprime}S$ or quantities defined on ${}^{\backprime}S$.

Now, let $\Phi{:}S \to {}^{\backprime}S \,|\, y^i = \phi^i(x^\beta)$ be a smooth but not necessarily invertible map. If $`f(`y^i)$ is a scalar-valued function defined on $`S$ (an element of $\Lambda^0(`S)$), then we define

$$f(x^\beta) = `f\circ\Phi = \Phi^*(`f) ; \tag{A7.5}$$

Thus Φ^* maps functions on $`S$ to functions on $S$. Now, let $`\omega(`y) = `w_i(`y^j)d`y^i$ be a 1-form on $`S$. The chain rule then gives

$$`\omega(`y) = `w_i(\phi^j(x^\gamma)) \frac{\partial\phi^i(x^\alpha)}{\partial x^\beta} dx^\beta = w_\beta(x^\beta)dx^\beta ,$$

so that we can map 1-forms on $`S$ onto 1-forms on S by

$$\Phi^*(`\omega) = \omega . \tag{A7.6}$$

The relation between these 1-forms is given by

$$\begin{aligned} `\omega &= `w_i(`y^j)d`y^i , \quad \omega = w_\alpha(x^\beta)dx^\alpha = \Phi^*(`\omega) \\ w_\alpha(x^\beta) &= `w_i(\phi^j(x^\beta)) \frac{\partial\phi^i(x^\beta)}{\partial x^\alpha} = (`w_i\circ\Phi) \frac{\partial\phi^i}{\partial x^\alpha} . \end{aligned} \tag{A7.7}$$

Now, any form can be made up from sums of products of elements of Λ^0 and Λ^1, and hence we can define the map Φ^* from $\Lambda^k(`S)$ to $\Lambda^k(S)$ by

$$\Phi^*(`\omega) = \omega . \tag{A7.8}$$

A straightforward set of calculations gives the following results.

If $\phi:S\to`S$ *is a smooth map, then* $\Phi^*$ *is a linear map of* k-*forms on* $`S$ *to* k-*forms on* S, *and*

$$\Phi^*(\omega\wedge\gamma) = (\Phi^*\omega)\wedge(\Phi^*\gamma) , \tag{A7.9}$$

$$\Phi^*(d\omega) = d(\Phi^*\omega) . \tag{A7.10}$$

Exterior forms thus have very nice behavior under mappings of their domain space.

The important thing to note here is that Φ^* *maps in the direction opposite to the direction of* Φ. Accordingly, if $\Phi: \mathbb{R} \to S | x^\alpha = \phi^\alpha(t)$, then Φ^* maps $\omega(x^\alpha)|_{x=\phi}$ onto a differential form on $\mathbb{R}$. Since the dimension of $\mathbb{R}$ is one, $\Phi^* f(x^\alpha) = f(\phi^\alpha(t))$, $\Phi^*(w_\alpha(x)dx^\alpha) = w_\alpha(\phi^\beta(t))\frac{d\phi^\alpha}{dt}dt$, and $\Phi^*(\omega(x^\beta)) = 0$ if $\omega \varepsilon \Lambda^k(S)$ for $k>1$.

A8. LIE DIFFERENTIATION

Let $V = v(x)^\alpha \, \partial/\partial x^\beta$ be a given vector field on S. This vector field can be used to generate the system of equations

$$\text{(A8.1)} \qquad \frac{\partial \gamma^\alpha(x^\beta;\lambda)}{\partial \lambda} = v^\alpha(\gamma^\beta(x^\mu;\lambda)) , \qquad \gamma^\alpha(x^\beta;0) = x^\alpha .$$

These equations define a flow in S that can be thought of as moving the points of S about by the map Ψ_λ that is defined by

$$\text{(A8.2)} \qquad \Psi_\lambda : S \to S | x^\alpha(\lambda) = \gamma^\alpha(x^\beta;\lambda) .$$

It is clear from (A8.2) that

$$\text{(A8.3)} \qquad x^\alpha(\lambda) = x^\alpha + \lambda v^\alpha(x^\beta) + o(\lambda) ,$$

$$\text{(A8.4)} \qquad \Psi_0 = \text{identity}.$$

Let $\omega(x)$ be a 1-form on S. Since exterior forms are scalar-valued, their values at different points can be compared *provided they are written in terms of the same arguments*. However, if $x^\alpha(\lambda)$ is the image of x^α under Ψ_λ, then $\Psi_\lambda^* \omega(x^\alpha(\lambda)) = \bar{\omega}(x^\beta)$, which expressed $\omega(x^\alpha(\lambda))$ in terms of the arguments x^β. Further, since $\omega(x^\beta) = \Psi_o^* \omega$, we obtain

$$\text{(A8.5)} \qquad \omega(x^\alpha(\lambda)) - (x^\alpha) = [(\Psi_\lambda^* - \Psi_o^*)\omega](x^\beta) .$$

We define the *Lie derivative* of ω with respect to the vector field V, written as $\pounds_V \omega$, by

$$(A8.6) \qquad (\pounds_V \omega)(x^\beta) = \lim_{\lambda\to 0} \left[\frac{(\Psi^*_\lambda - \Psi^*_o)\omega}{\lambda} \right] (x^\beta) \ ,$$

so that we have the linear interpolation formula

$$(A8.7) \qquad \omega(x^\beta(\lambda)) = \omega(x^\beta) + \lambda(\pounds_V \omega)(x^\beta) + o(\lambda) \ .$$

We now come to explicit calculation. Starting with $\omega(x^\beta) = w_\alpha(x^\beta)dx^\alpha$, then (A8.3) gives, for fixed λ,

$$\begin{aligned} \omega(x^\beta(\lambda)) &= w_\alpha(x^\beta + \lambda v^\beta + o)d(x^\alpha + \lambda v^\alpha + o) \\ &= w_\alpha(x^\beta)dx^\alpha + \lambda \left[\frac{\partial x_\alpha}{\partial x^\beta} v^\beta + w_\beta \frac{\partial v^\beta}{\partial x^\alpha} \right] dx^\alpha + o(\lambda) \ . \end{aligned}$$

Thus, the formula that evaluates $\pounds_V \omega$ is given by

$$(A8.8) \qquad \pounds_V \omega = \left(\frac{\partial w_\alpha}{\partial x^\beta} v^\beta + w_\beta \frac{\partial v^\beta}{\partial x^\alpha} \right) dx^\alpha \ .$$

Although we can calculate with (A8.8), it is not a very nice formula for it depends explicitly on the functions v^α and w_α that serve to define V and ω in terms of their natural bases. Further manipulations give

$$\begin{aligned} \pounds_V \omega &= \left(\frac{\partial w_\alpha}{\partial x^\beta} v^\beta + w_\beta \frac{\partial v^\beta}{\partial x^\alpha} + \frac{\partial w_\beta}{\partial x^\alpha} v^\beta - \frac{\partial w_\beta}{\partial x^\alpha} v^\beta \right) dx^\alpha \\ &= \left(\left(\frac{\partial w_\alpha}{\partial x^\beta} - \frac{\partial w_\beta}{\partial x^\alpha} \right) v^\beta + \frac{\partial}{\partial x^\alpha} (w_\beta v^\beta) \right) dx^\alpha \ . \end{aligned}$$

However, $V \rfloor \omega = \omega_\alpha v^\alpha$ and $V \rfloor d\omega = \left(\frac{\partial w_\alpha}{\partial x^\beta} - \frac{\partial w_\beta}{\partial x^\alpha} \right) v^\beta dx^\alpha$. Accordingly, we get the formula

$$(A8.9) \qquad \pounds_V \omega = V \rfloor (d\omega) + d(V \rfloor \omega) \ .$$

Now, (A8.9) is well defined for any exterior form of any degree since the operations $V \rfloor$ and d are well defined for any exterior form. Further when (A8.9) is applied to elements of Λ^0, we get

(A8.10) $$\pounds_V f = V \rfloor df ,$$

because $V \rfloor f = 0$, and hence the interpolation formula $f(x^\beta(\lambda)) = f(x^\beta) + \lambda(\pounds_V f)(x^\beta) + o(\lambda)$ remains valid. Thus, since $\pounds_V$ can be defined for 0-forms and 1-forms by (A8.9), it can be defined for all elements of all Λ^k by the same formula. Straightforward calculations then give the following results.

The operator $\pounds_V$ *is a map from* Λ^k *to* Λ^k *that is defined by*

(A8.11) $$\pounds_V \omega = V \rfloor (d\omega) + d(V \rfloor \omega)$$

and possesses the following properties:

(A8.12) $$\pounds_V(\omega+\gamma) = \pounds_V\omega + \pounds_V\gamma ,$$

(A8.13) $$\pounds_V(\omega \wedge \gamma) = (\pounds_V\omega) \wedge \gamma + \omega \wedge (\pounds_V\gamma) ,$$

(A8.14) $$\pounds_V d\omega = d\pounds_V\omega ,$$

(A8.15) $$\pounds_{fV}\omega = f\pounds_V\omega + df \wedge (V \rfloor \omega) ,$$

(A8.16) $$\pounds_{V_1+V_2}\omega = \pounds_{V_1}\omega + \pounds_{V_2}\omega .$$

These results completely delineate the action of $\pounds$ on exterior forms. The question now arises as to what happens when $\pounds$ acts on elements of $T(S)$. We note that $V \rfloor \omega$ is a well defined exterior form for any exterior form ω. Now, (A8.13) shows that $\pounds$ acts on exterior products as a derivation, so we can define the action of $\pounds$ on $V \in T(S)$ by requiring that $\pounds$ act as a derivation on $V \rfloor \omega$, that is

$$\pounds_U(V \rfloor \omega) = (\pounds_U V) \rfloor \omega + V \rfloor (\pounds_U \omega)$$

should hold for all exterior forms ω. This then gives the definition

(A8.17) $\quad (\pounds_{U}V)\rfloor\omega \;=\; \pounds_{U}(V\rfloor\omega) - V\rfloor(\pounds_{U}\omega) \quad$ for all ω .

In order to obtain an explicit evaluation of $\pounds_{U}\nu$, we proceed as follows. We take for ω the $n(n-1)/2$ 2-forms $dx^{\alpha_1}\wedge dx^{\alpha_2}$, $1\leq\alpha_1<\alpha_2\leq n$ that form a basis for Λ^2 . Direct calculations show that

$$V\rfloor(dx^{\alpha_1}\wedge dx^{\alpha_2}) \;=\; \nu^{\alpha_1}dx^{\alpha_2} - \nu^{\alpha_2}dx^{\alpha_1} ,$$

$$\pounds_{U}(V\rfloor(dx^{\alpha_1}\wedge dx^{\alpha_2})) \;=\; (U\rfloor d\nu^{\alpha_1})dx^{\alpha_2} - (U\rfloor dx^{\alpha_2})dx^{\alpha_1} + \nu^{\alpha_1}du^{\alpha_2} - \nu^{\alpha_2}du^{\alpha_1} ,$$

$$V\rfloor\pounds_{U}(dx^{\alpha_1}\wedge dx^{\alpha_2}) \;=\; (V\rfloor du^{\alpha_1})dx^{\alpha_2} - \nu^{\alpha_2}du^{\alpha_1} + \nu^{\alpha_1}du^{\alpha_2} - (V\rfloor du^{\alpha_2})dx^{\alpha_1} .$$

Thus,

$$\begin{aligned}(\pounds_{U}V)\rfloor(dx^{\alpha_1}\wedge dx^{\alpha_2}) &= (\pounds_{U}V)^{\alpha_1}dx^{\alpha_2} - (\pounds_{U}V)^{\alpha_2}dx^{\alpha_1}\\ &= \pounds_{U}(V\rfloor(dx^{\alpha_1}\wedge dx^{\alpha_2})) - V\rfloor\pounds_{U}(dx^{\alpha_1}\wedge dx^{\alpha_2})\\ &= (U\rfloor d\nu^{\alpha_1}-V\rfloor du^{\alpha_1})dx^{\alpha_2} - (U\rfloor d\nu^{\alpha_2}-V\rfloor du^{\alpha_2})dx^{\alpha_1}\end{aligned}$$

from which we conclude that

$$(\pounds_{U}V)^{\alpha_1} \;=\; U\rfloor d\nu^{\alpha_1} - V\rfloor du^{\alpha_1} \;=\; [U,V]^{\alpha_1} :$$

(A8.18) $\quad \pounds_{U}V \;=\; [U,V] \;=\; -\pounds_{V}U$.

Thus, $\pounds_{U}V$ is again an element of $T(S)$, and is given by the Lie bracket of U with V .

Now that we know that $\pounds_{U}V = [U,V]\varepsilon T(S)$, the question naturally arises as to how to compute $\pounds_{[u,\nu]}$. The secret is that

(A8.19) $\quad (\pounds_{U}V)\rfloor\Omega \;=\; \pounds_{U}(V\rfloor\Omega) - V\rfloor(\pounds_{U}\Omega)$

holds for every exterior form Ω. The definition of the Lie derivative gives

(A8.20) $$\mathcal{L}_{[u,v]}\omega = \mathcal{L}_{\mathcal{L}_u v}\omega = (\mathcal{L}_u V)\rfloor d\omega + d(\mathcal{L}_u V\rfloor\omega) .$$

Use of (A8.19) first with $\Omega = d\omega$ and then with $\Omega=\omega$ yields

$$(\mathcal{L}_u V)\rfloor d\omega = \mathcal{L}_u(V\rfloor d\omega) - V\rfloor(\mathcal{L}_u d\omega) ,$$

$$(\mathcal{L}_u V)\rfloor\omega = \mathcal{L}_u(V\rfloor\omega) - V\rfloor(\mathcal{L}_u\omega) .$$

When these results are substituted into (A8.20) and we use the fact that $d\mathcal{L} = \mathcal{L}d$, we finally obtain

$$\begin{aligned}\mathcal{L}_{[u,v]}\omega &= \mathcal{L}_u(V\rfloor d\omega) - V\rfloor(\mathcal{L}_u d\omega) + d[\mathcal{L}_u(V\rfloor\omega) - V\rfloor\mathcal{L}_u\omega]\\ &= \mathcal{L}_u(V\rfloor d\omega) - V\rfloor d\mathcal{L}_u\omega + \mathcal{L}_u d(V\rfloor\omega) - d[V\rfloor\mathcal{L}_u\omega]\\ &= \mathcal{L}_u[(V\rfloor d\omega) + d(V\rfloor\omega)] - V\rfloor d\mathcal{L}_u\omega - d[V\rfloor\mathcal{L}_u\omega]\\ &= \mathcal{L}_u[\mathcal{L}_v\omega] - \mathcal{L}_v[\mathcal{L}_u\omega] ;\end{aligned}$$

that is,

(A8.21) $$\mathcal{L}_{[u,v]} = \mathcal{L}_u\mathcal{L}_v - \mathcal{L}_v\mathcal{L}_u .$$

This result is very important. We know that vector fields form a Lie algebra under the multiplication $[u,v]$. It then follows from (A8.21) *the operators* $\mathcal{L}_u$ *form a Lie algebra under the multiplication* $\mathcal{L}_u\mathcal{L}_v - \mathcal{L}_v\mathcal{L}_u$. In particular, if we know that $\mathcal{L}_{u_i}\omega = 0$, $i=1,\ldots,r$ for r given vector fields $u_1,\ldots,u_r$, then every vector v that can be formed from the vectors $u_1,\ldots,u_r$ by the operations of addition, multiplication by numbers and formation of Lie brackets has the property that $\mathcal{L}_v\omega = 0$. It is customary to say that a form ω admits a vector field u as an *invariance vector field* and that ω *is invariant under transport by* u when $\mathcal{L}_u\omega = 0$. The above result then shows that *a form* ω *is invariant under transport by all vector fields that belong to the Lie algebra generated by any finite number of invariance vector fields of* ω *.* This is a very quick and simple method of generating many solutions to the equations $\pounds_u \omega = 0$ from knowledge of a finite number of solutions.

A9. THE THEOREMS OF FROBENIUS AND DARBOUX

We now come to the question of obtaining what may be called standard forms or representations of exterior forms. The first of these is given by the theorem of Frobenius.

Let ω^i , $i=1,\ldots,r$ *be* r *linearly independent* 1-*forms, so that*

$$(A9.1) \qquad \Omega_r = \omega^1 \wedge \omega^2 \wedge \ldots \wedge \omega^r \neq 0 .$$

There exists a nonsingular r-by-r *matrix of scalar-valued functions with entries* $A^i_j(x^\beta)$ *and* r *scalar-valued functionally independent functions* $u^i(x^\beta)$, $i=1,\ldots,r$, *such that*

$$(A9.2) \qquad \omega^i(x^\beta) = A^i_j(x^\beta)\, du^j(x^\beta)$$

if and only if

$$(A9.3) \qquad d\omega^i \wedge \Omega_r = 0 , \quad i=1,\ldots,r .$$

If these conditions are satisfied then the r 1-forms ω^i are said to be *completely integrable* and we have

$$(A9.4) \qquad \Omega_r = \det(A^i_j)\, du^1 \wedge du^2 \wedge \ldots \wedge du^r .$$

The reason for referring to such collections of 1-forms as completely integrable is because (A9.2) and the fact that $\det(A^i_j) \neq 0$ shows that the system of exterior equations $\omega^i = 0$, $i=1,\ldots,r$ that is completely integrable has the solutions $u^i(x^\beta) = \text{constant}$, $i=1,\ldots,r$.

It should be carefully noted that the matrix $((A^i_j(x^\beta)))$ and the functions $u^i(x^\beta)$ are not determined uniquely by the given forms ω^i, nor is the representation (A9.2) unique. The non-uniqueness of the representation is easily seen from the trivial manipulation

$$\omega^i = A^i_j\, du^j = d(A^i_j u^j) - u^j\, dA^i_j ,$$

in which case we obtain a representation in terms of an exact term $d(A^i_j u^j)$ and the remainder terms $u^j\, dA^i_j$. The nonuniqueness of the A's and the u's is seen as follows. Suppose that we set $f^1 = F(u^1)$ with $dF/du^1 \neq 0$, $f^i = u^i$ for $i \neq 1$, and consider the set of 1-forms $B^i_j df^j$. Now $df^1 \wedge df^2 \wedge \ldots \wedge df^r \neq 0$ since $du^1 \wedge \ldots \wedge du^r \neq 0$ and $dF/du^1 \neq 0$. We then have

$$B^i_j df^j = B^i_1 \frac{dF}{du^1}\, du^1 + \sum_{j=2}^{r} B^i_j du^j .$$

Accordingly, if we set $A^i_1 = B^i_1 df/du^1$, $A^i_j = B^i_j$ for $j \neq 1$ then we achieve $B^i_j df^j = A^i_j\, du^j = \omega^i$, and the matrix with entries B^i_j can be solved for from the above equations for a given nonsingular matrix with entries A^i_j. This amounts to the observation that the u's can be replaced by any other functionally independent collection of functions f^i such that f^i is constant on each hypersurface u^i = constant, $i=1,\ldots,r$.

We now turn to the question of how do we obtain an intrinsic characterization and a general representation of a general 1-form. As with most such problems, the first task is to obtain the characterization. To this end, we construct the following sequence of exterior forms of increasing degree:

(A9.5) $$I_1 = \omega , \quad I_2 = d\omega ,$$

(A9.6) $$I_{2k+1} = \omega \wedge I_{2k} = I_1 \wedge I_{2k} ,$$

$$I_{2k+2} = dI_{2k+1} = I_2 \wedge I_{2k} .$$

This sequence is finite for any given $\omega\varepsilon\Lambda^1$, for any form of degree higher than the dimension n of the space S is identically zero. We thus always have $I_{n+1} = 0$ on S. Further, for each point (x^β) in S or in some open n-dimensional subset of S we can find an integer $K(x^\beta)$ such that

$$\text{(A9.7)}\qquad I_{K(x^\beta)}(x^\beta) \neq 0\ ,\quad I_{K(x^\beta)+m}(x^\beta) = 0 \text{ for all } m>0\ .$$

The *class* of $\omega(x^\beta)$, written $K(\omega)$ is defined by

$$\text{(A9.8)}\qquad K(\omega) = \max_{x\varepsilon S} K(x^\beta)\ .$$

If $K(x^\beta) = K(\omega)$, then the point with coordinates (x^β) is said to be a *regular point* of the form $\omega(x)$, if $K(x^\beta) < K(\omega)$, the point with coordinates (x^β) is said to be a *critical point* of $\omega(x^\beta)$. The important thing to note here is that the class of a 1-form ω, its regular points and its critical points can be determined in a finite number of steps by simply constructing the sequence given by (A9.5), (A9.6). The *rank* of $\omega(x^\beta)$ is an even integer that is the largest even integer less than or equal to $K(\omega)$, and is written as $2\rho(\omega)$. Finally the *index* of $\omega(x^\beta)$ is defined by

$$\varepsilon(\omega) = K(\omega) - 2\rho(\omega)\ ,$$

so that $\varepsilon(\omega) = 1$ if $K(\omega)$ is an odd integer and $\varepsilon(\omega) = 0$ if $K(\omega)$ is an even integer. The Darboux theorem then gives the following representation.

Let $\omega(x^\beta)$ be a 1-form with class $K(\omega)$, rank $2\rho(\omega)$ and index $\varepsilon(\omega)$. There exist $\rho(\omega)$ scalar-valued functions $u_i(x^\beta)$ and $\rho(\omega) + \varepsilon(\omega)$ scalar-valued functions $v_i(x^\beta)$ that constitute a system of $K(\omega)$ functionally independent functions at every regular point of $\omega(x^\beta)$ such that

$$\text{(A9.9)}\qquad \omega(x^\beta) = \sum_{i=1}^{\rho(\omega)} u_i(x^\beta)\, dv_i(x^\beta) + \varepsilon(\omega)\, dv_{\rho(\omega)+1}(x^\beta)\ .$$

There are certain aspects of the proof of the Darboux theorem that are worth mentioning since they give a direct method for obtaining the functions u_i and v_i . One first establishes that there exists a scalar-valued function η such that

$$\text{(A9.10)} \qquad I_{K(\omega)} = d\eta \wedge I_{K(\omega)-1}$$

at all regular points of $\omega(x^\beta)$. Two classes of transformations are then constructed. The first consists of similarity transformations. A 1-form ω is said to undergo a *similarity transformation* to an image form ${}^{\backprime}\omega$ if

$$\text{(A9.11)} \qquad {}^{\backprime}\omega = \sigma\omega \ .$$

If we construct the sequence ${}^{\backprime}I_k$ from the image form ${}^{\backprime}\omega$, it follows readily that

$$\text{(A9.12)} \qquad {}^{\backprime}I_{2k} = \sigma^k I_{2k} + k\,\sigma^{k-1} d\sigma \wedge I_{2k-1} \ , \quad I_{2k+1} = \sigma^{k+1} I_{2k+1} \ .$$

Accordingly, if $K(\omega) = 2\rho(\omega)$, (A9.10) can be used to obtain

$${}^{\backprime}I_{2\rho({}^{\backprime}\omega)} = \sigma^k\, d\eta \wedge I_{2k-1} + k\,\sigma^{k-1}\, d\sigma \wedge I_{2k-1} \ , \quad k = \rho(\omega) \ .$$

The choice $\sigma = e^{-k\eta} = e^{-\rho(\omega)\eta}$ then reduces the class of the image form ${}^{\backprime}\omega$ to $2\rho(\omega)-1$. A 1-form ω is said to undergo a *gradient transformation* to an image form ${}^{\backprime}\omega$ if

$$\text{(A9.13)} \qquad {}^{\backprime}\omega = \omega + d\lambda \ .$$

This gives

$$\text{(A9.14)} \qquad {}^{\backprime}I_{2k} = I_{2k} \ , \quad {}^{\backprime}I_{2k+1} = I_{2k+1} + d\lambda \wedge I_{2k} \ .$$

Thus if $K(\omega)$ is an odd integer $= 2\rho(\omega)+1$, then use of (A9.10) gives

$${}^{\backprime}I_{K({}^{\backprime}\omega)} = I_{K(\omega)} + d\lambda \wedge I_{2\rho(\omega)} = d\eta \wedge I_{K(\omega)} + d\lambda \wedge I_{2\rho(\omega)} \ .$$

The choice $\lambda = -\eta$ thus reduces the class of ω to $2\rho(\omega)$. A succession of gradient transformations and similarity transformations will thus reduce the class of any form ω to zero. The functions u_i and v_i are then constructed from the σ's and the λ's by inverting the sequence of similarity and gradient transformations, on noting that a 1-form of class zero is a constant 1-form.

We note again the lack of uniqueness in the representation given by the Darbous theorem. In particular, since $u_i dv_i = d(u_i v_i) - v_i du_i$, (A9.9) can also be written as

$$\text{(A9.15)} \qquad \omega = -\sum_{i=1}^{\rho(\omega)} v_i du_i + d\{\varepsilon(\omega) v_{2\rho(\omega)+1} + \sum_{i=1}^{\rho(\omega)} u_i v_i\} .$$

There are thus many different representations of $\omega(x)$ that involve the same total number of functions, and for each, the function whose exterior derivative occurs with unity coefficient will be different.

A10. THE POINCARÉ LEMMA AND THE HOMOTOPY OPERATOR H

Let U be any open set in the Euclidean space E_n of n-dimensions. The set U is said to be *star-shaped with respect to the point with coordinates* (x_0^α) if and only if $(x^\beta)\varepsilon U$ implies $(\lambda x^\beta + (1-\lambda)x_0^\beta)\varepsilon U$ for all $\lambda\varepsilon[0,1]$. If U is star-shaped with respect to (x_0^α) , then a simple translation gives a new U that is star-shaped with respect to the origin. Let us assume that such a translation has been made. In this case we have $(x^\alpha)\varepsilon U$ implies $(\lambda x^\alpha)\varepsilon U$ for all $\lambda\varepsilon[0,1]$. Clearly, the specification

$$\text{(A10.1)} \qquad X(x^\beta) = x^\alpha \, \partial/\partial x^\alpha$$

defines a tangent vector field on U ; in fact, this tangent vector field is just the operator version of the radius vector of E_n considered as a vector space.

Define the homotopy operator H on $\Lambda^k(U)$ for $k \geq 1$ by

(A10.2) $$(H\omega)(x^\beta) = \int_0^1 X(x^\beta) \lrcorner\, \omega(\lambda x^\beta)\, \lambda^{k-1}\, d\lambda ,$$

where U is a star-shaped region with respect to the origin. We then have the following results.

The operator H *has the following properties:*

H_1: H *maps* $\Lambda^k(U)$ *into* $\Lambda^{k-1}(U)$ *for* $k\geq 1$ *and commutes with addition and multiplication by numbers;*

H_2: $dH\omega + Hd\omega = \omega$ for $\omega\epsilon\Lambda^k(U)$, $k\geq 1$, $Hdf(x^\beta) = f(x^\beta) - f(0^\beta)$;

H_3: $H(H\omega) = 0$, $(H\omega)(0^\beta) = 0$;

H_4: $HdH\omega = H\omega$, $dHd\omega = d\omega$;

H_5: $HdHd\omega = Hd\omega$, $dHdH\omega = dH\omega$, $(dH)(Hd)\omega = 0$, $(Hd)(dH)\omega = 0$;

H_6: $X\lrcorner H\omega = 0$, $H(X\lrcorner\omega) = 0$.

The first thing we obtain from these results is the Poincaré lemma: *Every closed form on* U *is exact.* This follows immediately from H_2 since $d\omega = 0$ if ω is closed, in which case we obtain $\omega = dH\omega$; that is $\omega = d\eta$ with $\eta = H\omega$.

There is significantly more information that can be gleaned from the above results by making use of the properties H_3 through H_6 . As before, we use $E^k(U)$ to denote the collection of all exact elements of $\Lambda^k(U)$. The following result is then an immediate consequence of H_2 through H_6 .

The operator dH *maps* $\Lambda^k(U)$ *onto* $E^k(U)$.

This result allows us to define a unique exact part of any exterior form ω for $k\geq 1$. The *exact part* ω_e of the form ω is given by

(A10.3) $$\omega_e = dH\omega .$$

It is then reasonable to define the other part that occurs in H_2 , namely $Hd\omega$, as the *antiexact part,* ω_a of ω :

(A10.4) $$\omega_a = Hd\omega , \quad \omega\epsilon\Lambda^k , \quad k>0 ; \quad f_a = f\epsilon\Lambda^0 .$$

It then follows immediately from H_3 and H_6 that the *antiexact part of any* $\omega\varepsilon\Lambda^k(U)$, $k>0$, *satisfies*

(A10.5) $$\omega_a(0^\beta) = 0 \ , \quad X\rfloor\omega_a = 0 \ .$$

This result provides us with a very simple characterization of the collection $A^k(U)$ of all *antiexact elements of* $\Lambda^k(U)$

(A10.6) $$A^k(U) = \{\omega\varepsilon\Lambda^k(U) \mid X\rfloor\omega = 0 \ , \quad \omega(0^\beta) = 0\} \ .$$

The operator Hd *maps* $\Lambda^k(U)$ *onto* $A^k(U)$ *and* $A^n(U) = 0$. The latter result follows on noting that $\Lambda^k \xrightarrow{d} \Lambda^{k+1} \xrightarrow{H} \Lambda^k$ and that $\Lambda^{n+1} = 0$.

We know that the exterior product of an element of $\Lambda^0(U)$ with an element of $E^k(U)$ is no longer a member of $E^k(U)$; that is, $E^k(U)$ is closed under multiplication by numbers but not under multiplication by functions. In this respect, the set $A^k(U)$ is much nicer than the set $E^k(U)$, for (A10.6) gives us the following immediate result.

The set $A^k(U)$ *is closed under addition and exterior multiplication.*

Thus exterior products of elements of $A^k(U)$ belong to $A^{k+\ell}(U)$, and, in particular, functions times elements of $A^k(U)$ belong to $A^k(U)$. This says that *the set* $A^k(U)$ *is a module over the associative algebra of* C^∞ *functions on* U . This result is instrumental in solving exterior differential equations when combined with the following:

The operator H *is the inverse of the operator* d *when* d *is restricted to* $A^k(U)$. This follows from noting that $\omega\varepsilon A^k(U)$ implies that $\omega_e = dH\omega = 0$ and H_2 yields $\omega = Hd\omega$. The set $A^k(U)$ also provides a unique representation of elements of ω . We first note that H_6 implies the following: *If* $\omega\varepsilon\Lambda^k(U)$ *satisfies* $X\rfloor\omega = 0$, $X\rfloor d\omega = 0$, *then* $\omega = 0$. This provides the uniqueness part of the following theorem.

Any $\omega\varepsilon\Lambda^k(U)$, $k>0$, *has the unique representation*

(A10.7) $\omega = dU_1 + U_2$

under the conditions $U_1 \varepsilon A^{k-1}(U)$, $U_2 \varepsilon A^k(U)$, *in which case we have*

(A10.8) $U_1 = H\omega$, $U_2 = Hd\omega$.

The simplest way of summarizing these results is by means of the following diagram in which $\Lambda(U)$ denotes the graded exterior algebra of forms, $E(U)$ denotes the graded exterior algebra of exact forms, and $A(U)$ denotes the graded exterior algebra of antiexact forms.

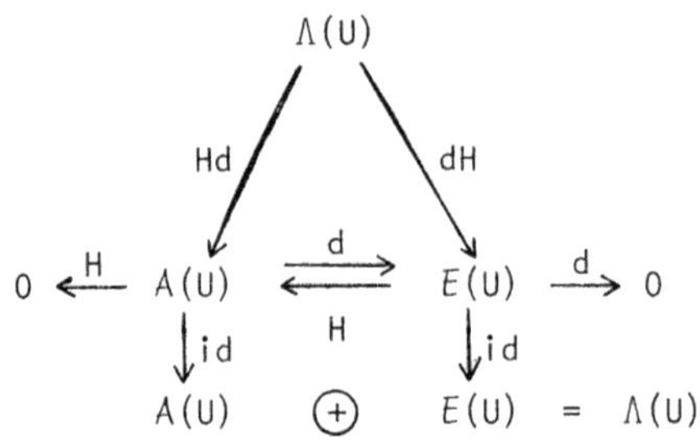

A11. SOLUTIONS OF EXTERIOR DIFFERENTIAL EQUATIONS

The purpose of this section is to show how the operator H and the unique decomposition of forms into an exact part and an antiexact part allows one to obtain explicit solutions of exterior differential equations. We start with a simple problem. Suppose that $\Gamma \varepsilon \Lambda^1(U)$, $\Omega \varepsilon \Lambda^{k+1}(U)$ are given forms and that we wish to obtain all forms $\omega \varepsilon \Lambda^k(U)$ that satisfy the exterior differential equation

(A11.1) $d\omega = \Gamma \wedge \omega + \Omega$.

Clearly, without further restrictions on the forms Γ and Ω, there can be no solutions, for exterior differentiation of (A11.1) yields

$$0 = d(d\omega) = d\Gamma \wedge \omega - \Gamma \wedge d\omega + d\Omega$$

$$= d\Gamma \wedge \omega - \Gamma \wedge \Omega + d\Omega .$$

We must therefore demand that Γ and Ω satisfy the integrability conditions

(A11.2) $$d\Omega = \Gamma \wedge \Omega - d\Gamma \wedge \omega .$$

The first thing we do is to introduce two new variables $\sigma \varepsilon \Lambda^0(U) > 0$ and $\gamma \varepsilon \Lambda^k(U)$ by the substitution

(A11.3) $$\omega = \sigma \gamma .$$

Since $\sigma > 0$, this substitution reduces (A11.1) to

(A11.4) $$d\gamma = (\Gamma - d \ln \sigma) \wedge \gamma + \sigma^{-1}\Omega .$$

The next thing to do is to use the unique decomposition to write

(A11.5) $$\gamma = \gamma_e + \gamma_a , \quad \Gamma = \Gamma_e + \Gamma_a$$

and to note that $\Gamma_e = dH\Gamma$, $\gamma_e = dH\gamma$, $d\gamma_e = 0$. Under this representation, (A11.4) reduces to

(A11.6) $$d\gamma_a = (\Gamma_a + dH\Gamma - d \ln \sigma)(\gamma_e + \gamma_a) + \sigma^{-1}\Omega .$$

This equation shows that a significant simplification can be achieved by choosing σ by

(A11.7) $$\sigma = \exp(H\Gamma) ,$$

in which case (A11.6) reduces to

(A11.8) $$d\gamma_a = \Gamma_a \wedge \gamma_e + \Gamma_a \wedge \gamma_a + \sigma^{-1}\Omega .$$

Now, $\gamma_a \varepsilon A^k(U)$ and $\Gamma_a \wedge \gamma_a \varepsilon A^{k+1}(U)$ since $\Gamma_a \varepsilon A^1(U)$. Thus, since H inverts d on $A^k(U)$ and $H(\Gamma_a \wedge \gamma_a) = 0$, (A11.8) yields

$$\text{(A11.9)} \qquad \gamma_a = H(\Gamma_a \wedge \gamma_e) + H(\sigma^{-1}\Omega) .$$

Collecting all of the results together and noting that $\gamma_e = d\rho$, we then obtain the general solution

$$\text{(A11.10)} \qquad \omega = e^{H\Gamma} d\rho + e^{H\Gamma} H(\Gamma_a \wedge d\rho) + e^{H\Gamma} H(e^{-H\Gamma}\Omega)$$

where ρ is an arbitrary element of $A^{k-1}(U)$. Conversely, if we use the integrability conditions (A11.2), it is easily shown that the form given by (A11.10) satisfies the original equation (A11.1) for any $\rho \varepsilon A^{k-1}(U)$.

A more representative class of problems consists of the following. Let $\underset{\sim}{\Gamma}$ denote an r-by-r matrix of 1-forms with entries Γ^i_j and let $\underset{\sim}{\Theta}$ represent an r-by-r matrix of 2-forms with entries Θ^i_j. The exterior product $\underset{\sim}{\Gamma} \wedge \underset{\sim}{\Theta}$ is then interpreted as the matrix product with exterior multiplication; that is $(\Gamma \wedge \Theta)^i_j = \sum_{k=1}^{r} \Gamma^i_k \wedge \Theta^k_j$. Our problem is to obtain the general solution of the system of exterior equations

$$\text{(A11.11)} \qquad d\underset{\sim}{\Gamma} = \underset{\sim}{\Gamma} \wedge \underset{\sim}{\Gamma} + \underset{\sim}{\Theta} , \quad d\underset{\sim}{\Theta} = \underset{\sim}{\Gamma} \wedge \underset{\sim}{\Theta} - \underset{\sim}{\Theta} \wedge \underset{\sim}{\Gamma} ,$$

which have no additional integrability condition (i.e.) the system (A11.11) implies that $d(d\underset{\sim}{\Gamma}) = 0$, $d(d\underset{\sim}{\Theta}) = 0$. If we use the unique representation theorem to write

$$\text{(A11.12)} \qquad \underset{\sim}{\Gamma} = \underset{\sim}{\Gamma}_e + \underset{\sim}{\Gamma}_a , \quad \underset{\sim}{\Gamma}_e = d\underset{\sim}{\gamma} , \quad \underset{\sim}{\gamma} = H\underset{\sim}{\Gamma} ,$$

then the first of (A11.11) becomes

$$\text{(A11.13)} \qquad d\underset{\sim}{\Gamma}_a = d\underset{\sim}{\gamma} \wedge d\underset{\sim}{\gamma} + d\underset{\sim}{\gamma} \wedge \underset{\sim}{\Gamma}_a + \underset{\sim}{\Gamma}_a \wedge d\underset{\sim}{\gamma} + \underset{\sim}{\Theta} .$$

If $\underset{\sim}{A}$ is a nonsingular r-by-r matrix of 0-forms, then we can change variables by the substitution

$$\underset{\sim}{\Gamma}_a = \underset{\sim}{A}\,\underset{\sim}{\bar{\Gamma}}\,\underset{\sim}{A}^{-1}\ ,\quad \underset{\sim}{\Theta} = \underset{\sim}{A}\,\underset{\sim}{\bar{\Theta}}\,\underset{\sim}{A}^{-1}\ , \tag{A11.14}$$

in which case (A11.13) becomes

$$\begin{aligned} d\underset{\sim}{\bar{\Gamma}} &= \underset{\sim}{A}^{-1}(d\underset{\sim}{\gamma}\,\underset{\sim}{A} - d\underset{\sim}{A}) \wedge \underset{\sim}{\bar{\Gamma}} + \underset{\sim}{\bar{\Gamma}} \wedge A^{-1}(d\underset{\sim}{\gamma}\,\underset{\sim}{A} - d\underset{\sim}{A}) \\ &\quad + \underset{\sim}{A}^{-1} d\underset{\sim}{\gamma} \wedge d\underset{\sim}{\gamma}\,\underset{\sim}{A} + \underset{\sim}{\bar{\Theta}}\ . \end{aligned} \tag{A11.15}$$

A significant simplification can be achieved in the above system of equations if we can choose $\underset{\sim}{A}$ so that $d\underset{\sim}{\gamma}\,\underset{\sim}{A} - d\underset{\sim}{A}$ belongs to $A^1(U)$, for in this event, $(d\underset{\sim}{\gamma}\,\underset{\sim}{A} - d\underset{\sim}{A}) \wedge \bar{\Gamma}$ and $\underset{\sim}{\Gamma} \wedge \underset{\sim}{A}^{-1}(d\underset{\sim}{\gamma}\,\underset{\sim}{A} - d\underset{\sim}{A})$ belong to $A^2(U)$. We therefore consider the system

$$d\underset{\sim}{A} = d\underset{\sim}{\gamma}\,\underset{\sim}{A} - \underset{\sim}{A}\,\underset{\sim}{\mu}\ ,\quad \underset{\sim}{\mu}\varepsilon A^1(U) \tag{A11.16}$$

where the unknowns are the r^2 variables A^i_j . Exterior differentiation of (A11.16) yields the integrability conditions

$$d\underset{\sim}{\mu} = -\underset{\sim}{A}^{-1}\ d\underset{\sim}{\gamma} \wedge d\underset{\sim}{\gamma}\,\underset{\sim}{A} + \underset{\sim}{\mu} \wedge \underset{\sim}{\mu}\ . \tag{A11.17}$$

Thus, since $\underset{\sim}{\mu}\varepsilon A^1(U)$ implies $\underset{\sim}{\mu} = Hd\underset{\sim}{\mu}$, $H(\underset{\sim}{\mu} \wedge \underset{\sim}{\mu}) = \underset{\sim}{0}$, we obtain

$$\underset{\sim}{\mu} = -H(\underset{\sim}{A}^{-1}\ d\underset{\sim}{\gamma} \wedge d\underset{\sim}{\gamma}\,\underset{\sim}{A})\ . \tag{A11.18}$$

When this is substituted into (A11.16) we see that

$$d\underset{\sim}{A} = d\underset{\sim}{\gamma}\,\underset{\sim}{A} - \underset{\sim}{A}\,\underset{\sim}{\mu}\ . \tag{A11.19}$$

Again, since $\underset{\sim}{\mu}\varepsilon A^1(U)$, $\underset{\sim\sim}{A\mu}\varepsilon A^1(U)$, while the scalar nature of the entries of $\underset{\sim}{A}$ yields $\underset{\sim}{A} = \underset{\sim}{A}(0^\beta) + Hd\underset{\sim}{A}$. Thus, the entries of the matrix $\underset{\sim}{A}$ satisfy the system of integral equations

$$\underset{\sim}{A}(x^\beta) = \underset{\sim}{A}(0^\beta) + H(d\underset{\sim}{\gamma}\,\underset{\sim}{A})(x^\beta)\ . \tag{A11.20}$$

Standard existence theorems for this system of linear integral equations establishes existence of solutions in some neighborhood of (0^β) in U . Further, it follows from (A11.19) and $\underset{\sim}{\gamma}(0^\beta) = 0$

(i.e. $\underset{\sim}{\gamma} \varepsilon A^0(U)$) that

$$\det(\underset{\sim}{A}) = \det(\underset{\sim}{A}(0^\beta))e^{tr(\underset{\sim}{\gamma})} , \tag{A11.21}$$

where $tr(\underset{\sim}{\gamma})$ denotes the trace of the scalar-valued matrix $\underset{\sim}{\gamma}$. Nonsingular solutions of (A11.19) thus exist for any choice $\underset{\sim}{A}(0^\beta)$ such that $\det(\underset{\sim}{A}(0^\beta)) \neq 0$.

We now use any such matrix $\underset{\sim}{A}(x^\beta)$ in (A11.15) that is determined by solving (A11.16). This gives

$$d\underset{\sim}{\bar{\Gamma}} = \underset{\sim}{\mu} \wedge \underset{\sim}{\bar{\Gamma}} + \underset{\sim}{\bar{\Gamma}} \wedge \underset{\sim}{\mu} - d\underset{\sim}{\mu} + \underset{\sim}{\mu} \wedge \underset{\sim}{\mu} + \underset{\sim}{\bar{\Theta}} . \tag{A11.22}$$

However, $\underset{\sim}{\bar{\Gamma}}$ and $\underset{\sim}{\mu}$ belong to $A^1(U)$ so that $\underset{\sim}{\bar{\Gamma}} = Hd\underset{\sim}{\bar{\Gamma}}$, $\underset{\sim}{\mu} = Hd\underset{\sim}{\mu}$, $H(\underset{\sim}{\Gamma} \wedge \underset{\sim}{\mu}) = H(\underset{\sim}{\mu} \wedge \underset{\sim}{\Gamma}) = H(\underset{\sim}{\mu} \wedge \underset{\sim}{\mu}) = \underset{\sim}{0}$, and (A11.22) yields

$$\underset{\sim}{\bar{\Gamma}} = - \underset{\sim}{\mu} + H(\underset{\sim}{\bar{\Theta}}) . \tag{A11.23}$$

Combining all of the above substitutions, we finally obtain

$$\underset{\sim}{\Gamma} = (d\underset{\sim}{A} + \underset{\sim}{A}\, H(\underset{\sim}{\bar{\Theta}}))\underset{\sim}{A}^{-1} . \tag{A11.24}$$

This solution determines $\underset{\sim}{\Gamma}$ in terms of the 2-forms $\underset{\sim}{\Theta}$ and the matrix $\underset{\sim}{A}$ of 0-forms that satisfy (A11.20).

It now remains to solve the second set of the system (A11.11), namely

$$d\underset{\sim}{\Theta} = \underset{\sim}{\Gamma} \wedge \underset{\sim}{\Theta} - \underset{\sim}{\Theta} \wedge \underset{\sim}{\Gamma} . \tag{A11.25}$$

Under the substitution (A11.14), this system becomes

$$\begin{aligned} d\underset{\sim}{\bar{\Theta}} &= - \underset{\sim}{A}^{-1}\, d\underset{\sim}{A} \wedge \underset{\sim}{\bar{\Theta}} + \underset{\sim}{\bar{\Theta}} \wedge \underset{\sim}{A}^{-1}\, d\underset{\sim}{A} + \underset{\sim}{A}^{-1}\{\underset{\sim}{\Gamma} \wedge \underset{\sim}{\Theta} - \underset{\sim}{\Theta} \wedge \underset{\sim}{\Gamma}\}\underset{\sim}{A} \\ &= H(\underset{\sim}{\bar{\Theta}}) \wedge \underset{\sim}{\bar{\Theta}} - \underset{\sim}{\bar{\Theta}} \wedge H(\underset{\sim}{\bar{\Theta}}) \end{aligned} \tag{A11.26}$$

when (A11.24) is used. If we now use the unique decomposition to write

$$\underset{\sim}{\bar{\Theta}} = d\underset{\sim}{\theta} + \underset{\sim}{\bar{\Theta}}_a , \quad \underset{\sim}{\theta} = H\underset{\sim}{\bar{\Theta}}\ \varepsilon A^1(U) \tag{A11.27}$$

we obtain

$$\text{(A11.28)} \qquad d\underset{\sim}{\Theta}_a = -\, d(\underset{\sim}{\theta} \wedge \underset{\sim}{\theta}) + \underset{\sim}{\theta} \wedge \underset{\sim}{\Theta}_a - \underset{\sim}{\Theta}_a \wedge \underset{\sim}{\theta} \ .$$

Thus, since $\underset{\sim}{\theta} \wedge \underset{\sim}{\Theta}_a$ and $\underset{\sim}{\Theta}_a \wedge \underset{\sim}{\theta}$ belong to $A^3(U)$, (A11.28) yields

$$\text{(A11.29)} \qquad \underset{\sim}{\Theta}_a = -\, \underset{\sim}{\theta} \wedge \underset{\sim}{\theta} \ ,$$

so that (A11.27) gives

$$\text{(A11.30)} \qquad \bar{\underset{\sim}{\Theta}} = d\underset{\sim}{\theta} - \underset{\sim}{\theta} \wedge \underset{\sim}{\theta} \ .$$

Accordingly, when we combine the above substitutions, we obtain the general solution

$$\text{(A11.31)} \qquad \underset{\sim}{\Gamma} = (d\underset{\sim}{A} + \underset{\sim}{A}\,\underset{\sim}{\theta})\underset{\sim}{A}^{-1} \ , \quad \underset{\sim}{\Theta} = \underset{\sim}{A}(d\underset{\sim}{\theta} - \underset{\sim}{\theta} \wedge \underset{\sim}{\theta})\underset{\sim}{A}^{-1} \ ,$$

where $\underset{\sim}{\theta}$ and $\underset{\sim}{A}$ are determined in terms of $\underset{\sim}{\Gamma}_e = d\underset{\sim}{\gamma}$ by

$$\text{(A11.32)} \qquad \underset{\sim}{A}(x^\beta) = \underset{\sim}{A}(0^\beta) + H(d\underset{\sim}{\gamma}\ \underset{\sim}{A})(x^\beta) \ ,$$

$$\text{(A11.33)} \qquad \theta = H(\underset{\sim}{A}^{-1}\underset{\sim}{\Theta}\underset{\sim}{A}) \ .$$

A12. INTEGRATION OF EXTERIOR FORMS AND STOKES' THEOREM

Let Y_r be a closed, r-dimensional region of r-dimensional numbers space E_r and let ∂Y_r denote the boundary of Y_r. We know that the dimension of $\Lambda^r(E_r)$ is one, so that any element ω of Λ^r can be written uniquely as

$$\text{(A12.1)} \qquad \omega(y^i) = w(y^i)\ dy^1 \wedge dy^2 \wedge \ldots \wedge dy^r$$

in terms of the natural Cartesian coordinate functions $(y^1,\ldots,y^r)$ of E_r. Further, any two elements of Λ^r can only differ in the function $w(y^i)$ and in the order in which the $dy^1,\ldots,dy^2$ are multiplied. Thus, if we choose the order so that we have the simple elementary volume element

$$W = dy^1 \wedge dy^2 \wedge \ldots \wedge dy^r , \qquad \text{(A12.2)}$$

then we fix all simple elements of Λ^r to within a factor of ± 1. Choice of the plus sign, that is, choice of W given by (A12.2) thus fixes things completely. This choice of the order in which the factors occur in the simple elementary volume element is referred to as a choice of *orientation*, while making this choice assigns an orientation to Y_r. It is then a trivial matter to assign an orientation to ∂Y_r, for ∂Y_r is of dimension $r-1$ and all $(r-1)$-forms on an r-dimensional space are also simple. We simply take a basis for the simple $(r-1)$-forms at any point on ∂Y_r and complete it to a basis for r-forms at that point. If the resulting r-form is a positive multiple of W at that point, then the element of ∂Y_r is said to have a *positive induced orientation.* If the resulting r-form is a negative multiple of W at that point then ∂Y_r is said to have a *negative induced orientation.* It is clear geometrically that we can always assign a positive induced orientation to ∂Y_r provided the set Y_r is simply connected with a sufficiently smooth boundary.

Let

$$\omega(y^i) = w(y^i) dy^1 \wedge dy^2 \wedge \ldots \wedge dy^r \qquad \text{(A12.3)}$$

be a given element of Λ^r. We define the integral of ω over the set Y_r by

$$\int_{Y_r} \omega(y^i) = \int_{Y_r} w(y^i)\, dy^1 dy^2 \ldots dy^r \qquad \text{(A12.4)}$$

where the integral in the right-hand side of (A12.4) denotes the r-dimensional Riemann integral over the region Y_r. It is now a direct, but messy calculation to establish Stokes' theorem.

If $\omega \in \Lambda^{r-1}(E_r)$ *is defined over a closed arcwise connected, simply connected* r*-dimensional region* Y_r *of* E_r *and* Y_r *has a smooth boundary, then*

$$\text{(A12.5)} \qquad \int_{Y_r} d\omega = \int_{\partial Y_r} \omega ,$$

when the orientation of ∂Y_r *is that induced from the orientation of* Y_r .

Suppose now that we have an element $\omega(x^\beta)$ of $\Lambda^r(S)$. If we construct a smooth map

$$\text{(A12.6)} \qquad \Psi : Y_r \xrightarrow{\text{onto}} S_r \subset S | x^\beta = \psi^\beta(y^i)$$

then $\Psi^*(\omega)$ becomes an r-form on the r-dimensional set Y_r . We then define the integral of ω over S_r by

$$\text{(A12.7)} \qquad \int_{S_r} \omega(x^\beta) = \int_{Y_r} (\Psi^*\omega)(y^j) .$$

This definition, when combined with (A12.5) gives the general version of Stokes' theorem: *If* $\omega(x) \in \Lambda^{r-1}(S)$, *then*

$$\text{(A12.8)} \qquad \int_{S_r} d\omega = \int_{\partial S_r} \omega ,$$

provided $S_r = \Psi(Y_r)$, $\partial S_r = \Psi(\partial Y_r)$. A more useful way of writing this is

$$\text{(A12.9)} \qquad \int_{Y_r} \Psi^*(d\omega) = \int_{Y_r} d(\Psi^*\omega) = \int_{\partial Y_r} \Psi^*\omega .$$

REFERENCES

I. EXTERIOR FORMS

1. Slebodzinski, W.: *Exterior Forms and Their Applications* (Polish Scientific Publishers, Warsaw, 1970).

2. Cartan, H.: *Differential Forms* (Houghton Mifflin Co., Boston, 1970).

3. Choquet-Bruhat, Y.: *Géométrie Différentielle et Systèms Extérieurs* (Donod, Paris, 1968).

4. Sternberg, S.: *Lectures on Differential Geometry* (Prentice-Hall, Englewood Cliffs, 1964).

5. Loomis, L. H. and S. Sternberg: *Advanced Calculus* (Addison-Wesley, Reading, Mass., 1968).

6. Lovelock, D. and H. Rund: *Tensors, Differential Forms, and Variational Principles* (John Wiley, New York, 1975).

7. Flanders, H.: *Differential Forms* (Academic Press, New York, 1963).

8. Matsushima, Y.: *Differentiable Manifolds* (Marcel Dekker, New York, 1972).

9. Vaisman, I.: *Cohomology and Differential Forms* (Marcel Dekker, New York, 1973).

II. GENERAL

10. Eisenhart, L. P.: *Continuous Groups of Transformations* (Dover Publications, New York, 1961).

11. Ovsjannikov, L. V.: *Group Properties of Differential Equations* (Siberian Sect. Academy of Sciences U.S.S.R., 1962).

III. MECHANICS

12. Goldstein, H.: *Classical Mechanics* (Addison-Wesley, Cambridge, Mass., 1951).

13. Whittaker, E. T.: *A Treatise on the Analytical Dynamics of Particles and Rigid Bodies* (Dover, New York).

14. Abraham, R. and J. E. Marsden: *Foundations of Mechanics* (W. A. Benjamin, New York, 1967).

15. Corbin, H. and P. Stehle: *Classical Mechanics*, Second Ed. (John Wiley, New York, 1960).

16. Liapounov, M.: *Probleme General de la Stabilite du Movement* Princeton Univ. Press, Princeton, 1949).

17. Pars, L.: *A Treatise on Analytical Dynamics* (John Wiley, New York, 1965).

18. Wintner, A.: *The Analytical Foundations of Celestial Mechanics* (Princeton Univ. Press, Princeton, 1941).

INDEX

V

W

LIST OF FREQUENTLY USED SYMBOLS

C	Configuration space	3
d	Exterior derivative	265
E	Event space	4
$E(t;q^{\beta};y^{\beta})$	Total energy function	103
F	Fundamental form	139, 210
G	Graph space	200
$\tilde{G}$	Union graph space	248
H	Homotopy operator	284
$I(K)$	Identity class of K	215
J	Fundamental form	103, 210
j	Fundamental form	138
K	Kinematic space	5, 200
$\tilde{K}$	Union kinematic space	247
L	Lagrangian function	56, 61
N	Non C-H form	137
$N(W)$	Noetherian vector fields	219
$N(W;B)$	Noetherian vector fields	219
T	Kinetic energy function	44
$T(K)$	Tangent space of K	9ff
W	Work and balance form	45, 210
Z	Linear operator	10
Z_i	Linear operators	202
δ	Variation	35ff
$\Delta(V)$	Extended variation	42
π_i	Surface forms	201
Φ^*	Pull back of Φ	273
Φ_*	Pull forward of Φ	272
ω	Volume form	200
ω^i	Constraint forms	80
$\Omega(V)$	Noetherian form	212

MIX
Papier aus verantwortungsvollen Quellen
Paper from responsible sources
FSC® C105338

If you have any concerns about our products,
you can contact us on
ProductSafety@springernature.com

In case Publisher is established outside the EU,
the EU authorized representative is:
Springer Nature Customer Service Center GmbH
Europaplatz 3, 69115 Heidelberg, Germany

Printed by Libri Plureos GmbH
in Hamburg, Germany